The Hydraulic State

The Hydraulic State explores the hydraulic engineering technology underlying water system constructions of many of the ancient World Heritage sites in South America, the Middle East and Asia as used in their urban and agricultural water supply systems.

Using a range of methods and techniques, some new to archaeology, Ortloff analyzes various ancient water systems including agricultural field system designs known in ancient Peruvian and Bolivian Andean societies, water management at Nabataean Petra, the Roman Pont du Garde water distribution *castellum*, the Minoan site of Knossos and the water systems of dynastic (and modern) China, particularly the Grand Canal and early water systems designed to control flood episodes. In doing so the book greatly increases our understanding of the hydraulic/hydrological engineering of ancient societies through the application of Complexity Theory, Similitude Theory and Computational Fluid Dynamics (CFD) analysis, as well as traditional archaeological analysis methods.

Serving to highlight the engineering science behind water structures of the ancient World Heritage sites discussed, this book will be of interest to archaeologists working on landscape archaeology, urbanism, agriculture and water management.

Charles R. Ortloff is the director of CFD Consultants International and Research Associate in Anthropology at the University of Chicago.

Routledge Studies in Archaeology

Dwelling
Heidegger, Archaeology, Mortality
Philip Tonner

New Perspectives in Cultural Resource Management
Edited by Francis P. Mcmanamon

Cultural and Environmental Change on Rapa Nui
Edited by Sonia Cardinali, Kathleen Ingersoll, Daniel Ingersoll Jr., Christopher Stevenson

Making Sense of Monuments
Narratives of Time, Movement, and Scale
Michael J. Kolb

Researching the Archaeological Past through Imagined Narratives
A Necessary Fiction
Edited by Daniël van Helden and Robert Witcher

Cognitive Archaeology
Mind, Ethnography, and the Past in South Africa and Beyond
Edited by David Whitley, Johannes Loubser and Gavin Whitelaw

Archaeological Networks and Social Interaction
Edited by Lieve Donnellann

The Hydraulic State
Science and Society in the Ancient World
Charles R. Ortloff

For more information on this series, please visit www.routledge.com/Routledge-Studies-in-Archaeology/book-series/RSTARCH

The Hydraulic State

Science and Society in the Ancient World

Charles R. Ortloff

LONDON AND NEW YORK

First published 2021
by Routledge
2 Park Square, Milton Park, Abingdon, Oxon OX14 4RN

and by Routledge
605 Third Avenue, New York, NY 10017

First issued in paperback 2022

Routledge is an imprint of the Taylor & Francis Group, an informa business

Publisher's Note
The publisher has gone to great lengths to ensure the quality of this reprint but points out that some imperfections in the original copies may be apparent.

British Library Cataloguing-in-Publication Data
A catalogue record for this book is available from the British Library

Library of Congress Cataloging-in-Publication Data
A catalog record for this book has been requested

ISBN: 978-0-367-50238-6 (pbk)
ISBN: 978-0-367-85808-7 (hbk)
ISBN: 978-1-003-01519-2 (ebk)

DOI: 10.4324/9781003015192

Typeset in Times New Roman
by Apex CoVantage, LLC

Lolie

with me throughout time

Author biography

At the midpoint of the author's 50-year research and development career concentrating on aerodynamic and hydrodynamics analysis applications for major national and international corporations and government agencies involved in defense, nuclear reactor, petroleum, chemical and ocean engineering subject areas, he developed an interest in the technical accomplishments of ancient water engineers and the theoretical base underlying their urban and agricultural water system designs and structures. This led to a second career involving collaboration with university scholars and international archeological institutes which led to projects and travels to archaeological sites in South America, the Middle East and Southeast Asia to analyze water system designs by using modern hydraulic engineering methodologies. His 2010 Oxford University Press book *Water Engineering in the Ancient World* summarized many new discoveries on this topic; the present book expands further on the hydraulic engineering knowledge base discovered by ancient world engineers. The author is currently Director of CFD Consultants International and Research Associate in Anthropology at the University of Chicago. His work continues to highlight the use of modern engineering methodologies in gaining new perspectives on historical and scientific aspects of ancient societies.

Contents

Figures

Tables

Acknowledgments

With a 50-year research and development career in aeronautical/mechanical engineering for major US and foreign companies, together with 40 years in archaeological research and travels, there have been many opportunities to work with leading research specialists in both engineering and archaeology. This provided opportunities to contribute research discoveries related to ancient engineering practice. Although many references describe the water control structures of the ancient world, chapters of this book go further to uncover the ancient science and engineering discoveries underlying the design and function of these structures and thus discover if elements of modern hydraulic engineering theory have earlier precedents worthy of revisions to the history of hydraulic science. Work started in archaeology in the early 1980s with participation in Dr. Michael Moseley's ancient irrigation project on the north coast of Peru, (*Programma Riego Antiguo*) centered at the Moche Valley site of Chan Chan, the capital city of the Chimu empire dating from AD 1000–1432. From exploration of this area over the next five years, data from the 75-km-long Chicama-Moche Intervalley Canal was used for CFD analysis; results indicated that Chimu water engineer's use of indigenous hydraulic technologies duplicated aspects of modern hydraulic engineering practice as described in my 2009 Oxford University Press book *Water Engineering in the Ancient World*. Continuance of work in the Moquegua Valley of far-south Peru with Dr. Michael Moseley and Dr. Robert Feldman after the north coast project provided insight into the Tiwanaku-Wari interface site of Cerro Baul and Wari canal technology. With the thought that there were undiscovered revelations about the engineering capabilities of Andean civilizations that had eluded previous researchers, there was an invitation by Dr. Alan Kolata, Chair of Archaeology at the University of Chicago leading to a further project that was initiated shortly thereafter to participate at the site of Tiwanaku (AD 400–1100) in Bolivia. Tiwanaku excavation had been denied to foreign researchers before this project initiated in the mid-1980s, so the opportunity for initial discoveries of water technologies used for both urban and agricultural use provided a unique opportunity for discovery – here Chapter 5 summarizes the hydraulic/hydrological technologies of Tiwanaku civilization of equal level to that found in modern engineering practice. The transfer of early 1930s aerial photos of urban Tiwanaku by Dr. Alan Sawyer and later cooperation with Dr. John Janusek (Vanderbilt University) proved vital discovering discovering hidden truths about the water system of Tiwanaku city and its famous "moat." What emerged from six years of site research was a methodology to rapidly dry infiltrated rainfall using moat water collection and drainage capabilities – this provided hygienic health benefits to city inhabitants. A further discovery related to the moat and its supply canals ability to maintain the groundwater level constant through seasonal changes in rainfall – this provided structural stability for the many large monuments as well as continued production of specialty crops within city limits. Results of this research are given in Chapter 5 of this book. A later invitation by Dr. Ruth Shady in the 1990s to participate in research at the newly discovered preceramic site of Caral (2500–1600 BC) in the Supe Valley of north-central Peru yielded much new information about the early development of hydraulic engineering for water supply to the main city temple compound area; research associates Dr. Moseley (University of Florida), Dr. Dan Sandweiss (University of Maine) and Dr. Robert Keefer (US Geological Survey) combined efforts to broaden perspectives on geophysical landscape changes that affected Caral's historical development and ultimate demise as described in Chapter 7 of this book. A continuous stream of hydraulic engineering advancements from early preceramic (2600 BC) to late Inka times (AD 1400–1532) as demonstrated at the Inka site of Tipon (Chapter 6) reveals 3500 years of hydraulic engineering discoveries worthy of a revision to the history of science through water engineering discoveries made by Andean water engineers. While New World revelations of Andean hydraulic science add a new dimension to the history of science, the Old World empires of Rome, Greece and Nabataean Jordan demonstrate many elaborate water control structures in the form of aqueducts, channels, canals, siphons and pipelines but not the underlying water engineering technologies behind their design, construction and use. To investigate technologies underlying such constructions, an invitation to join with Dr. Phillip Hammond in early 2005 to participate in his investigation of the Temple of the Winged Lions at Petra with subsequent field investigations with Dr. Talal Akasheh, Director of the Queen Rania Jordanian Cultural Heritage Foundation, led to revelations about the hydraulic engineering accomplishments of the Nabataeans at Petra (Chapter 8). Further investigation of the Pont du Garde aqueduct and *castellum* built by Roman engineers revealed use of a knowledge base familiar to modern water engineering practice (Chapter 9) and more recent excursions to Minoan Crete and the Knossos palace site revealed further pipeline water transfer technologies not previously reported in the literature (Chapter 11). Dipl. Ing. Gilbert Wiplinger and site director Professor Dr. Krinzinger of the Austrian Archaeological Institute provided an introduction to the urban water systems at Ephesus through projects over the next five years; association with Dr. Dora Crouch during this period with exploration of many ancient Roman and Greek sites in Turkey led to several joint publications made possible through her sharing of information on ancient world sites. Further projects developed through invitation of Dr. Alan Kolata to explore the Cambodian site of Angkor Wat in early 2005; research done on the massive water reservoirs produced many new discoveries described in my earlier Oxford University Press book. To all these key associates, my profound gratitude for their gift of knowledge and opportunities to explore the water technologies of the ancient New and Old Worlds. With many individual journeys to South and Central America (Mexico, Peru, Bolivia, and Guatemala), the Middle East (Egypt, Israel, Turkey, Jordan, Lebanon) and Asia (China, Viet Nam, Laos, Cambodia, Taiwan) to investigate their ancient water systems, a comprehensive picture is beginning to emerge of engineering technologies discovered and described for the first time in both the earlier OUP book now augmented with further discoveries in the current Routledge book. With 40 years thus far devoted to this discovery and analysis endeavor, I am only beginning to reveal what I hope future generations of scholars will continue to investigate to bring forward the true nature of ancient science and engineering practice.

I

Hydraulic engineering in pre-Columbian Peru and Bolivia

1 Origins and development of water science in the Andean world and societal development according to Modern Complexity Theory

Andean water science in Preceramic times

Wetland agriculture in Preceramic times (10,000–4000 cal BC) in the form of raised fields at the site of Huaca Prieta-Paredones (Dillehay 2011, 2017; Dillehay et al. 2005) located in the Peruvian Chicama Valley coastal zone is, to date, the earliest indication of organized agriculture to support the local population. The Paredones complex was apparently the main farming area of the site whose agriculture was supported by high groundwater levels from rainfall runoff interception/infiltration and Chicama River input into the aquifer supporting the farming area phreatic zone. This site likely represents the earliest use of groundwater technology for farming known at present and signals the start of water control technology in coastal Peru. From recent research (Dillehay 2011:214:Figure 11.9; Dillehay et al. 1997, 2005), early Preceramic Period canals in the Nanchoc area of the Zaña Valley have been discovered associated with early settlements. The canal sequence consists of the earliest canal at the lowest depth with several later canals following the same path in the sequentially debris filled-in canals at higher elevations. The earliest canal dates from ~7500 BP with the latest canal dating to ~850 BP. The canals demonstrate increasing canal cross-section width and increased cross-section area profiles from early to late times. The debris infilling of early canals and the placement of later canals above earlier ones represent episodes of mass infilling due to erosion/deposition mass wasting effects over time and subsequent episodes of site abandonment. Apparently the spring water resource supplying the canal sequence over time was repeatedly reused for limited (or seasonal) agricultural purposes by site occupiers. This staging required origination and excavation of a new canal designs placed into the accumulated soil layers that covered the debris buried, earlier canal versions. The lowest and earliest canal (~7500 BP) was at a depth of 3.5 m from the present surface with a large width and low slope; about 80 cm above this canal was a canal dated to ~6100 BP with a depth of ~30–40 cm. The third later canal is about 1.0 m above this canal and dates to ~4900 BP. From data reviewed from Dillehay et al. 2005, the slope of the earliest canal is less than those of later canals. Presuming all canals originated from the same spring source and proceeded to a flat agricultural area, all canal versions had the same flow rate (Q = V/A, where V is the average water velocity and A is the canal cross-section area), and all canal versions shared the same final destination to a lower agricultural area. The hydraulic significance of this multi-bedded canal arrangement (with all canal versions originating from the same spring location and ending at the same field location) sharing the same path is that to maintain the same spring flow rate in each canal version, the canal cross-sectional area of later canals at lower slopes must increase. For the same flow rate Q in all canal versions, for velocity V low in the latest shallower slope canal, then A is prescribed to be larger to produce the constant Q flow rate value. For the earlier (deeper) canal versions, the canal slopes were higher resulting in higher water velocity values – but to maintain the same constant Q value, the canal cross sections must be lower. A progression exists of larger canal cross-section areas from the higher slope earlier and deeper canals to later canals to carry the same spring flow rate. This vertical sequence of different shape canal cross-sections superimposed along the same path and made over a long time period between redesigns incorporates an early form of hydraulic engineering knowledge derived from common sense observations of to how water flow rates (the volume of water flowing in a canal over a given time period) can be controlled by canal cross-sectional area shaping. Although this process was derived from observations by the ancient site occupants using trial-and-error observations involving canal shape changes, its significance in early use of water control can best be explained using modern hydraulic science notation. From the Manning equation (Morris and Wiggert 1972; Van Te Chow 1959; Henderson 1966), the steady flow rate (Q) in a canal of mild slope is predicted by $Q = 1.49\ R_h^{2/3}\ S^{1/2}\ A/n$ where R_h is the hydraulic radius (wetted canal cross-section area A divided by the canal wetted perimeter P) and S is the canal slope, A is the cross-sectional area of the water stream and n is a measure of channel roughness using English units. The slope S is defined as the ratio of declination distance d from horizontal over a given length L, i.e., d/L, and n is a measure of the canal wall roughness. From modern hydraulic principles, for the same flow rate Q in different shape canals, as the slope S increases, $R_h^{2/3}$ must decrease. Practically, this is accomplished by increasing the wetted perimeter P leading to a wider canal having an area A. This result is then consistent with the observed field data where the oldest, most deeply buried canal at a steeper slope is of smaller width than later canals at lower slopes that demonstrate higher widths at lower slopes where water velocity is low. The earliest lower placed canals can be set at high slopes while sequential debris infilling over time limited later, higher elevation canal to lower slopes. This empirical observation from ancient field data is an early indication of an empirical observation related to hydraulic science put into practical use by ancient canal designers – albeit this conclusion can only be advanced in approximate form given the limited excavation data at one excavation trench. As an early common sense application of water control using canal shaping and canal slope change to regulate water flows necessary for irrigation agricultural use, the observed field data serves as a start of a chain of observations of water behavior and control which was of vital importance to later developing societies. From this early example, later Andean societies developed and integrated more reliable agricultural practices vital to their developing economies that relied on higher levels of hydraulic knowledge for their canal irrigation networks. The development chain of hydrological knowledge thus had a beginning in the form of empirical observations of canal geometry change effects on water flow rates; this early beginning laid the foundation for later knowledge developments in codified form that underwrote development of later hydraulic societies. In subsequent sections related to later phases of Andean hydraulic practice, the increase in hydraulic engineering sophistication used for agricultural development is apparent to the point where it is incorporated and celebrated in religious practices elaborated with ceremonial devices to codify their mastery of water science. The control of water for urban and agricultural use was vital to Andean society's prosperity, sustainability and development and is causal to higher forms of social structure and ceremonial elaboration over time as later sections demonstrate.

For other early preceramic sites, the site of Caral (2600–1500 BC) located ~20 km inland in the Supe Valley appears to have developed an *amuna* system to support floodplain agriculture (details discussed in Chapter 7). Here a series of highland lakes still provide an abundance of water to valley bottomlands through a valley geologic fault line (although locals say that an underground tunnel exists!) – the net result is a high water table only a few feet from the ground surface that persists on a year-round basis with multiple springs, reservoirs and large intravalley lakes that supported agriculture in early BC centuries – and even to the present day. Figures 1.18 and 1.18a shown later in the text indicate the presence of water reservoirs and lakes in low valley bottom areas that penetrate the high water table.

The abundance of surplus water in the lower Supe Valley farming area is demonstrated by the presence of a large drainage canal adjacent to the entry road that regulated the groundwater height of adjacent field systems. The presence of a low slope, contour canal (Ortloff and Moseley 2012) taking water from a high elevation intersection with the Supe River and directing it to the Caral urban center and the nearby site of Chupacigaro along contour structures built along the cliff side wall indicates that an accurate surveying practice was well developed at that time (Figure 1.19).

While the valley bottom took its agricultural water supply from *amuna* supplied groundwater (presently the water table is less than a few feet from the ground surface), the surrounding elevated plateau above the valley bottom area was the location of the Caral urban center as well as ten contemporaneous nearby sites distributed along both sides of the Supe River. As such, their water supply came from terraced contour canals originating from the Supe River to deliver water to the higher elevation plateau sites (as a later section on Caral describes). Based on the abundance of water from the *amuna* source together with careful management of groundwater level for agricultural use and creation of intravalley lakes for local water table control through evaporation, the Supe Valley was a natural choice of early inhabitants due to its abundance of water resources. An additional factor elevating the groundwater level is a mid-valley natural subterranean choke point formed between closely separated elevated banks of the Supe River. As the groundwater flow through the porous aquifer encounters the subterranean choke contraction, the height of the groundwater interface behind the choke location increases in order to develop higher hydrostatic pressure necessary to overcome the flow resistance in subterranean soils. This hydrologic condition produces sufficient hydrostatic pressure to permit aquifer flow through the natural subterranean choke. The backup of the groundwater in the aquifer necessary to increase hydrostatic pressure raises the groundwater level behind the choke thus improving the farming potential of upvalley lands. While traces of preceramic canals are to be found elsewhere in La Galgada, El Paraiso and the upper reaches of the Zaña and Santa Valleys and other areas (Moseley 2008:109, 128–133; Greider et al. 1998:20–23, Dillehay et al. 2005), little technical data is available on these canals to extract the level of technology that existed with their construction and use. Later Formative Period coastal sites (Moseley 2008:134) saw the rise of irrigation and corporate reclamation farming areas; in the Moche Valley alone the Caballo Muerto site located near the Moche Valley neck was the forerunner of later period (Early and Late Intermediate Period) north- and south-side irrigation networks that included practically all arable land area in the Moche Valley. The Early Intermediate Period Moche Valley canals have been analyzed in detail (Ortloff 2009) given the technical data available for their construction and use. With the advancement of hydrological and hydraulic science from early Preceramic to Formative Period to Early Intermediate Period to Middle Horizon to Late Intermediate Period and finally to late Horizon times, supportive rituals and ceremonies elaborated the success of corporate agriculture practice that provided the food supply and economic sustenance base of many Andean societies located in different ecological zones. Essentially rituals and ceremonies reflected the acknowledgment of beneficence of celestial and worldly deities to transfer water control knowledge from nature observations to support the advancement of Andean societies from family related groups to statehood and finally to empire status. The following sections describe ceremonial aspects related to the control of water for agricultural and urban use to illustrate the discovery and dependence of water science related to water rituals in Andean societies.

Water ritual and symbols in the Andean world: *huacas* and *pacchas*

In Andean societies, ritual objects denoted as *huacas* (Bray 2015a, 2015b; Quilter 2014; Cachot 1955; Glowacki and Malpass 2003; Sherbondy 1992) convey the concept of power associated with objects and the concept that with proper rituals and ceremonies, this inherent power can be liberated to benefit and advance societal progress. The Inka viewed flowing water as an animating force (*camay*) having the capability to animate *huacas*. Here *huacas* of many forms that incorporated water flows possessed supernatural powers once given their *camay* life force – here the importance of water for sustaining life-sustaining agriculture was a prime objective of rituals and ceremonies. Many different *paccha* types are associated with the major ancient civilizations of Peru and Bolivia as well as for later Spanish colonial societies. The reciprocal relationship of liberating forces in objects was based upon worshipping sacred objects, idols and geographic features (such as sacred mountain places, pillars, rock outcroppings as well as ancestor mummy bundles), so that the deities and spirits associated within the objects and sites would oversee, protect and provide agricultural land fertility vital for the sustainability of the society devoted to the object (Moore 2005). Such associations continue into modern society as witnessed by inanimate religious icons given reverence to liberate their inner spiritual power through prayer and ritual to benefit the needs of worshipers – here many current religions maintain icons and rituals operating in this regard. A special class of *huacas* denoted as *pacchas* deserve special attention in pre-Columbian Andean societies due to their association with life-giving water vital for sustaining agricultural fertility. While *pacchas* come in many different forms (wooden and ceramic figurines, rock carvings and ceremonial sites) they essentially serve as props for the celebration of the importance of water to life.

Use of *pacchas* in ancient South American societies (Cachot 1955) has been associated with ceremonial rituals that relate the importance of water to the fertility of agricultural and pastoral resources that helped maintain the sustainability of urban and rural societies. In many cases, rituals involving *pacchas* are related to the agricultural cycle as exemplified by their ritual use in the Ayacucho area involving the cleaning of irrigation canals tied to the September equinox, the upcoming rainy season and the time for planting (Isbell 1985:138–139). Further associations of *paccha* ritual devices used in ceremonies and ritual processions in this regard involve festivals in Yura (Bolivia) related to plowing, planting and harvest as well as Aymara rituals tied to the agricultural cycle (Bastien 1985:51–53, 61; Moore 2005) among others. Further use of *pacchas* related to the pouring of *chicha* into canals in the Cusco area to promote crop fertility (Meddens et al. 2010:50; Rowe 1946) indicate continued use of ritual devices and ceremonies into Late Horizon Inka times. The *paccha* association with water, irrigation, plant and animal representations derives from Andean mythology when animals could speak and springs came forth as a result of rivalries between *huacas* for dominance (Rostworowski 1999). Additionally, many Andean ethnic groups claimed origins from lakes, springs and the ocean giving further association with ceremonial representations of water and its prominent use in rituals. One aspect of the Andean concept of reciprocity (denoted as *tincuay*) involved giving back to the earth what the earth had given to sustain societies; rituals involving water returned to the earth through elaborate water conveyance ceramics, wooden objects and stone carvings (Figures 1.1 to 1.6) denoted as *pacchas*, were key to ceremonies performed by priests designated to perform sacred rituals. From the mythological consciousness of deities that control nature's forces for either the benefit or harm to societies, ritual *pacchas* that transmitted prayers directed toward favorable deity intervention was a necessary function of the rituals. As water for agriculture and soil fertility to produce bountiful crop yields was a main concern of ritual supplications, any sign interpreted as spiritual habitation within water, as indicated by certain patterns in water flow through *paccha* channels, served as an indication that the deities were signaling a response through this means.

Figure 1.1 Pre-Columbian water ritual involving a *paccha* (lower part of figure).
Source: R. C. Cachot 1955.

The recreation of such surface flow patterns in *pacchas* therefore may have meaning related to observed irrigation channel flow patterns associated with bountiful harvests and are duplicated in ritual supplications in rites performed by the priests. Figures 1.4, 1.8 and 1.9 indicate the correspondence between surface wave patterns and their interpretation as animal snake spirits coming forward on the water surface. Figures 1.10 and 1.11 indicate the presence of undulation surface waves at critical flow conditions. Other interpretations of *paccha* flow patterns follow from the mixing of two fluid streams originating from two different sources in intersecting flow passageways in *pacchas* indicating that the gift of life and agricultural bounty proceeds from combining and mixing sources of divine origin with human endeavor in a cooperative manner.

Figure 1.2 Different *paccha* examples – top row labels a, b; second row labels c, d; third row labels e, f; fourth row labels g, h.

Source: R. C. Cachot 1955.

Typical *paccha* devices of this type are illustrated in Figures 1.2, 1.7, 1.12 and 1.12a as well as in stone carved channels (*huacas*) shown in Figures 1.3 and 1.6. From ethnographic sources (Bray 2013, 2015b; Sherbondy 1992; Hyslop 1990; Kaulicke et al. 2003), the importance of water to life continuity was manifest through water rituals; this emphasis continues from ancient to contemporary times in the Andean world. The concept of *paccariscca* (MacCormack 1991; Bray 2013) was associated with founding sites related to ancestral groups that provided the physical link between the descendant population and the territories they claimed through land and water rites; here ancestor worship through sacrifices and rituals bonded the past with the present. The creator god Viracocha provided the origins of societies through association with Lake Titicaca and its sacred island site as the source of the human race (Betanzos 1996 [1551–1557]; Sarminto 2007 [1572]) as understood by the Bolivian Middle Horizon Tiwanaku society and especially the Late Horizon Inka society. Additionally, among the contemporary Aymara of altiplano Bolivia drawing and continuing ceremonial

Figure 1.3 Multiple dual zigzag channel pattern on carved boulder surface – Inka origin.

beliefs from ancient ancestral sources, water has different magical and functional properties according to its source – rainfall water, channeled water from snowmelt sources, water from springs, water from streams and rivers or from groundwater seepage – each individual source has different ritual uses. Anointment of the newborn with water from all such sources, through ritually designated *pacchas*, then guarantees success, health, prosperity and prominence in life according to local custom and tradition. A further example within the ancient Andean world derives from rituals performed by the Inka at the confluence of the Saphi and Tullu Rivers to form the Huatanay River in Cuzco – the *purapucy* festival purification ceremonies performed at the confluence junction node of the river branches provided absolution from past sins. Here the junction node of rivers had special religious significance as *tincuay* symbolism was key to the success of rituals performed

Figure 1.4 Stylistic drawing of the Fuente de Lavapatas carved boulder, San Austin, Columbia, showing animal spirits interacting with water paths originating from reservoirs A and C and multiple water inlets.

Source: R. C. Cachot 1955.

at these sites – again, the multiple separation and confluence of water streams in certain *paccha* types (Figure 1.2) represented symbolically the ritual importance of these devices to confer life, success and fortune benefits to all when used in community ceremonies. While water was frequently used in this manner, ceremonies that involved *chicha*, well known for social bonding communion in Andean community rituals from ancient times to the present day, would have yet further ceremonial significance in celebratory functions. Specific examples of *pacchas* are found in practically all the ancient societies of Peru and Bolivia including the Nazca, Tiwanaku, Puquina, Inka, Chimu, Huaylas, Chavin and Recuay societies, among others as well as for colonial period Andean societies.

Figure 1.5 Single zigzag channel on carved boulder surface – Inka origin.

One class of *pacchas* consisted of ceramic or wooden vessels incorporating elaborate internal water passageways combined with zoomorphic figures, representations of agricultural crop staples and agricultural field tools as well as representations of anthropomorphic and mythic deities that signify the intimate connection between water, animal husbandry, crop yields and nature deities (Figures 1.1, 1.2, 1.20 and 1.21). A subclass of this *paccha* type consists of multiple nested bowls each having individual inlets and flow passages to separate bowl segments. Each separate nested bowl contains a different liquid and drinking from each in turn brought the blessings of each liquid type.

Other classes of *pacchas* are massive carved boulders (*huacas*) elaborated with numerous stylized animal figures and anthropomorphic deities with carved networks of water channels (Figures 1.3, 1.4 and 1.5) that reinforce the intimate connection between fertility and zoomorphic messenger deities, water and

Figure 1.6 Inka double zigzag channel at the site of Samaipata.

wildlife native to the Andean world. In many cases, zoomorphic representations are highly stylized derivatives of recognizable wildlife species (Figures 1.20 and 1.21). Incorporated into zoomorphic figure representations are the mythic roles these figures represent to ancient societies. Well-known examples in this category are the Lavipatas Fountain at San Augustín in Columbia (Figure 1.5), the carved boulder at Saihuite in Peru (Doig 1973:518) and the Inka double zigzag channels down the Samaipata rock outcrop top surface (Hyslop 1990:124) shown in Figure 1.6 as well as Recuay examples (Doig 1973:332–333) and further Inka examples shown in Figure 1.3. A further class of *pacchas* are elaborately carved water channels (Figures 1.14, 1.15 and 1.16), some associated with petroglyphs, that show channel paths with sequences of right angle bends clearly intended for ritual use as the channel elaboration in such systems clearly exceeds that required for practical agricultural irrigation purposes and, in many cases, no agricultural

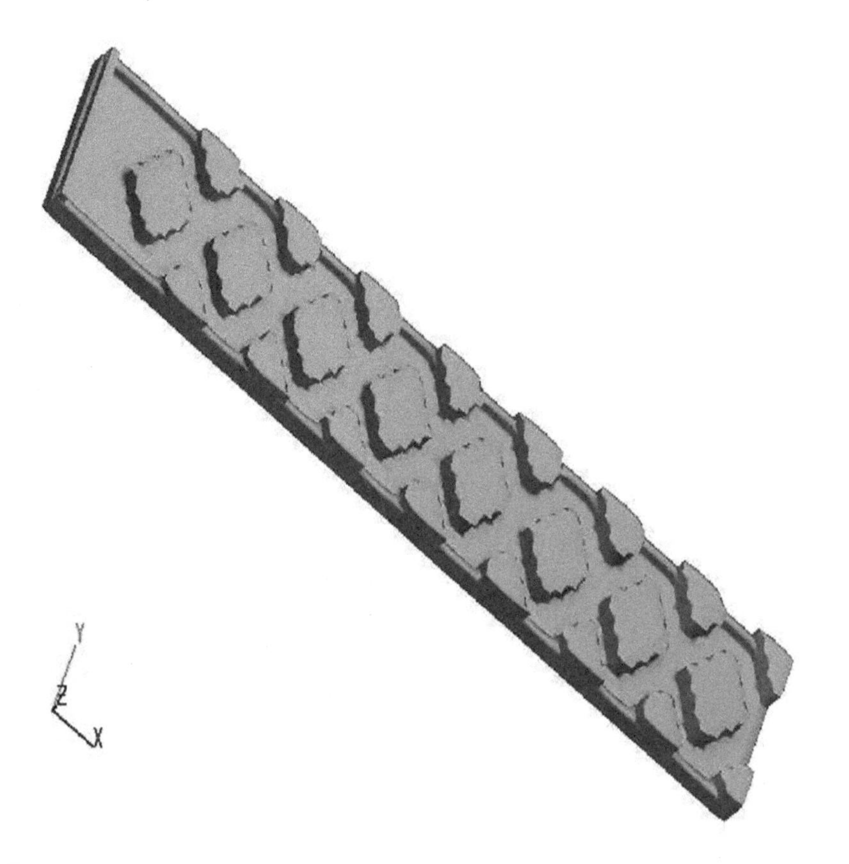

Figure 1.7 FLOW-3D CFD typical scale model of Figure 1.2 *pacchas* b, d and e with multiple branching/converging flow paths.

Figure 1.8 Two-dimensional surface standing wave pattern induced by a bottom obstacle for critical flow conditions.

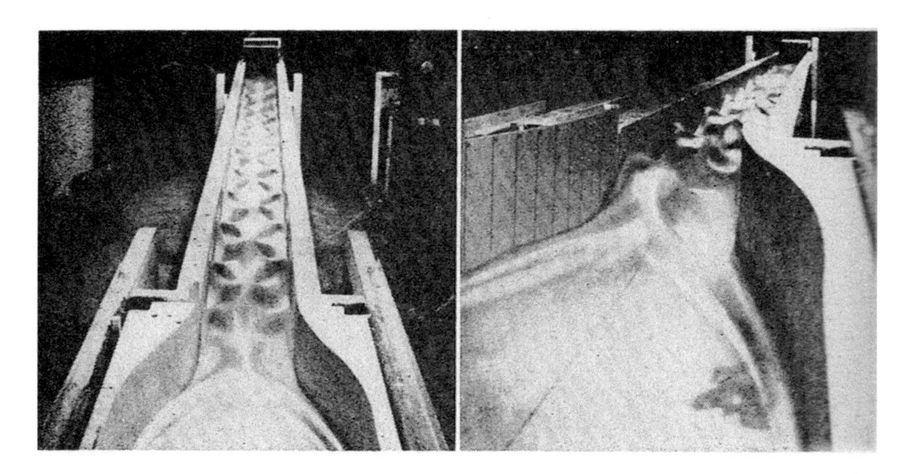

Figure 1.9 Area transition in a flume causing critical flow (Fr = 1) standing surface wave patterns.

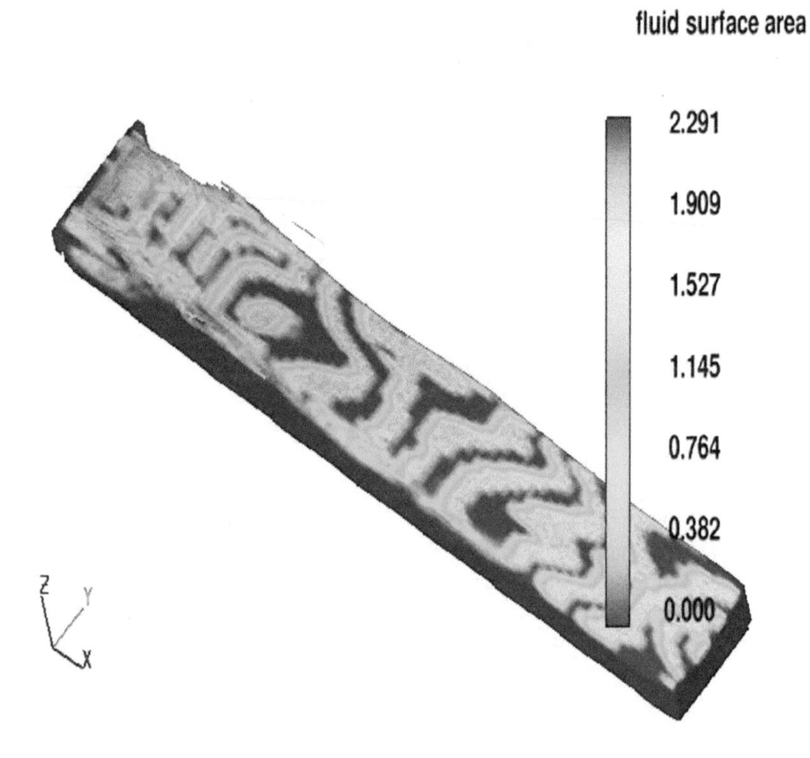

Figure 1.10 FLOW-3D computer results for critical flow in a straight channel. Variations in surface area patterns are indicative of surface waves.

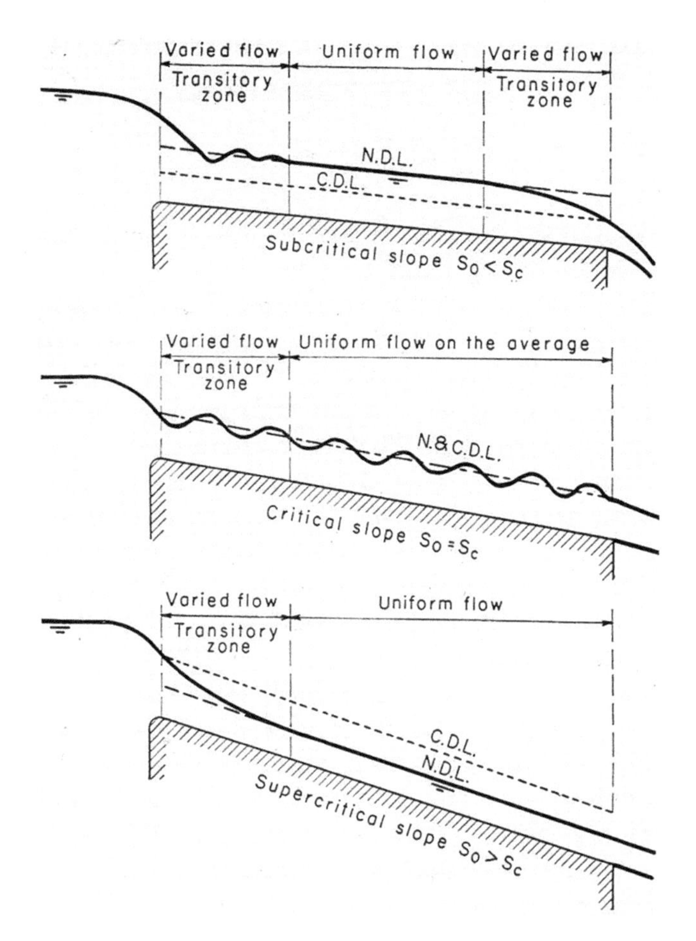

Figure 1.11 Influence of channel slopes on asymptotic surface flow geometry. N.D.L. is the normal asymptotic depth; C.D.L. is the asymptotic depth line; S is the channel slope. Note that N.D.L. and C.D.L. coalesce at Fr =1 critical flow conditions at critical slope S_c causing an oscillatory surface wave pattern.

fields are associated with canals of this type. An example of this type of water system is the Cumbemayo Canal (Doig 1973:245) located in the northern sierra region just outside of Cajamarca. Channels with sinusoidal carved channels (Figures 1.3 and 1.5) that constitute a subclass of this large-scale channel type. A further subset of this category are the elaborate fountains and reservoir systems served by canals found at Pachacamac (Rostworowski 1999), as well as water conveyance structures found at Kenko, Tambomachay, Vilcashuamán, Ollantaytambo/Incamisana (Wright et al. 2016) and Winaywayna, among others, found throughout Peru and mainly of Inka origin (Villafana 1986) with the exception of the Cumbemayo Canal thought to have been constructed about ~1300 BC (but

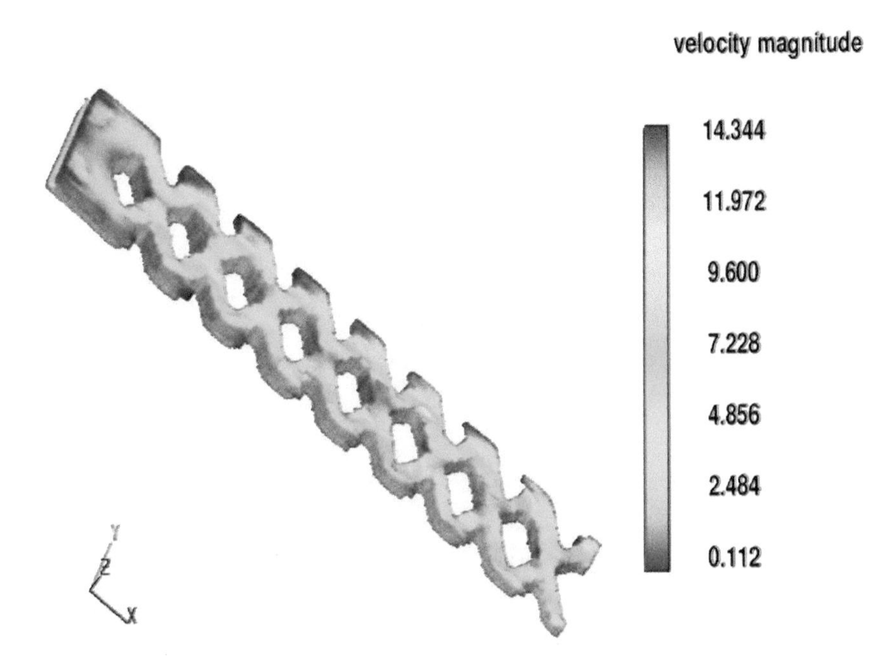

Figure 1.12 CFD results for surface velocity magnitude from the Figures 1.6 and 1.7 FLOW-3D model. Note that near constant surface velocity denotes near complete mixing of water streams originating from different inlet channels.

dating is conjectural). The elaborate Inka site of Moray (Wright et al. 2011; Figure 1.10) is now proven to have a ceremonial function and, as such, is now classified as a later version of ceremonial canals. A highly elaborate class of *pacchas* is found in carved stone deity monuments, such as the Lanzon monument at Chavin, where an interior water passageway integral to the deity monument exists. Given the religious practices that intend communication with nature gods, many *paccha* and *huaca* types include zoomorphic and anthropomorphic forms native to the Andean world to oversee the passage of water through the water channels carved into *pacchas*. Water itself was regarded as the messenger and conveyor of prayers to higher spiritual domains and figures in many of the Andean creation myths. The surface flow patterns produced in convoluted sinusoidal *paccha* channels by different flow conditions would confirm to ritual participants that a hidden life existed within the water that could be interpreted as animal spirits coming forward and confirming the presence of animal messengers within the flowing water. As an example, when snake-like undulations were manifest in surface water patterns in *paccha* channels, then to observers the water is exhibiting the animistic soul characteristics of a snake messenger and thus deity communication is confirmed by messenger spirits within the flowing water in the *paccha* channel (Figures 1.4 and 1.9 for example). Likewise, for different animal messengers represented on

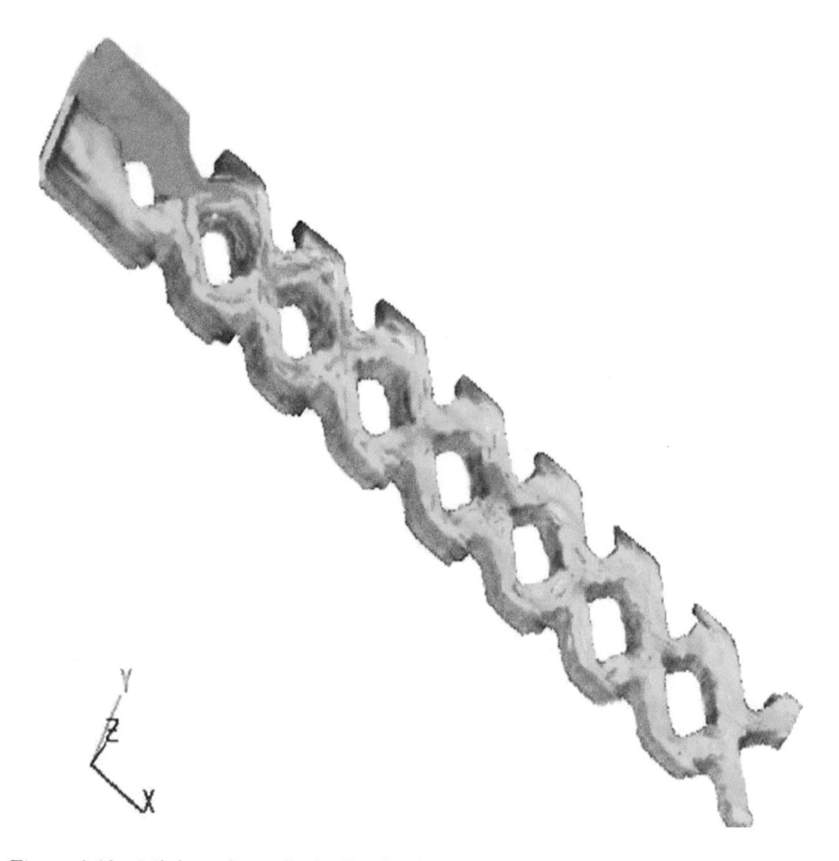

Figure 1.12a Mixing of two dissimilar fluid streams illustrating the *tinquay* concept.

pacchas (Doig 1973:332–333), when water flow patterns imitate some characteristic trait of an animal, the communication would be verified as having been successful. Water flow patterns of this type can be demonstrated to have merit through recreation of these patterns within *pacchas* by Computational Fluid Dynamics (CFD) methods to produce a deeper understanding of the importance of *pacchas* as communication instruments using zoomorphic and mythic animal representations. Here CFD methods permit calculation of flow patterns in models of ritual devices from the basic equations governing fluid flow (Flow Science 2018). Given the multiplicity of sinusoidal and intersecting channel arrangements (Figure 1.7 for example), and the differences that water depth (D) and inclination angle (S) may have on water surface flow patterns, the task of relating these patterns to animal characteristics is next investigated.

A further class of water-related rituals are centered on *wak'as* – mythical human and subhuman deities (Weismantel 2018; Bray 2015a, 2015b) that interact with societies providing social behavioral oversight according to its moods –here an

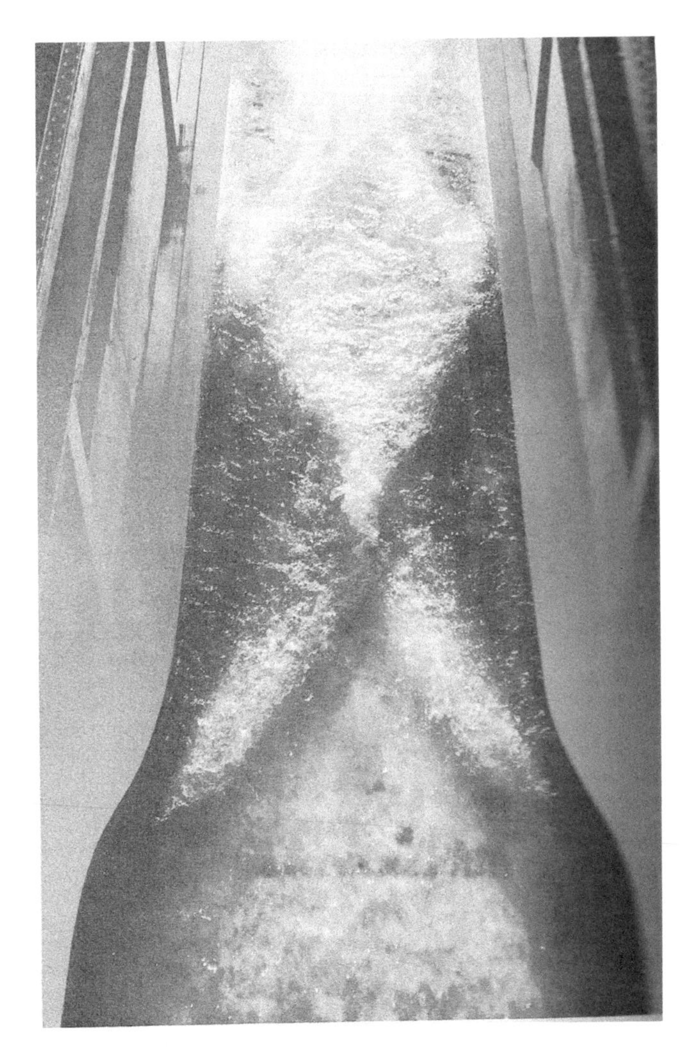

Figure 1.13 Channel contraction inducing supercritical (Fr >> 1) downstream flow showing consecutive reflecting waves off channel walls.

El Niño flood event is an expression of dissatisfaction with destructive societal development proclivities while periods of productive farming, absence of external and internal conflict and cooperation between elements of a society are rewards from the overseeing deity. Ritual devices such as *pacchas* have many controlling deities that represent water physical phenomena given the personification of nature's patterns.

Use of Computational Fluid Dynamics (CFD) is made to compute surface flow patterns observed in convoluted *paccha* channels. This permits the observance of flow patterns observed in ancient rituals and provides a connection to the intention and use of different *paccha* types shown in Figure 1.2. A CFD model (Figure 1.7) is made of a typical *paccha* device (Figure 1.2) involving sinuous, stream-crossing, water flow paths. The computation of the flow pattern then duplicates the flow pattern made by flowing (real) water in convoluted *paccha* channels as would be observed in rituals and ceremonies. Use of CFD in this manner provides the route to bring forward flow patterns that support anthropomorphic beliefs that animal spirits reside in flowing water. For example, computed surface water patterns in channels (Figures 1.10, 1.12) demonstrate undulatory or standing surface wave motion that mimics the undulating motion of snakes thought to be messengers of the gods – here anthropomorphism to ancient observers was verified through water patterns that displayed an internal spirit living within moving water. The rock carving shown in Figure 1.4 emphasizes the belief that multiple snake representations associated with water flow patterns in channels reside in water motion. Figure 1.10 shows an undulatory surface wave pattern such as would exist in critical flow (unit Froude number flow) in a rectangular channel as Figure 1.11 indicates. In a *paccha* device with an upstream reservoir (Figure 1.2) and a straight channel, this flow pattern is achieved simply by varying the inclination of the *paccha* device to a critical slope that brings forward the surface wave patterns. The sudden appearance of the undulatory surface flow pattern in the *paccha* would have the dramatic effect of bringing forward an animated life spirit that suddenly appears in the water surface flow pattern. Figure 1.4 verifies this association given the presence of reptiles and snakes associated with water flow patterns. The appearance of undulatory surface flows correspond to the highest flow rate per unit width in a device that can be achieved; this connection has profound significance in the design of channels as the appearance of snake-like surface waves corresponds to the maximum flow rate that a channel can achieve given its cross-sectional geometry. In terms of modern hydraulic theory, this critical flow condition is associated with unit Froude number given as ($Fr = 1$) (Morris and Wiggert 1972) where the Froude number (Fr) is defined as $Fr = V/(g\ D)^{1/2}$, where V is water velocity, g the gravitational constant (32.2 ft/sec^2) and D the hydraulic depth defined as the ratio of the wetted canal cross-section area divided by the wetted perimeter. This modern nondimensional term (Fr) is used repeatedly in subsequent discussions to bring forward its physical significance in describing channel flow physics.

With respect to flow patterns computed for the Figure 1.7 *paccha*, Figure 1.12 CFD results reveal that the surface velocity of intersecting water streams is about constant. This indicates that dual stream separation and reconnection produces a complete mixing of similar fluid streams indicative of *tinquay* combined reciprocity of giving and receiving to earth spirits for the benefit of society. Figure 1.12a shows the *tinquay* mixing of two fluid entry streams in the Figure 1.7 *paccha*; the different fluid types are denoted by orange and blue streams of different viscosities, densities and flow rates where the orange stream has a lower flow rate than the blue stream. The forced mixing of the streams caused by the intersecting fluid paths is evident and likely represents the combining of different social groups or classes into a single collective society as represented by the totally mixed fluid appearing at the outlet location. This indicates a shared common goal and beliefs and values, as represented by the *paccha* flow pattern, necessary for societal cohesion and continuity. Other interpretations related to the mixing of different fluid streams may arise from the joining of human and revered deities united in common purpose to promote the welfare of a society. Additional use of wooden *pacchas* at the Inka site of Pachacamac (Eeckhout 2019) are consistent with the observations of the early Spanish conquistador Francisco de Jerez who observed "in all streets of this city and the main gates, and around the temple of Pacacamac, there were many wooden idols," this observation is consistent with and verifies the use of *pacchas* in religious ceremonies involving communication with ancestors and praying for water and fertility to maintain the economic basis of Inka society. Similarly, wood and ceramic *paccha* idols are found at Wari, Moche and Chimu sites conducive of their continued religious use from earlier to later periods of Andean societies.

Additional hydraulic flume channel photographs (Henderson 1966:244; Ven Te Chow 1959:468) shown in Figures 1.9 and 1.13 show additional hydraulic water surface displays that can be reproduced by *paccha* devices that show a channel width reduction that transforms subcritical flow ($Fr < 1$) in the wider channel to supercritical ($Fr < 1$) flow in the contracted width channel. The supercritical flow shown in Figure 1.9 produces surface ripple patterns associated with near critical flows; the flow pattern shown in Figure 1.13 indicates V-shaped wave patterns that continue downstream. The difference in these patterns is due to the input Froude number in the wide channel and the contraction ratio of this width to the downstream contracted width. As observed, different flow patterns emerge in downstream contracted and expanded width channel regions that originate from upstream expanded and contracted width channel sections. As what results from channel width change depends upon the Froude number classification (sub- or supercritical) of the flow in the upstream channel section to decide the flow pattern and Froude number in the downstream channel section, clearly some degree of complexity is involved. Figure 1.17 (Bakhmeteff 1932) presents results derived from the Euler continuity and momentum equations to help resolution of this issue – here an input Froude number (1) term represents conditions in the upstream channel section and the output Froude number term (2) the conditions in the downstream channel section. The w2/w1 ratio curves represent contraction ($w2/w1 < 1$) and expansion ($w2/w1 > 1$) in channel widths given the input Froude number (1) value to determine the output Froude number term (2). Although somewhat complicated using modern hydraulic theory, all results related to channel expansion/contraction effects can be obtained by carefully controlled experimental observation. In a later Chapter 6 section related to the Inka site of Tipon, some form of test observations related to channel expansion/contraction effects underlie the design and function of the waterfall section albeit in an Inka methodology and format yet to be discovered.

One conclusion is that there must have been an observation and awareness of the many patterns of water behavior in nature that could be modified and used for the design of canal irrigation systems necessary for agriculture. As lessons of beneficial water behavior patterns were extracted and subsequently applied to optimize water delivery to agricultural field systems through optimized water delivery systems, the celebration of this achievement was represented in some fashion in *pacchas*. The *pacchas* were then used as representative flow duplication devices that represented in some fashion the beneficial flow patterns used for canal irrigation designs. The religious aspects celebrated by *paccha* use related to water deities that brought forward into use and continue to oversee canal irrigation use as a vital component population food supply. The ritual *paccha* devices acknowledged these connections vital for the agricultural sustainability of their societies. *Paccha* devices thus reproduce and demonstrate hydraulic phenomena useful for canal designs used for canal water transport used for irrigation agriculture; this connection appears in celebratory rituals that acknowledge its importance for the successful continuance of a society.

Irrigation canal designs mirror hydraulic phenomena observed from *paccha* flow patterns

For a *paccha* incorporating only a straight channel (Figure 1.3), water flows from a reservoir bowl through the channel then exits in free overfall. If the *paccha* is held at different angles, then different fluid surface patterns occur. If, for example, in terms of modern hydraulic science notation, the *paccha* channel is maintained at a mild or near horizontal slope (slope $S < S_c$, S_c the critical slope), then the flow is uniformly laminar past an entry transition zone. A further declination slope increase is next made so that the slope approaches the critical slope critical depth D_c. Here the Froude number is close to unity for this slope (Morris and Wiggert 1972; Ven Te Chow 1959; Woodward and Posey 1941) and surface wave undulations appear. If the *paccha* declination slope is advanced to a yet steeper angle ($S > S_c$), then the flow depth within the *paccha* channel approaches the asymptotic normal depth D_n on an S-2 profile if $D > D_n$ and on an S-3 profile if $D < D_n$. For flow angle $S = S_c$, then a major change in surface flow pattern occurs as the angle for critical flow is approached; here the Froude number approaches unity and $D_n = D_c$. Typical unstable flow patterns then exist, as shown in Figure 1.11 for critical flow conditions. Thus simple inclination angle changes of a *paccha* channel can bring forward vastly different surface flow patterns indicative of useful hydraulic applications in canal designs to control water flows. The central figure in Figure 1.11 indicates a sinusoidal depth variation pattern within the channel that resembles undulations typical of a moving snake or amphibian as suggested from Figure 1.3 representations. The ancients would, of course, without knowledge of modern hydraulic principles in this detail, observe that angle changes of the straight channeled *paccha* led to a sudden animation of the water flow surface pattern at a particular inclination slope (critical slope S_c) as if an internal spirit within the water became alive. The unit Froude number ($Fr = 1$, $S = S_c$ = critical slope) case corresponds to the *maximum flow rate* in the channel achievable for a given water reservoir head per unit width of the channel – a fact of vital importance that must have been noticed by ancient engineers designing water channels to transport the maximum amount of water to field systems. The maximum flow rate could be easily observed from the rapid emptying of a supply water bowl when the *paccha* channel was set at a given slope; slopes more than or less than this (critical) slope would have longer emptying times. If a rod were to be inserted into the critical flow water surface, it would be observed that no disturbance pattern is created around the rod; this simple observation yields a criterion consistent with critical flow. Ancient engineers could simply utilize this observation to design a transport water channel slope to achieve the channel's maximum flow rate. These criteria would prove useful to design irrigation channels that could transport water at the highest possible rate. If a V-shaped wave originated behind the rod with a half angle θ, then according to modern hydraulic science, $\theta = \sin^{-1}(1/Fr)$ and the flow would be supercritical with $Fr > 1$. For no apparent upstream influence on the water surface caused by the rod, then flow would be subcritical, with $Fr < 1$. For either $Fr > 1$ or $Fr < 1$, the *paccha* reservoir would empty slower than that occurring for $Fr = 1$ critical flow. Such indications of different surface flow patterns related to the declination angle of a *paccha* would certainly be noticed by ancient engineers as the Fr number of the flow determines not only a maximum flow rate condition but also how a channel width expansion (or width contraction) drastically alters the exit flow velocity due to the change in channel width geometry under different Froude number conditions. The *paccha* surface patterns and reservoir emptying rate had a practical use for canal design engineering practice. It may be speculated that larger versions of ritual *pacchas* existed for test purposes vital for irrigation channel design although no such devices (or experimental facilities) have thus far been identified. For sinusoidal *paccha* channels (Figure 1.2), similar water surface instabilities exist as for straight channel types for unit Froude number but now internal water flow patterns contain rotational effects that modify the oscillating surface disturbances. Again, for *pacchas* that have multiple intersecting and dividing channels (Figures 1.2 and 1.7) reference to the Aymara concept of *tinkuy* likely had relevance. Here the concept of intersecting complementary halves (channels) coming together to create a combined flow (for example, intersecting streams to form a river, man and woman to produce progeny, rain and soil to produce crops, leaders and community to produce a harmonious society, ritual warfare among Inka moieties to promote the spirit of vigorous group rivalry similar to the competitive spirit of a sports event) may have representation in the comingling fluid streams of this *paccha* category. Figure 1.12a shows a computation of flow in a *paccha* of this type – the different property streams originating from separate channels are seen to coalesce at intersecting junction points, separate once again and coalesce again at a series of intersecting nodal junctions. At each nodal junction, flow from the two separate inlets is mixed and proceeds downstream to further mix at downstream cross-over junctions. At the exit, streams of different fluids starting from separate inlets are totally mixed together and indistinguishable from each other – an indication that cooperation rather than rivalry is the path to a better society. This concept is later explored by Complexity Theory (Chapter 1) and Similitude Theory (Chapter 2) to demonstrate how societal cooperation has economic benefits to promote the advancement of a society to higher levels of sustainability. The multiplicity of intersecting fluid streams at junction nodes therefore reinforces the number of positive, complementary joining outcomes induced by use of this type of *paccha* as applied to a specific ritual ceremony. Accompanied by carved and painted animal forms on the *paccha*, the communication of positives by zoomorphic messengers is promoted.

Water channel geometry variations underlie development of hydraulic science used to optimize agricultural field system designs

For the Cumbemayo Canal located near the highland Peru city of Cajamarca, Figures 1.14, 1.15 and 1.16 detail complex channel shapes characterized by sequential right angle channel path changes and curved turns.

A further feature of this canal observed by the author in exploration of this canal is the penetration of a solid stone outcrop ~5 m long by a channel with a carved, smooth, interior surface of rectangular cross section with dimensions of ~0.8 m in width and ~0.5 m in height indicating the ability to excavate hard stone through a narrow passageway. From personal exploration of the upper reaches of the Cumbemayo channel, some of the canal extensions are rectangular flow passageways cut through stone outcrops with widths of ~0.5 m and heights of ~0.5 m over several meters in length. The internal walls of these passageways are smooth. It remains a mystery as to how such channels were cut through hard stone given that the small size of rectangular openings hardly permits a worker to perform the stone-cutting and internal wall-smoothing tasks.

While current reporting of this channel system characterizes localized channel geometry changes as "resistance elements" to slow channel water velocity, the elaborate nature of the many channel geometric variations associated with petroglyph inscribed rock faces, temple structures and elaborate burial areas suggests associated ceremonial reasons for the convoluted channel geometry sections other than for agricultural purposes. As this channel still supports water flow despite its ancient origin (the channel is carved through rock ensuring its permanence), examination of the hydraulic behavior induced by the convoluted channel geometry sections indicates that flow is largely subcritical ($Fr < 1$) with occasional supercritical ($Fr > 1$) flow regions induced by channel contractions.

The presence of a single (Figure 1.14) or multiple (Figure 1.16) right angle channel turns induce vortex motion in bend corners that have the effect of narrowing the streamtube flow width thus increasing local velocity in the streamtube through the channel. This results in a flow pattern equivalent to flow in a sinusoidally curved channel.

While some localized surface undulations and wave structures appear throughout convoluted channel sections due to induced supercritical flow (as the flow path is effectively narrowed from the effect of induced corner vortices), some similarity to surface flow patterns observed in handheld and carved rock sinuous channel designs (Figures 1.1 to 1.6) occur. Given the similarity of convoluted

Figure 1.14 Cumbemayo channel showing sequential right angle and curved channel geometry; post-rainy season flow in channel appears low subcritical ($Fr \ll 1$) so flow resistance is negligible – only at higher flow rates during rainy season is low resistance capable of lowering the flow velocity.

channel flow paths in early Formative Period times, as seen in the Cumbemayo channel, and the continued use of convoluted *paccha* channels in rituals (Figures 1.1 and 1.2) throughout later period times, some ritual codification and connection of ritual beliefs involving water flows in convoluted channels and

Figure 1.15 Cumbemayo rectangular channel constriction.

pacchas is apparent. The continuous use of convoluted water channel geometries for religious and cultic ritual purposes in *paccha* form was apparently passed along through the centuries with specific embellishments provided by different societies to conform to their religious belief patterns. As for the Cumbemayo channel system, although the convoluted channel shapes induce flow resistance, this effect is minor for subcritical flows. Given the association of the ritual use of curved channel *pacchas* and the association with water patterns associated with convoluted channel geometries, clearly some association between practical use and ritual devices is apparent.

It is noted that although water phenomena induced by channel shaping is discussed with relation to religious and ritual observations and practice, the Cumbemayo channel may be an early indication of hydraulic engineering practice through use of channel cross-section shaping change along channel lengths. The use of

Figure 1.16 Cumbemayo right angle channel constriction.

channel cross-sectional shaping change to control the Froude number between $0.8 < Fr < 1.2$ to produce stable flows is well documented in the later Late Intermediate Period Chimu Intervalley Canal (Ortloff 2009) as well as in the Late Horizon Inka Tipon Principal Waterfall system described in Chapter 6). Froude numbers slightly above or less than $Fr = 1$ ensure stable flow and flow rates close to the maximum value associated with unstable $Fr = 1$ flows. Early origins of open channel flow hydraulic technology may have a *paccha* device memory path to later stages of use in Andean societies. In summary, the Cumbemayo channel may be an early ceremonial *huaca* with later precedents shown in later period (Inka) rock carvings shown in Figures 1.3, 1.5 and 1.6. A speculative reason for this channel's existence may lie in its capability to observe flows in convoluted channel shapes as a source of experimental data later to be used in both irrigation system design as well as ceremonial display water devices. Certainly later developments of the Chimu using shaped channel cross sections to control water flows (Ortloff 2009) must have had some original inspiration from earlier work although this technical transfer path is presently little explored in the current literature.

From this discussion, several uses of the *paccha* are suggested that tie together its ritual and practical usages based upon similarities in fluid motions observed in channels and paccha ritual devices. For ritual use, the presence of zoomorphic animated motions typical of messenger spirits and/or fluid mechanical analog representations of social principles and catechisms basic to societal continuity are apparent. For practical use, observation of *paccha* flow patterns as dependent upon declination angle, cross-section geometry, convoluted zigzag channel shaping and flow rate are transferrable to irrigation canal designs. While only a few classes of *paccha* designs are discussed, the multiplicity of designs clearly suggest that water and its passage through complex passageways through and over stone, ceramic and wooden ritual objects were expressions of the importance of water control; this parity between practical and ceremonial use is seen by the ancients as a blessing that deities bestow that grant the gift of life.

The *paccha/huaca* concept: ritual and practical usage

Many ceremonial and ritual usages of *paccha* devices and antecedent convoluted channel geometries are apparent from the prior discussion. The effects on surface water flow patterns from channel geometry changes vital for ceremonial purposes may have originated ideas that a practical use of channel geometry changes to control water flow characteristics had practical application for water transport control vital for canal irrigation agriculture. From an observational point of view of the archaeological record, irrigation canals of different slopes and cross-section geometries existed, which must have proceeded from ancient water engineer's designs and calculations given the water requirements of agricultural field areas, different crop types and the periodic water supply times for irrigation as well as the intermittent river source providing water to field systems. Such variations on canal geometry may have proceeded from trial *paccha* flow devices and experiments. As no recognizable Andean water flow laboratories permitting investigation of open channel flow characteristics have been identified in the archaeological record thus far, it may be permitted to say that the source of hydraulic knowledge comes from some observational form of flow trials proceeding from trial canal geometry variations that may have originated from variations of *paccha* counterparts. Of course, it is well known that chance discoveries occur from otherwise well-planned research efforts – perhaps this is best said by Shakespeare in the *Tempest* as "by accident most strange can come bountiful fortune." Given the importance of water engineering to agriculture, and the necessity for irrigation for water-dependent crops such as maize and cotton in dry Peruvian coastal areas, sophisticated hydraulic techniques were developed out of necessity to sustain life and likely celebrated in rituals that guaranteed connection to the favor of deities. While acknowledgment of divine association characterized many Andean societies through complex ceremonies and rituals involving ritual *paccha* use that established a ritual hierarchy as part of government, the demands of increasing population size and secure food supply generated the need for higher levels of hydraulic technology to bring water from distant sources to complement local supplies. For coastal irrigation based on canals emanating from intermittent valley rivers dependent upon mountain runoff and infiltration, water supplies were seasonally variable; the need to transport water from larger rivers in adjacent valleys through lengthy intervalley canals required a quantum jump in hydraulic technology from canals derived from intravalley river sources. As canal cross-section geometry and canal slope play a central role in irrigation water control systems and knowledge of how to transfer water efficiently over long distances to bring new water to desiccated fields in adjacent valleys was a survival option during extended drought conditions, sophisticated hydraulic engineering must be in place to accomplish design goals. As before, ritual *paccha* devices not only celebrate but encode indications of the level of hydraulic engineer knowledge. As an example of the engineering knowledge base available to hydraulic engineers in Late Intermediate Period times, an example case is next presented originated by Chimu hydraulic engineers.

From examination of a section of the Chimu Intervalley Canal it is apparent that the cross-section geometry of the ~75-km-long transport canal varies substantially along its length. Here detailed analysis of the effect of lengthwise canal cross-section geometry changes (Ortloff et al. 1985; Ortloff 2009) indicate that the Froude number (Fr) is maintained between $0.8 < Fr < 1.2$. This effect, as practiced in modern transport canal design, permits the maximum flow rate per unit width of canal to occur for a given head while limiting the occurrence of surface flow instabilities characteristic of unit Froude number ($Fr = 1$). Further, the canal design makes a specific flow rate the maximum flow rate with this flow rate designed to activate a supply canal in an adjacent valley to reinstitute agriculture. Instabilities associated with $Fr = 1$ conditions are to be avoided in channel designs due to accelerating erosion effects on unlined channels due to agitated flows enhanced from turbulence and transient undulatory flow effects. Here the maximum flow rate in the LIP Chimu Intervalley Canal coincides by design with the acceptance flow rate into the main feeder canal in the Moche Valley (the Vinchansao Canal) to resupply the desiccated agricultural field systems surrounding Chan Chan due to extensive 10th- to 11th-century drought. Essentially, the hydraulic knowledge transfer from observation of channel geometry changes encoded in *paccha* rituals serves as the basis to regulate flow instabilities. As open channel flow determination is affected by channel slope, channel wall roughness, channel cross-section geometry and, most importantly, the different effects of inlet subcritical ($Fr < 1$) and supercritical ($Fr > 1$) flows on channel contraction and expansion on outflow velocity and Froude number, obviously more observational detail had been added in later years to refine open channel flow science as practiced by the Chimu. While refinement of open channel water control technology proceeded by codified observations over the centuries and technical transfer between different societies, the Chimu must be given credit for technical innovations particular to their extensive intervalley water transfer project.

Paths to technical knowledge: Andean and western society similarities

Similar to this technical evolution path in the Andean world, in western science Christian beliefs manifest in the Enlightenment Period (Hampson 1968; Outram 1995; Sloan 2003) focused on the perfection of a supreme deity to perform nature's tasks in an optimum manner with the least amount of energy involved. Translating this religious belief into practical science, mathematical theorems related to minimizing the energy expenditure in physical processes flourished in the Enlightenment period as they provided a practical mathematical path toward engineering solutions to practical problems. One such development is captured by Hamilton's Theorem which postulates minimization of mathematical expressions of kinetic and potential energy involved in solid body dynamics problems; other applications codified in Castigliano's Minimum Strain Energy formalism used in structural analysis problems derives from the notion that incorporates a supreme deity's optimum, least energy expenditure way of performing nature's processes and translates this idea into mathematical formalisms that provided answers to practical engineering problems. These rules were put into calculus format then available from 17th century innovators (primarily Newton and Liebnitz) and are in use in modern engineering practice. Here their celestial origins somewhat mimic the connection between what ancient Andean societies accomplished by their observations and codification of hydraulic phenomena as having divine origins. While European societies had the advantage of sophisticated mathematics and sponsored research in water science, it appears that Andean societies accomplished complex water management tasks solely on a codified observational basis. As experimental work serves this end in both ancient and modern science, Andean water engineers appear to have mastered the tasks given them with equal technical knowledge albeit in a format as yet undiscovered by modern researchers. Here what a supreme deity uses to construct the universe's physics underlies the source of western science – this brought about through experimental observations or theoretical investigations. In the Andean world, manifestations of water movement in convoluted water channels caused by underlying spirit motions and manifestations of animal behavior served as enlightenment paths to bring forward water flow patterns by manipulating shapes of water channel paths to understand water movement patterns. In this manner, a water control science was originated that ultimately translated and transformed into a practical irrigation science as religious practices transform into secular uses. Here early western and Andean science appear to share similar origins and paths based upon initial observations of physical phenomena controlled by deities with rituals and ceremonies performed to acknowledge the presence, favor and protection of deities. This path is followed by ways to control physical phenomena and then followed by ways to codify physical phenomena in a predictive manner. In western science this is done by mathematical formalisms; in Andean science by predictive formalisms not yet fully understood but hinted at by developments discussed in subsequent chapters.

The Andean worldview (Quilter 2014:47–48) incorporates complementary forces that cause movement characterized by the concept of *ayni*. This is expressed as reciprocal interchanges between competing and cooperative forces with reciprocal interchanges that govern how the universe works. Thus forces that are larger and more forceful than others and out of equilibrium balance cause movement and make events happen. Divisions of water from one source stream into two streams that separate and combine are important to represent in ritual device paths as exemplified by Figures 1.2, 1.3, 1.6, 1.7 and 1.12. For Figure 1.12a, two different fluid types that combine through multiple separation and reuniting paths in a *paccha* clearly represent two forces can combine to produce something different from each. The *quecha* concept of *camay* represents life forces associated with water; for example, water in mountainous highlands is linked to ocean water through return cycles of rain and snow from clouds. In ritual *paccha* devices such as those shown in Figures 1.3, 1.4 and 1.5, such downhill sloped return paths of collected water led through zigzag channels can be interpreted as elements in this cyclical waterr return path. The dual pairing of forces may be *sami*, or light forces or *hucha* as dense, heavy and sporadic forces; such forces reside in an upper world of celestial beings (sun, moon, ancestors) while *kay pacha* is the middle world where humans live and *ukhu pacha* is the underworld dissimilar in all ways to the structured forces of the former categories. As the condor, puma and snake are the animal representatives of these cosmological worlds in descending order, Figure 1.4 in some way associates and unites these concepts in a ritual *paccha*, or as a *huaca*. While priests and shamans perform rituals to maintain a balance between cosmological forces at all levels, illness, political instability and climate/weather disruptions are the consequence of imbalances between heavy and light forces. Ritual devices represent cosmological beliefs and their use in ceremonies are but a part of shamanistic practice to maintain balance and unison between competing unstable forces as Figure 1.1 demonstrates.

From a broader perspective, Comte's positivism (Lepenieas 2010, Levy-Bruhl 1899) the thesis that, in those fields of knowledge where humankind has been able to make progress, a transition from an initial stage (theological) in which the explanation of observed physical phenomena was the causal action of supernatural entities was followed by a successive stage (metaphysical) in which an explanation was found in fundamental principles of reality (such as the principle of causality) to interpret the rational order of the world), to a third and final stage (positive) in which scientists produce an accurate description of physical phenomena regularities that are in the form of natural laws. To Comte, the universe existed to give lessons to mankind (Cohen 1977). Once scientific rigor in astronomy, physics, chemistry and physiology was established and verifiable, belief in the principles established in these sciences is undeniable. This historical understanding of the course of human knowledge was achieved from one stage to the subsequent stage as the result of a struggle that is never finished as theological and metaphysical efforts of interpreting reality continue to occur in societies when leading edge science uncovers new, unexplained phenomena that lends itself to interpretation as having a theological source (Laudan 1981; Lepenieas 1988; Levy-Bruhl 1899; Macherey 1989). Even as science advances to provide rational answers it is important to continue the dialog between religion and metaphysics as the mysteries answerable initially only from a religious perspective (theological) provide topics that open investigative inquiries (metaphysics) seeking guiding physical principles that sequentially progress to formal verifiable scientific answers (positive) in a final stage. A different perspective is offered by Toynbee (1972:349) for which concepts of "coming of age" of a society is improper as aging and maturing are the experiences of organisms and neither a species nor a society is a set of organisms but rather a set of relations; this thought comes closer to Routledge's (2013:1, 157) "concern from what the state was to what the state does." Contrary to neo-evolutionary thoughts on state formation, Routledge (2013) poses the question of how states arise, continue in progress and guide political life by asking "if states are not things, but the effect of practices, discourses and dispositions, how and why do these effects generate an apparent collective agent or instrument?" The discussion to follow in later chapters advances scientific progress as a catalyst toward societal complexity and progress as evidenced by the historical development of western societies and, to a certain degree, the progress of Andean societies that depend heavily on a stable economic base generated by increases in agricultural productivity. While modern 21st-century societies continue to advance the development of science and engineering knowledge derived from induction based observation and experiment as the source of future progress (as formulated in Bacon's 15th-century *Novum Organum* part of his *Magna Instauratio*), further scientific progress laid in the pre-Enlightenment and Industrial Revolution periods treat nature as an empirical source that underlies theories of its behavior (Shapin 1996; Jardine 1999). The Andean world had its survival rooted in understanding nature's ways and accommodating its cities and agricultural systems to what nature's lessons had provided to exploit. As such, while science in western societies developed formulaic recipes to encode natural phenomena, Andean societies remained closer to observing and expanding upon what nature provided as lessons to improve the sustainability of its cities and agricultural systems. The persistent use of ritual objects, ceremonies and rituals integrated with technical applications of nature's lessons. In the Andean world viewing *paccha* examples as elements of theology with the later branching of hydraulic observations into advanced practical use in open channel hydraulics practiced by the Chimu in Intervalley Canal construction, the final Comte stage (positive) appears as hydraulic and hydrological science. A dual reliance on religious aspects of water behavior through rituals and ceremonies celebrated deeper learning of hydraulic technology as more was revealed by nature observation and put into practical use in urban and agricultural settings. Later chapters indicate that application of hydraulic principles advanced to optimum field system designs that maximized food production verifying that an active agricultural science was present in several of the coastal and highland Andean societies.

Hydraulic societies revisited

Unlike well-researched ancient far eastern societies dominating large continental areas where a supreme ruler and his selected council advisors determined societal norms and practices, the diversity of Andean South American societies occupying different isolated ecological niches in different time periods using different agricultural systems prevents a unified generalized verifiable conclusion regarding the connection between state control, the political economy and the relation of management structure of water systems for urban, agricultural and ritual use. While it is clear that successful design and management of urban and agricultural water systems underlie and provide the foundation for a stable society, it is unclear if such systems were initially generated by elite, top-down management or evolved from bottom-up origins (or a mixture of both). At later societal stages that involve organization of vast labor resources to implement large-scale irrigation and water supply projects, it is apparent that centralized control originating from elite control centers was in place to provide project engineering resources, project design specification, labor resources, logistical support and project management and oversight to implement major water supply projects. Although knowledge of an elite societal structure performing and guiding these tasks is assumed from the presence of segregated royal and administrative compounds isolated from secular housing and workshop areas, what happens within these elite structures as to the process of originating and managing water projects is somewhat conjectural given lack of archaeological knowledge of the degree of the subdivision of elite control between technical projects and religious, ceremonial duties. Here the archaeological literature provides little guidance as to technical management capabilities and support groups augmenting ceremonial tasks and rituals. The water infrastructure projects show that advanced technical achievements were prevalent in many of the Andean societies (as subsequent chapters detail) and, for the example cases examined, show the near optimum deployment of land, water, labor and technology for large-scale projects originating from elite management sources as determined from Similitude Theory (Chapter 2). An advisory council is suggested within governing elite structures able to improve field system designs to maximize comestible output. Additionally, from ritual objects demonstrating religious significance and evidence of an involved ceremonial life as evidenced by architectural evidence in the form of religious pyramid, temple, ritual plaza and ceremonial compounds, a practical administrative branch addressing worldly issues through design and oversight of a successful agricultural system provided the basis for ceremonial activities that acknowledge and celebrate divine intervention that guides daily life. As such, a similarity to modern governmental practice exists where technical advisory groups, both in government, academic, political and commercial spheres are inherent to governance success to conceive and perform major technical projects whose success often is interpreted through societal conditioning and controlled political information implanting in a society susceptible to divine presence oversight and influence. On this basis, many of the Andean world's elite ruling class practices resemble those of successful industrial age modern societies where worldly commercial and social control activities have a religious companion.

The narrative presented thus far is based on observation of water flow patterns that can be induced by channel shaping, either man made or occurring naturally in nature, as a source of learning of how water can be controlled. Early Peruvian preceramic coastal societies utilized water resources to organize an egalitarian societal structure where the benefits of cooperation between different groups involved exploiting marine and terrestrial land use for agriculture. For example, no elite management structures were apparent at Huaca Prieta-Paradones in Early Preceramic times (Dillehay 2017). The nature of this developing existence pattern generated the desire to improve to a more predictable and stable existence pattern by transforming from natures' vagaries controlling human existence to more stable existence controlled by human endeavor. Once this mindset was in place, then control over nature's ways followed with technical improvements that depended upon understanding of hydraulic and hydrologic water control used for agricultural field systems (as Comte's theories would envision). The agricultural resource base could be stabilized, exploited and expanded upon in one area enhancing population growth. As water control was vital for predictable agricultural output that would provide for societal stability and ultimately lead to more complex societal governance patterns, controlled experiments in the manipulation of water flow patterns became vital to improve and optimize food output from given land and water resources to develop irrigation agriculture to its full potential. Water then proved to be a controllable resource as predictable water patterns emerged in a predictable manner when subject to imposed variations in main factors that influenced water flow rates to field systems. These controllable factors include channel cross-section geometry, wall roughness and channel slope effects on water flow rates. What previously had been thought outside of human control was now seen as controllable through vestiges of an emerging water science. Convoluted channel geometries that produced dramatic water pattern effects celebrated the transition from what nature chose to bestow to a controllable resource that could now be utilized to guarantee a stable, sustainable existence through planned agriculture. This transition from uncertainty to certainty likely was interpreted as a reward bestowed from beneficent deities whose rituals with *pacchas* and *huacas* in various forms acknowledged. In this sense, the observation that physical laws govern water flows (in the form of conservation of mass, momentum and energy) that came later in western science had predecessors in the Andean world through observation of flow patterns in early convoluted channels (as the Cumbemayo Canal would appear to indicate). Whereas western science could codify water motion through solution of governing equations, the Andean world likely used observations of water motion as the basis of their empirical science. How such empirical data was recorded, stored and codified remains a research topic given known recording (*quipus*) and calculation (*yupana*) devices that likely played a role on codification. From Ortloff (2009, Figure 1.35), a *yupana* diagram of a solution for a canal surveying problem is presented as an example of codification of a technical problem solution in canal design; overlaying *quipu* strings over the *yupana* columns then could represent the same solution in terms of knot sequences that represent the "dots" in *yupana* boxes. As the existence of Peruvian coastal societies depended upon knowledge of the administration and control of water resources as well as the ability to mobilize and control labor forces capable of maintaining the existing irrigation systems and building new ones

(Ortloff 1993, Ortloff and Moseley 2009, 2009), such codification and scalability of land, water, labor and technology related to water control for irrigation was a vital administrative resource to develop their societies to more advanced levels (Marcus and Stanish 2006). Rostworowski (1999) amends these precepts with two additional features of coastal hydraulic societies. These include local control of valley water sources and canal intakes and freedom from highland domination. Without these additional conditions, it was impossible for ancient Peruvian coastal dwellers to ensure control over their water resources and with it, their autonomy. Thus some measure must be given to coastal-sierra highland rivalries (Rostworowski 1999:214–219) for control and distribution of water resources as an element influencing technical advances in irrigation technology as limited water resources implies use of sophisticated water control techniques to obtain the most value from a limited resource. For example, use of shortest possible length Fr ~ 1 critical flow canals designed to transport water at a maximum flow rate was vital to speed up water distribution times to multiple sections of field systems designed for different crops with different water requirements to limit evaporation loss and ground seepage loss. Given groundwater resources from infiltrated highland rainfall as well as rainfall runoff into coastal rivers delivered to coastal farming areas, use of *hoyas* (sunken fields) were utilized by the coastal Chimu to augment river sourced canals that irrigated fields to increase food production. Thus scarcity of water sources led to engineering efficiency to extract maximum food production per unit arable land area – this in turn resulted from observance of alternates to river sourced canal irrigation systems through observance of a further water resource available from pit excavation to groundwater levels to sustain agriculture. An example lies in a Chimu sequence of pits (*wacheques*) adjacent to their capital city of Chan Chan that are excavated to the groundwater phreatic level used as a supplemental food resource base when intermittent rivers experience a decline in their flow rates. As the water table declines during the dry season (or especially during extended drought), new pits closer to the ocean edge appear to intercept the available water table to sustain agricultural production from this source.

It is of interest to note that in a compacted sandy coastal environment, nature's energy efficiency dictates that a flood event will carve a maximum flow rate channel. This is usually in the form of a wide channel eroded from the sandy soil. While good from a rapid water removal standpoint, this shallow channel cross-section shape is not useful for irrigation purposes which require daughter distribution canal arms to distribute water to different field plots. Here shallow, wide daughter channels would promote rapid evaporation and ground seepage water losses which would not be efficient for large scale agriculture. Previously, early Chimu Canal versions of this type were later replaced by deeper depth, controlled cross-section geometry canal versions indicating attempts to improve irrigation agriculture practice beyond nature's sometimes negative example lessons. Hence the technical challenge to design lined canals to limit seepage loss combined with canal geometry variations that take a prescribed amount of water from a river source to field systems required knowledge of canal slope, cross-section geometry, wall roughness and associated Froude number effects on flow heights and velocity in canals. As an example of the modern counterpart of effects of canal geometry expansion/contraction effects as shown in Figure 1.17 to predict velocity and water height changes as dependent upon the Froude number of the incoming flow, the Chimu version (albeit in some prescientific format) demonstrated and used in a section of the Chimu Intervalley Canal (Ortloff 2009:51) gives indication of the complexity of hydraulic science. Thus some version of Figure 1.17 resided in the Chimu knowledge base to permit the intentional canal cross-section modifications observed in the Intervalley Canal albeit in a format yet to be discovered. While such effects are codified in modern graphic format, it remains a research topic

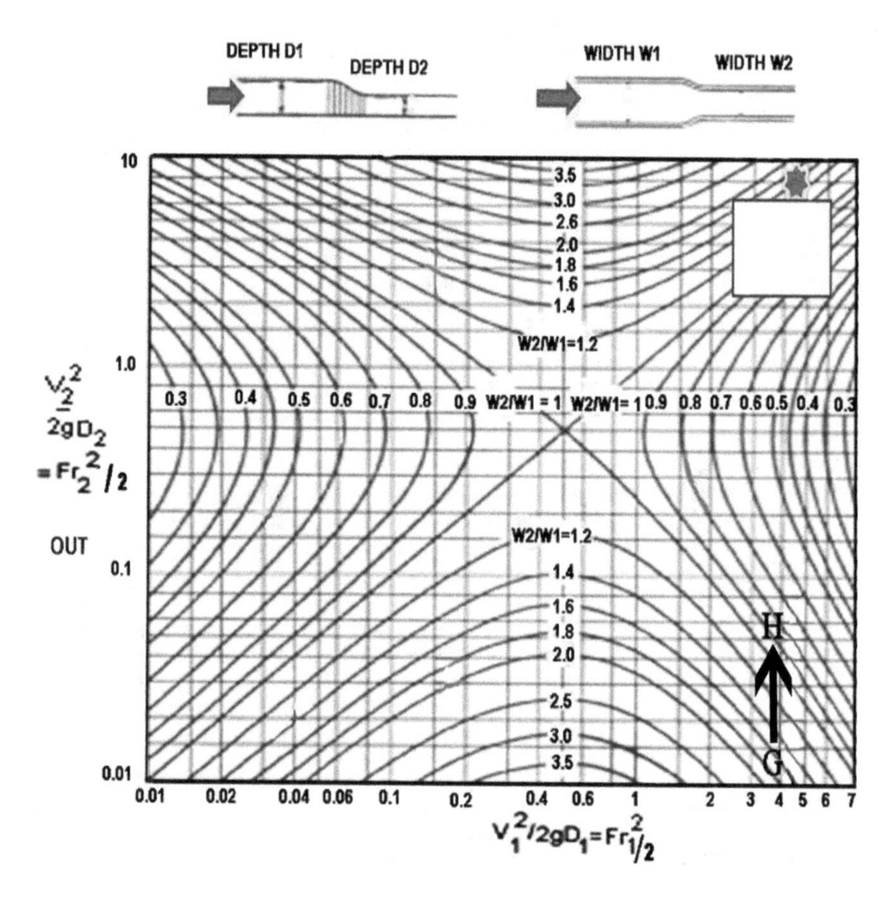

Figure 1.17 Flow diagram indicating the complexity involved in determining channel post expansion (E) or contraction (C) velocity and water height changes given inlet Froude number Fr_1, outlet Froude number Fr_2 and depth changes D1, D2 from a width (W2/W1) change in a channel.

Source: From Woodward and Posey 1941:122

Figure 1.18 Large reservoir in the Supe Valley originating from an *amuna* (highland lake water transfer via geologic fault and/or surface or subsurface channel to valley bottomlands) that supplied high groundwater supporting agriculture for the preceramic site of Caral.

as to how ancient Andean societies recorded observational water phenomena by their *quipu* (or other) recording devices.

Later development stages of Andean hydraulic science

Inka society had the advantage of incorporating the lessons learned from prior societies' hydraulic accomplishments they had contact with as well as unique discoveries derived from specific challenges that their environment posed regarding water availability and landscape complexity. The Tipon Chapter 7 summarizes Inka mastery of hydraulic science principles with several known precedents from

Figure 1.18a Reservoir in the Supe Valley made by excavation into the highwater table.

Figure 1.19 Elevated Plateau and Supe Valley bottomlands – a canal built on a supporting terrace along the side face of the cliff brought water supplies to the Caral and Chupacigaro urban centers on the elevated plateau areas.

earlier societies indicating a possible borrowing of earlier society's discoveries together with independent research and development activity devoted to urban and agricultural water distribution applications and aesthetic displays involving fountains and ceremonial water cascades that had *paccha* counterparts. As an example of Inka hydraulic science in practice and the application of accumulated learning, the Principal Fountain at the site of Tipon is supplied by a spring and several subsidiary channels that capture water flow en route to a four-channel waterfall display (Chapter 6). What is unique about this display is its dependence upon a series of contracting channels upstream of the waterfall. Analysis shows that there is a deliberate plan in this design: flow goes from subcritical ($Fr < 1$) in the first wide channel accepting spring water to critical ($Fr = 1$) and slightly above critical ($Fr > 1$) in the sequentially contracted width downstream channels. What is the purpose behind what Inka hydraulic engineers had in mind with this design? As achievement of an aesthetic waterfall display requires a stable and equal water flow to all four waterfall channels, how does the creation of critical flows by sequential channel width contractions serve this purpose as envisioned by Inka hydraulic engineers? The answer lies in Inka knowledge that critical ($Fr = 1$) and supercritical ($Fr > 1$) flows in channels have no upstream influence on flow in first wide spring supplied channel – therefore no disturbances from downstream flows can propagate upstream to cause instabilities and unsteadiness in the supply flow en route to the waterfall that would disturb its steady flow pattern. While this feature is built into the design, it indicates that the Inka had developed observational knowledge that differences in water velocity V, depth D and some equivalent to the modern Froude number notation of an incoming flow from a wide channel to a contracted width channel can have a vastly different effect on the flow velocity, depth and Froude number in the widened channel. Again, the insertion of a rod into a flow and observation of upstream surface height change resulting from rod insertion ($Fr < 1$) as well as water surface pattern changes from critical ($Fr = 1$) and supercritical flows ($Fr > 1$) can be used to classify different flow regimes as previously discussed. Figure 1.11 illustrates that, depending upon the Froude number of the incoming flow from a wide channel and the contraction ratio to a smaller width channel, surface waves are produced when the Froude number equals unity in the contracted channel section. Further, for a high-speed, low-water height supercritical flow ($Fr > 1$) in a contracted channel, insertion of a rod into the flow does not propagate any surface waves in the upstream direction but only in the downstream direction – here a V-wave pattern develops from the rod apex. The upstream flow before the rod does not see influence of the presence of the rod as a resistance source. This simple demonstration would serve as a test for high Froude number flows that cannot propagate disturbances in the upstream direction. Although described in modern hydraulic engineering parlance, some equivalent Inka nomenclature must have existed to describe Froude number effects albeit in an as yet discovered format. For the Tipon waterfall, use of the channel design indicated in Tipon (Chapter 6) guarantees stable flow to the waterfall to produce the desired aesthetic display. Modern hydraulic theory explains these phenomena but its presence and use within an Inka hydraulic system indicates an advanced state of hydraulic knowledge not previously noted in the history of the hydraulic sciences. While Tipon's hydraulics can be understood and described by use of modern hydraulic engineering methods, observation and recording of effects of channel cross-section variation and streamwise channel expansion/contraction effects on water velocity and depth change by Inka water engineers underlies the application seen at Tipon.

Aspects of water technology in the old and new worlds

In retrospect, it appears that terracotta and lead pipeline flow technology has its origins at Roman, Greek and other Middle Eastern Levant sites given the preponderance of pipeline remains found at their many urban sites used to supply baths, fountains, urban housing, ceremonial structures as well as cisterns and reservoirs. Among the earliest (~800 BC) pipelines are the ones at the site of the Temple of Artemis (Ortloff 2009:248) consisting of ~400-pound lead pipelines connected by stone toroidals. Descriptions of pipeline and inverted siphon systems in early independent Middle East kingdoms, the Turkish interior and south and west coasts and later Roman occupied territories in the Europe and the Levant (Pergamun, Cadiz, Smyrna, Methymna, Alatri, Lugdunum, Ephesus and Segobriga, among many others) are described in the vast literature on this subject (Nikolic 2098). For Andean sites, only Inka Cuzco's pre-colonial and colonial fountains show traces of (pressurized) pipeline use while water supplies to other pre-Late Horizon urban sites derive mainly from aqueduct supplied open channel canals, wells penetrating into the groundwater level and channeled river sources. For dynastic Chinese sites (Chapter 12), hydraulic technology appears to be directed toward controlling rivers to limit flooding, use of offshoot canals from rivers for irrigation agriculture purposes, river transport of agricultural and commercial goods as well as enhancing communication and central control between sites controlled from a central dynastic site. Here early dynastic versions of China's north to south running Grand Canal served to connect the west to east running Yellow and Yangtze Rivers to fully integrate both east-west and north-south water transport and communication to all parts of the country. It appears that pipeline flows were only used within Asian royal and ceremonial structures such as the Beijing Forbidden City. Further examples of short pipeline usage are evident at the Neak Pean site in Angkor (Ortloff 2009:371–373) for ceremonial water displays but not found elsewhere within city precincts. Given the scarcity of water in Middle Eastern sites compared to the abundance of water at Asian sites, it is clear that a high level of hydraulic engineering applied to water control and conservation by sophisticated pipeline systems was vital for urban centers to flourish. On this basis, much experimental work likely provided the foundation and codification of hydraulic engineering principles applied by Roman, Greek and Levant society water engineers and seen in their many water structures.

Of interest are the many words in the Quecha language used by the Inka related to water (Brundage 1967:379–380). Many relate to the hydraulic technology that was prevalent in Late Horizon times such as *pincha*, a water pipe; *rarca*,

Figure 1.20 *Paccha* typical of Figure 2 (LRHS) ceramic types from the Mochesociety: water poured into bowl goes through internal piping to exit at the base of the ceramic figure.

an irrigation ditch; *patqui*, a channel; and *chakan*, a water tank – while other words describe water motion such as *pakcha*, water falling into a basin; *huncolpi*, a water jet; and *yaku*, water flows guided by an internal agency (perhaps anticipating later century concepts of mass, momentum and energy laws governing fluid motion?). While these words were commonplace among the general public given the dependence upon water for agricultural and urban use, there must have been a

Figure 1.21 Ceramic *paccha* with Inka origins; water entry at top of ceramic exits at base.

more complex vocabulary to describe more complex hydraulic phenomena used by Inka hydraulic engineers involved in the design the Tipon water system given the complex technology used in that system.

Toward Complexity Theory views of societal development

The transition from observance of physical phenomena in nature to the creation of explanations for the source of such phenomena has occupied theologians and philosophers over the centuries resulting in theories on the origin of religious beliefs. For the present discussion, as religion and ritual play a vital role in Andean history, the mechanism behind transfer of nature observations to scientific principles is of philosophic interest as the work on noted scholars on this subject reveals.

Friedrich Creuzer (Shalaeva 2014) offered a theory of symbolism in which the myths and religious belief systems revealed that which otherwise would be nearly impossible to express. A physical symbol, often expressed in the form of a natural event or observation from nature, concealed how it was formed or used to convey meaning to people; as such, it was deities who revealed themselves in symbols.

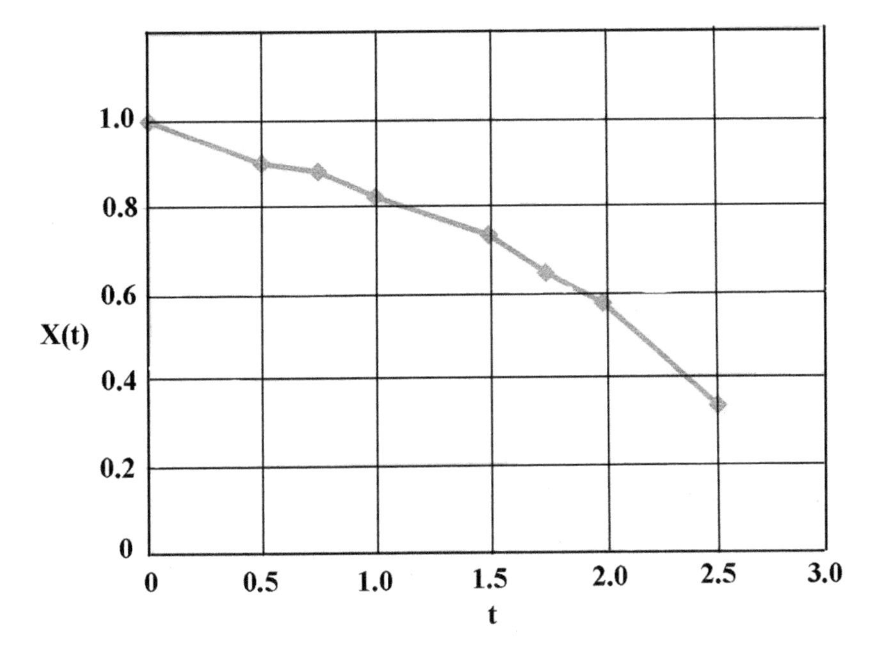

Figure 1.22 Numerical solution of Equation (5): $X_\alpha(t)$ vs. time plot showing decline of $X_\alpha(t)$ over time resulting from extensive drought.

David Hume held that all humans as rational beings would have arrived at a belief in a single powerful being through a contemplation of the size and order of the universe. Out of experiences as to the origins of natural phenomena, polytheistic deities served as explanatory agents. Creuzer adds that gods revealed themselves through symbols – in this case particularly through observation of water phenomena so vital for life and agriculture. Modern theorists (Benyus 1997) reprieve natures' examples as inspiration for solutions to practical problems in engineering. The modern study of *biomimicry* (Kurin 2016) has demonstrated practical applications derived from observation of nature's many efficient solutions. In recent years these efforts have become more intensive, with researchers seeking rules, concepts and principles of biology to inspire new possibilities in materials, mechanisms, algorithms and fabrication processes as well as to provide insight into human societal structure developments that follows biological developments of self-organizing structures.

This discussion expresses ideas to as to origins of deities through observation of the sources of natural phenomena. The transition from observance of water flow patterns in nature as messages from benign deities to realization that such patterns could not only be duplicated through water engineering efforts for practical use in urban and agriculture usage but represented an evolutionary course to replace ecclesiastical explanations with predictable science methods. Similar to modern advances in engineering and science through *biomimicry* where examples of nature's implementation to organize more complex systems from elementary components to prevail against outside challenges and threats to survival, the Andean world appears to have derived similar lessons that a society acting collectively with common beliefs, purpose and goals provides the key to societal advancement. While in nature some form of collective unity prevails to ensure survival of its component members by changes in individual structure and collective organization of its component members (Wieczorek and Wright 2012; Nicolis and Prigogine 1989), similar lessons from nature influenced the societal development of the Andean world through organizational changes that promoted societal continuance and sustainability. Similar to this path is the development of hydraulic and hydrological science and the managerial structures that promote collective unity and survival continuity.

Comte's staged evolutionary progression of societal development leads to transition from polytheistic deities giving symbolic messages through natural phenomena to understanding through a form of science to provide useful tools for advancement of civilizations. Again, this is an early form of *biomimicry* characterizing Andean societies that combine belief in nature as animate and approachable through ritual and ceremonies combined with practical usage of nature's lessons.

Comte's original three stages of societal development (Levy-Bruhl 1899):

(theological) → (metaphysical) → (positive)

as applied to Andean versions of a hydraulic society has a notable variation as follows:

(practical science/engineering applications)
(theological) → (metaphysical)↓ → (positive) ↕
(continued ritual practices to connect and acknowledge beneficent deities as a source of knowledge)

in that ritual beliefs persist along with practical applications of engineering knowledge. The difference is that in modern societies, progressive technical development and science are separate from classical religious belief patterns which concentrate on moral behavior patterns and ecclesiastical mysteries beyond logical explanation. For Andean societies, the development of science/engineering is accompanied by reverence through ritual to the deities that bestowed gifts of knowledge upon worshippers and devotees.

Later century European Enlightenment Period views project a mechanical universe governed by physical laws was opposed by the Christian religious hierarchy condemning this view as heresy. Here the Andean world appears to be respectful of sources of knowledge giving deity connections as the source of knowledge and paying respect to these sources through complex rituals. Objects in *paccha* and *huaca* form demonstrate the duality to reinforce ecclesiastical connections to show appreciation and respect for the deities that gave gifts to mankind. The nature of Andean hydraulic societies then appear to simultaneously incorporate Comte's Three Stage theological, metaphysical and positive phases as ritual life continues simultaneously with practical applications of water technology throughout Andean history. The extension of Comte's chain of history to the present time in western society may yet produce a next phase related to technical achievements creating their own religion. The management of genes to produce genetic advantages, artificial intelligence and robotics to guide decision making and replace human labor, machine-human mind coalescence and subdivision of societies into rulers and subjects based upon economic advantage are but a few considerations that may shape the future. In a world where laws are subjective to certain elements of society and power and control are the aims of a few in positions of power, the vision of Orwell may come to pass as media control by an authoritative government used to form the opinions, actions and noncontestive thoughts in the populace may yet evolve to produce a passive society content with state supplied physical comforts as the achieved final purpose to be delivered by a government.

Returning to the Andean world, as Late Intermediate Period Chimu society manifested societal and administrative control over subjects by elites as evidenced by the eight royal compounds in Chan Chan (Figure 1.23) each of which was devoted to an individual leader and his retinue in perpetuity.

Similarly, the Tiwanaku had a ceremonial core of religious and royal compound structures controlled by aristocratic patrician lineages (Kolata 1992:172–176). In similar fashion, the Inka royalty centered in Cuzco controlled many aspects of the lives of their subjects. The case for elite groups in societies in charge of intelligent

Figure 1.23 Walled compound structures of the Chimu capital city of Chan Chan.

planning using engineering principles to guide their decision making for urban and agricultural field system design and function, given the optimum nature of these designs observed from the archaeological record (Chapter 3), is evident. The elite class dominance of the Chimu, Wari, Tiwanaku and Inka societies thus served as the development agency of large scale applications of water science, as indicated by the many labor intensive water structures in the archaeological record requiring large-scale planning and massive labor input.

Self-organization in hydraulic societies: Complexity Theory

Analysis presented thus far of the urban and agricultural systems of the Inka, Tiwanaku, Wari and Chimu societies reveal complex knowledge of hydraulic and hydrological science put into practice. Celebratory rituals and ceremonies involving *pacchas* and *huacas* of different types emphasize the importance of water to Andean societies for both practical and religious purposes. Discussion thus far has emphasized scientific progress through lessons provided from nature observations translated into practical extensions useful for water control for urban and agricultural settings. While nature serves as a guide in this manner, the social structure of a society requires a mechanism to absorb these lessons and provide a management structure that can translate nature observations into practical use for the betterment of society through advances in agricultural development, urban lifestyle security and hygienic improvements. This involves management of land, water, labor and technology resources in different ecological environments. While ideas about the origins of hydraulic science and related social structure developments have been advanced in the preceding sections and summarized by many authors in the literature (Marcus and Stanish 2006; Stanish 2006:364–366; Kolata 1996a; Koestler 1964; Bartlett 1958), questions arise as to the Andean societal structure that originates, encourages, supports and puts into practice empirical knowledge given the relative isolation of major Andean societies in time and distance from each other. For the north coast Chimu society, although complex technology is demonstrated in canal designs, earlier societies occupying the same areas exhibited lesser levels of technical acumen indicating that the later, more advanced Chimu society required more predictability and certainty about food resource production for their larger population and therefore required a higher level of technical inventiveness to address a more complex problem set than their Moche predecessors. This process included a series of inventive and defensive solutions for water supply and control problems not addressed by earlier societies. A further question arises as to the degree to which observations from nature were translated to practical use to guide and influence technological development of agricultural and urban water control systems so that rise to optimum performance levels. Given the unpredictability of nature's climate and weather variations and their effect on agriculture and societal sustainability, nature as a benevolent guide may seem doubtful as an always reliable path that can yield benefits to human society. Given the limits of translating all of natures' lessons into positive societal outcomes, a further obstacle to societal cooperation to coordinate activities toward a positive, universally accepted goal lies in acceptance by the populace of leadership dictates.

Societal development according to Modern Complexity Theory

To further penetrate the mechanism behind societal cooperative efforts, Modern Complexity Theory (Nicolis and Prigogine 1989) provides insight into how technical advances can be the catalytic source behind strategic decisions from among K alternative strategies that lead to success of a singular decision. Provided the attractiveness of a particular strategic decision put forward by elite management can be translated into economic benefits for individuals, then cooperation is assured within the populace to willingly perform tasks involving labor investment from all society members. This process will involve informal discussions between individuals and groups where advocates of different points of view, not necessarily those of elite management, vie for acceptance of their point of view. When majority population acceptance of a proposed concept coincides with official policy (or vice versa), then the means for universal cooperative effort from all society members is assured. The steps toward this end involve dictates that are accepted by all the populace (or accepted by passive non-response) and not necessarily ordered by elite decree but rather accepted by the apparent economic benefits to individual society members.

In terms of agriculture, different experiments and/or insights put into practice together with evaluation of results over time to increase food productivity are effective means to verify that a chosen decision agricultural strategy over another candidate strategy was justified. Here the role of leadership to guide or enable decisions (Plowman et al. 2007; Spencer 1993) is an important factor but given incomplete and preliminary notions about the structure of Andean societies' management decision making capability that resides within elite groups to guide technical decisions, the origins of progress are still somewhat subjective given the additional factor that their science depth is only assayed from preliminary analysis of existing archaeological remains of water systems. However, some degree of self-organization leading to progress, whether from a top-down elite management structure or from bottom-up localized groups, must prevail to arrive at the optimized water control systems for urban and agricultural use as observed from the archaeological record. The decision processes involved in the path to a cooperative union rely on economic advantages apparent to all – in Marxian terms (Durant and Durant 1968:51–57) "there is a contest among individuals, groups, classes and states for food, materials and economic power that are rooted in economic realities – in such cases, motives of leaders may be economic but the result is largely determined by the passions of the masses." In the discussion to follow, the origins for choices of an agricultural strategy are based upon the apparent benefits to all participants in terms of decreased labor leading to increased agricultural productivity. As agriculture is basic to the growth and survival of a society – particularly in formative Andean societies emerging from *allyu* and multi-generation community roots – it is of interest to examine societal complexity development based upon selection of a common agricultural strategy – among various choices put forward by different elements of society that ultimately form the common bonding structure so that there is a shared interest among all society members to move forward on rules and practices accepted by all society members. Only on the basis of cooperation toward common goals can society unite and progress to higher levels of complexity and state formation. Of course, consensus to unite elements of a society toward a commonly accepted trajectory is preconditioned by population elements conditioned by subsistence shifts, class inequality, notions of superiority among different kin groups, newness of the urbanism experience, uncertainty of economic realities, man/woman visions of what is important, differences in religious and behavior standards, previous sociopolitical trajectories and the usual vagaries of human nature that at times go against realities. In many cases, a non-egalitarian society is the preferred outcome of choice and projects are conceived and executed by common consensus without formal elite class management. For the present discussion, a path toward a consensus sociopolitical structure that characterizes many of the major Andean societies that develop an elite management structure is to be examined by modern Complexity Theory to examine possible self-organizing paths toward increased societal complexity based upon organized development of competing ideas about agricultural systems that underlie societal sustainability.

In mathematical terms useful to explore self-organizing societies and the means to their progress, for two choices of an agricultural strategy α, β, then A_α represents the benefit of choice α and X_α represents the number of people having adopted strategy α at a given time. The relative number of individuals X_β wishing to consider an alternate path to choice A_β is related to the number of those that have adopted some other choice (like A_α) multiplied by the relative attractiveness of the β strategy – this is $A_\beta/(A_\alpha + A_\beta)$. In practical terms, A_α may represent strict population conformance to an elite class-imposed labor tax inherent to a collective agricultural system while A_β may represent different independent communities controlling their agricultural lands without elite management oversight. Here Routledge (2013:27) would express this division as "how sovereign power is carried out against all odds and against the material interests of at least some of those involved" – a route by which A_α is given dominance by a small X_α elite leadership group over X_β opposition dissidents. While this state of affairs ultimately characterizes many examples that history provides (Toynbee 1972), a trend toward ultimate acceptance of an A_α strategy based upon demonstrated successful economic and societal stability advantages promoted by the X_α elite minority can ultimately convince the X_β minority toward acceptance of elite management control. In terms of Complexity Theory, those individuals wishing to leave choice β in favor of strategy α will be proportional to the number of people X_β multiplied by the relative attractiveness of strategy α which is $A_\alpha/(A_\alpha + A_\beta)$. Given the population N of a society is $X_\alpha + X_\beta = N$, then for a = recruitment rate per individual and t = time, the differential equation governing the rate of change of the X_α population is:

$$dX_\alpha / dt = aX_\alpha \left\{ \left[A_\alpha X_\beta / \left(A_\alpha + A_\beta \right) \right] - \left[A_\beta X_\alpha / \left(A_\alpha + A_\beta \right) \right] \right\} \quad (1)$$

or, after $X_\alpha + X_\beta = N$ substitution, following Nicolis and Prigogine 1989:239,

$$dX_\alpha / dt = aX_\alpha \left[N\, A_\alpha / \left(A_\alpha + A_\beta \right) - X_\alpha \right] \tag{2}$$

The equation gives the rate of change of the number of people selecting option α. For cases where the attractiveness of a particular strategy is evident over time due to positive economic results enjoyed by all classes of society, i.e., A_α is large compared to A_β, then a greater number of the population participates in its use as is evident by larger values of dX_α/dt. Here $0 < A_\alpha < 1$, $0 < A_\beta < 1$. Nondimensionalizing Equation (3) with $X_\alpha/N = X_\alpha'$, $X_\beta/N = X_\beta'$ so that $X_\alpha' + X_\beta' = 1$, and with $t/t_0 = t'$ and assuming $t_0 = 1$ yr and then solving Equation (2) after dropping the ' notation,

$$X_\alpha \sim e^{\Omega a\, t} / \left(e^{\Omega a\, t} - 1 \right) \tag{3}$$

where $\Omega = [1 - (A_\beta / A_\alpha + A_\beta)]$. Here the $t >> 0$ solution just shown is implicit in interpreting Equation (3) as the main interest is to observe the long-time effect of a favorable, positive, large A_α strategy to increase the population X_α that endorses this strategy over a long time period. From Equation (3), as $t >> 0$, $X_\alpha \rightarrow 1$ indicating total dominance of strategy α. Here the population segment X_α endorsing the A_α strategy ultimately dominates in time the minority X_β opposition as the process of cooperation over coercion takes time to accomplish given the success of the recruitment rate (a) from the increasing X_α population segment to overcome opposition of X_β individuals involved in the decision process. In other words, X_α becomes the majority population involved in implementing strategy A_α over increasing time. This approach is evolutionary and implies some democracy in the decision process; decisions and strategies may evolve from the outcome of trial-and-error agricultural experiments, observations from nature interpreted as insights to produce sustainable crops and from leadership roles in the management structure that correlates past agricultural experiments, good or bad, to encourage the most attractive A_i option to go forward. Such societies exhibiting some form of democratic decision making are here noted as $\Omega < 1$ societies. Note that if A_β regains popularity for some reason, then from Equation (3), the X_α population decreases as expected. Note also that the recruitment rate (a) influences the conversion rate to new thinking – this implies existence of leadership personnel that can convincingly influence or command large segments of the N population or alternatively, command conformance by edict such as would be expected from a dominant hierarchical egalitarian management structure. If a command decision is made by an autocratic, egalitarian minority in power to enforce A_α, then $\Omega \rightarrow 1$ as $A_\beta \rightarrow 0$ and the conversion to X_α is instantaneous as Equation (3) predicts; this mode of decision making by wise leadership, particularly in a fast evolving crisis situation, may prove to facilitate population survival by command decision leadership that replaces argumentative democratic processes that take more time than is available to develop consensus. The interpretation of the Equation (4) result is that as an agricultural strategy prevails from early bottom-up acceptance of a mutually beneficial situation for all participating labor groups, then some semblance of a managerial group acceptable to all members is put in charge of implementing the A_α strategy. That an individual multi-generation kin group *allyu* X_α possessing capability to generate a successful agricultural strategy A_α meeting the demands for a specific agricultural terrain and ecology has been demonstrated by Erickson (2000); here the X_β population element has blended into the X_α group based upon economic benefits of cooperation rather than competition using a common agricultural system. The results developed thus far appear to follow many of the precepts developed by Feinman and Neitzel (1984). Here new traditions are reworked, new rationalizations are developed and new social norms are established by consensus of a population to the logic of a managerial leadership class. This may take the form of reorganizing communal activities in the form of new building projects, new moral and ethics codes, new communal festive bonding feasts and wider scope activities such as new religious practices, implementing new agricultural practices, establishing trade routes, expansion policies by military conquest or trade relations with resource rich societies. While elements of a society may not readily conform to change from established norms and practices promoted by the leadership group, provided economic benefits of a more secure existence prevails, then community consensus will ultimately prevail as long as society members believe they are benefiting from the new organization (Giddens 1984, 1979). The exploration of multi-party collaboration and joint decision making directed toward a common goal in water governance practice has been investigated (Magnuszewski et al. 2018) using game-based methodology that explores the role of individuals in a collaborative group and how they weigh their personal advantages following group dictates. While modern democratic societies emphasize personal freedoms of speech and collective action directed toward a common goal, little is known about the presence of similar individual freedoms of expression in hierarchical society's characteristic of the ancient Andean world necessary to relate modern conventions to those of other societies with different governance practices. Further research (Stanish 2017; Carballo et al. 2014; Spencer 1993) emphasizes collective ritual behavior in stateless societies as the common bonding A_α strategy element of self organization. In this respect, *paccha* and *huaca* rituals and ceremonies are key elements of this process as observed in Andean societies. Provided leadership guiding acceptance of an A_α strategy with mutual economic and social benefits is apparent to a majority X_α body of the population, then, as Equation (3) indicates, elements of a collective, cooperative society appears on the road to statehood. A further option arises in times of conflict where communities may yield to absolutist forms of leadership authority (Spencer 1993; Routledge 2013:17) or, in cases of extended drought or flooding emergencies inducing collapse of agricultural systems, cede total control to authority figures that provide the basis for collective action to organize labor to begin restoration of damaged water supply systems. In such cases, A_α is the only option and the X_α majority is under total immediate control of authority figures. Outside of convincing logic and coercion methods relevant to self-organizing societies lie societies that resist establishment of hierarchical leadership and proceed with household organized labor that resisted the emergence of local leaders (Langlie 2018); in this case, there is only one A_α strategy but X_α remains fixed and not subject to change in time. Note that from Equation (3), that for a society that defers egalitarian rule and prefers community consensus government, that for $t >> 0$, only X_α exists confirming coherent society rule by consensus rather than by an elite class.

Equation (1) is most applicable to complex, pre-state societies which are characterized by hierarchy without hereditary rank where community, rather than individuals or elite groups, are created to provide focus for community rituals and initial materialization of a leadership class. This social structure is often found in late preceramic societies of the Peruvian north coast and central highlands where complex architecture and agricultural system appear as a result of community cooperation (Stanish 2001:45–47; Moseley 1975). In later coastal Late Preceramic and Early Horizon sites (Caral in particular, Chapter 7), trade of marine resources available from fishing communities for agricultural products available from inland sites becomes the start of more complex agricultural system development relying upon irrigation technologies and enlarged field systems under a central management system as trade comes with traders versed in commercial dealings. Associated with the evolution of these societies are complex architectural developments that indicate the creation of worship sites and administrative centers – all of which are steps toward states that arise in Middle Horizon times. As different Andean communities exploit local ecological conditions to devise their agricultural systems, a wide range of agricultural technologies arise to provide the stable foundation for more complex societal development seen in later periods.

Returning to the state formation discussion of Andean societies (characterized as hydraulic societies), reliance by individual Tiwanaku *allyu* and clan based groups exploiting unique terrain and ecology conditions requiring individual agricultural strategies is assumed to underlie and translate into cooperative labor reliance to effectively manage raised field agriculture over a vast scale (19,000–54,000 hectares). This cooperative strategy relies on recruitment of large labor resources guided by a centralized, top-down control that managed a comprehensive overall agricultural strategy. A single agricultural strategy guided by a central authority productive to all elements of the Tiwanaku society appears necessary to provision Tiwanaku's large city population rather than reliance on individual groups with individual strategies only accountable to their individual welfare. For the Late Horizon Inka, established elite management controlled all aspects of land resettlement, labor assignments and land allocation for populations under their control (Baudin 1961); here population consensus to hierarchical control was achieved by the success of the agricultural programs that Inka management instituted through mastery of the technical aspects of different agricultural systems for different environmental conditions.

As the archaeological record shows for later stages of major Andean societies, vast arrays of ceremonial structures and corporate administrative compounds exist within large urban compounds signaling the arrival of top-down decision making authority; this is consistent with the $\Omega \rightarrow 1$ strategy. Although different decisions to implement different strategies exist at the top management level, it is likely that elite class management had evolved to become a command structure to lower orders of the population that accepted rule that brings stability, predictably and a degree of prosperity to all levels of the social order. Archaeology theory, as applied to early Peruvian and Bolivian societies, is largely nonconclusive when it comes to agreement on how early societies transitioned from independent familial clan groups to an advanced state of organization of cooperative groups that provided the infrastructure necessary for the building of large ceremonial and administrative complexes. Was there an initial form of authoritative coercion involved in the decision processes ($\Omega < 1$) involving accession to by societal leaders to organize labor resources according to Equation (2) or did a $\Omega \rightarrow 1$ governing hierarchical structure originate from democratic consensus or from assumption of power by a hierarchical elite to rapidly advance progress toward more efficient use of land, water, labor and technology resources? As only architectural structures of administrative and ceremonial function remain in the archaeological record together with cultural remains as a source of interpretation of the societal structure that originated them, many questions remain unanswered as to the transition process of early Andean societal structure. However, the results of Complexity Theory indicate at least one path toward consensus and cooperation through recognition and acceptance of economic benefits that can be experienced by all society members – a logic which in any society has the strength of consensus building.

Examination of the later Late Horizon Inka society (AD ~1400–1532), for which many accounts are available from Spanish chroniclers related to the social organization of that society, indicate a well ordered and well organized society with many rules, rites and rituals accepted as orthodoxy by a population controlled by the top-down Inka hierarchy. Other major societies, such as the Chimu, Wari and Tiwanaku, apparently follow this same path with acceptance of the stability, prosperity and predictability that top-down management had accomplished. As populations increased in these societies over time, the archaeological record shows that top-down management was vital for labor intensive projects in the form of water resource management projects for agriculture and urban use. The management skills of logistics organization and planning, organizing communal labor participation, resource management together with technical analysis to ensure the success of major projects and advanced hydraulic/hydrological means to accomplish complex water engineering tasks, could only be directed by a central authority. The presence of colonies with their agricultural and resource bases as well as conquest of other resource rich societies to expand dominance and control from a central urban center can be seen as a further top-down management function common to most major Andean societies. While societal structure variations exist in the path toward empire, many of the major Andean societies dependent upon agricultural success appear to follow a variation of Wittfogel's vision of a hydraulic society.

From theory to reality

In later applications on similitude analysis that originate the source and progression of the Andean path to statehood and empire, the case is made for the basis for conversion from A_β to A_α strategies based upon economic improvement for their populations. While only two strategies are considered for the present discussion, further generality and extension to more competing strategies and strategy variants involving different supporting segments of the population are summarized in Nicolis and Prigogine 1989:240; Eq. 6.15 by extension of Equation (1) to multiple population groups X_i supporting different A_i strategies with different intensity and success levels of recruitment methodologies. Although the idea of systematic planning evolves from the preceding discussion and this ultimately leads to development of planners within a top-down societal management structure, perhaps a less theoretical and more realistic path consistent with human endeavor is that a trial agricultural system produces marginal benefits over time and is then replaced by a new trial innovation whose A_i benefit are only apparent over an extensive trial run period. This progress path requires a stable environment over time to evolve – but given nature's climate and weather vagaries, new solutions may need to proceed from new environmental and ecological conditions affecting the agricultural and social structure environment. The notion of global planners attempting to optimize a particular strategy valid over long time periods may more realistically be replaced by localized in time and location stable agricultural and management patterns that have localized improved results. The conclusion is that agricultural progress derives from a form of planning from early societal group activity to later, more evolved and complex corporate planning and that this evolutionary path is vital to consider in any assessment of societal management structures. Here the hydraulic and hydrological technology available to the Tiwanaku and other major Andean societies (as later chapters describe) served as the infrastructure to progress to a more complex social structure capable of more complex hydraulic and hydrological projects serving the population with improved economic and security benefits.

Despite infrequent but major duration drought periods in Andean history, stable agriculture patterns apparently could develop for long time periods under stable climate conditions as the archaeological record indicates. Thus a gradual emergence of an organized agricultural pattern over long time periods (as Equation (3) predicts) with an administrative and top-down management center evolved for major Andean societies with supportive organizations, perhaps religious in nature, to influence the deities that influence the human-deity connection to bring continuity, sustainability and progress to a society. As noted in later chapters, significant technical advances evolve in the hydraulic and hydrological sciences as observed in Tiwanaku, Wari, Chimu and Inka archaeological history – this likely indicates an active imbedded branch of government in these societies focused on higher levels of technical improvement much in the same manner that modern societies exhibit. While planning and agricultural productivity progress evolves from stable ecological conditions as the Complexity Theory model predicts, rapid change in environmental and societal conditions can invalidate progress made in previous time periods under different ecological and societal structural conditions. As Andean history is characterized by several infrequent, but long-lasting, unstable climate periods, many defensive strategies have evolved to protect agricultural continuity (as Figure 1.3 describes); these strategies result from recording and memory of past destructive climate events. Complexity Model equations can be made more complex with destructive nonlinear climate and social disintegration effects added on to Equations (1) and (2). These effects include change in behavior patterns, bifurcation phenomena, symmetry breaking and chaotic behavior typical of nonlinear equation solutions that can result from major events that challenge societal stability. These mathematical features, typical of nonlinear equation solutions may have counterparts in demonstrating societal collapse mechanisms noted in certain Andean societies under climate duress as societal norms are compromised by collapse of sustaining agricultural systems. As such, these deviations from predictability, as represented by nonlinear model behavior, have counterparts in history and often tell the story of rapid changes in societal structure, both positive and negative, from unforeseen events both from random nature effects and human sources.

Tiwanaku societal collapse: drought according to Complexity Theory models

From the preceding discussion and details of Tiwanaku social organization (Kolata 1986, 1993, 2003; Ortloff and Kolata 1993) it appears that an initial *allyu* societal structure prevailed in early versions of raised field agriculture. From results of the subsequent Similitude Theory chapter, it can be demonstrated that cooperation and consolidation of initial *allyu* groups led to reconfiguration and reorganization of raised field systems on a larger, more integrated scale that promoted economic gains in the form of increased agricultural productivity for lower labor input that benefited participants from all *allyus*. From the Complexity Section discussion, this process suggests that a cooperative agricultural strategy advantage A_α predominated over any individualistic previous A_β strategies so that the X_α majority conversion prevailed to universal acceptance. As demands of a more secure food supply and provisions for an increasing population became evident ($\Omega < 1$), the need for a higher level of management expertise prevailed ($\Omega \rightarrow 1$) to expand raised field productivity that incorporated hydraulic and hydrological science advances to better the lives of city inhabitants. As hierarchical rule governed more complex water management projects involving groundwater height control and labor organization to carry forward complex water control projects, consensus from all *allyu* groups was achieved resulting in a higher level of elite control to secure the future of Tiwanaku society. With this came amplification and elaboration of the ritual and ceremonial structures to assure cooperation of local deities to oversee progress and prevail against weather and climate catastrophes. Celebratory rituals that acknowledged success of the use of scientific agricultural advances together with deity assurance and community bonding rituals (as Comte's theories would dictate) now became part of city life. With respect to the Tiwanaku being classified as a hydraulic society, some aspects provide confirmation of the Wittfogel hypothesis (Wittfogel 1955, 1956, 1957a, 1957b) of vast irrigation systems underlying the emergence of a state structure. The same conclusion is warranted for the Chimu society where complex irrigation systems in the form of river supplied canal networks required a high level of technology to implement. For the Inka society, a complex royal hierarchy and government structure prevailed to control all aspects of agricultural productivity in a $\Omega \rightarrow 1$ manner from their administrative center of Cuzco. Control of Inka agricultural systems extended from local Urubamba *andenes* to a vast array of sites obtained from conquest that were farmed under imported *mit'a* labor overseen by Inka royalty subordinates. Basic to this evolution was the gradual accumulation of hydraulic science that was inspired from nature observations and codified into water rituals and *pacchas* that provided the foundation for later agricultural productivity and societal structure advances on the way to a state administrative societal structure. As climate change can interrupt the path toward societal progress, Equation (1) is next modified to ascertain societal structural change resulting from these destructive effects.

From Ortloff 2009:180 and results from the Similitude discussion, under extreme drought from the 10th to 12th centuries AD, the change in Tiwanaku city population N is such that $N \sim K \exp\{-kt^2\}$ where K and k are constants and t is time. Here k is an intensification constant related to the onset time of drought. For $\Lambda = a K A_\alpha / (A_\alpha + A_\beta)$

then Equation (2) can be written as

$$dX_\alpha / dt = \Lambda X_\alpha exp\{-kt^2\} - aX_\alpha^2 \tag{4}$$

where $\Lambda = a K A_\alpha/(A_\alpha + A_\beta)$. The solution to Equation (4) is

$$X_\alpha(t) = exp\{\Omega\, erf(t)\} / \{(a/k^{1/2}) \int_0^t exp\{\Omega\, erf(t)\}\, dt + [X_\alpha(0)]^{-1}\} \tag{5}$$

where $\Omega = (\Lambda/2)(\pi/k)^{1/2}$ and where $X_\alpha(0) > 0$ is the X_α population at the start of the precipitous drought at time $t = 0$. The erf(t) term is the standard error function given by

$$erf(x) = (4/\pi)^{1/2} \int_0^x exp(-s^2)\, ds \tag{6}$$

Note that at $t = 0$, the initial condition $X_\alpha(0)$ is recovered on both sides of Equation (5). The $X_\alpha(t)$ population consists of most all members of society participating in given societal norms and mutually agreed upon agricultural practices – here a small number of the population consists of leadership personnel and, as such, share the fate of the entire society subject to the variable climate and weather conditions affecting agricultural production and its role in influencing societal stability and coherence. Equation (5) indicates that the majority X_α population diminishes exponentially (Figure 1.22) by minus time squared under extreme drought duress; this may result from food shortages or dispersal migration to local near city areas with local water supplies or external sites that can support agriculture for smaller clan groups.

While the total population decreases, the majority $X_{\alpha}(t)$ segment of the population, which includes the elite leadership class, diminishes rapidly but in a more damaging way as the disappearance of leadership elites disassembles the main guidance and leadership figures of the elite ruling class. Here rituals and celebratory ceremonies under elite sponsorship no longer are performed and trust in leadership's authority and association with benign deities is compromised. Under these conditions, social fragmentation and political unrest follows with abandonment of trust in deities and their connection to the elite class (Janusek 2004). From personal observation after many years research at Tiwanaku, excavated deity statuary was defaced and, in one particular instance, a deity representation carved into a stone block was found as part of a building wall with the deity face reversed into the interior part of the wall. The occurrence of human sacrifices in main areas of Tiwanaku is also noted as offerings to deities to restore stability. Apparently, these signs represent recognition of the abandonment by the deities of the welfare of the Tiwanaku population and resentment of the deities for their role in the destruction of organized city life.

Figure 1.22 provides the numerical evaluation of Equation (5) for $K = 1$, $k = 1$, $a = 1$ and $\Omega = \pi^{1/2}/2$. Drought timing and intensity may be regulated by increased values of K and k; the $a = 1$ notation denotes full involvement of all members of the population subject to drought conditions. The negative effect from Equation (5) is, according to the archaeological record of the Tiwanaku society, accompanied by the dispersal of small groups abandoning the city to areas with localized water resources to permit a limited form of agriculture. The lure of city life with the promise and advantage of unity and prosperity derived from sharing common goals together with managed land areas that guaranteed adequate food supply provided the draw for rural populations to initially band together to form a city – as drought intensified, advantages of city life receded as the stable food supply diminished. As population from the city center decreased over time and affected all segments of a society, key management and government personnel were no longer available making the command and leadership decisions and no longer able to manage labor and resources effectively. Social fragmentation and political discontent would logically follow with abandonment of formerly beneficial deities and rituals no longer thought effective to continue societal harmony and prosperity (Janusek 2004). It has been observed that ritual statues were defaced in this difficult time period as drought severity deepened and, from personal observation, some carved blocks with deity representations were reused as building blocks for emergency house construction with the deity faces reversed from view. With respect to Tiwanaku's extended drought and how it affected population levels, the first defensive moves led to abandonment on near-lake edge raised fields and the transfer of farming to more distant raised fields from the Titicaca lake edge where the water table still remained high to support vestiges of raised field agriculture and limited open field farming utilizing rainfall infiltration. Thus the decrease in

Titicaca lake height from extended drought (Binford et al. 1996) affected near lake raised fields directly as the groundwater profile declined with the lake height decrease. Distant groundwater levels supplied by rainfall infiltration into large collection plains near mountainous areas distant from the Lake Titicaca remained stable over long time periods to sustain distant groundwater levels vital for *cocha* farming and shallow swale raised fields to continue. Essentially, the near-lake groundwater profile and raised field swale water heights followed the lake shore height decline more rapidly than at distant locations from the lake edge forcing abandonment of productive near lake raised fields and use of smaller area raised fields distant from the lake edge. The presence of *cochas* excavated down to the aquifer phreatic zone supporting lower population groups is also noted as drought intensified in the 11th and 12th centuries AD consistent with dispersal of city population into smaller groups at multiple *cocha* sites and local areas where the high water table could support limited agriculture (Kolata 1996:266). In final drought stages, the Tiwanaku urban center was largely abandoned by AD ~1100 and the return of rainfall norms in later times then formed the basis of the next phase of population development characterized by multiple, dispersed defensive sites that likely guarded precious water resources.

Complexity theory allows for multiple segments of society to play individual roles under dominating state governance. According to Lumbreras (1987, 1974) views on Andean society, social formation is the structural entity out of which culture emerges. From Complexity Theory, the process of social formation that permits different views to compete for dominance (A_α, A_β, . . .) by different competing elements of the population (X_α, X_β, . . .) ultimately defines the culture that emerges. This model likely characterized early Tiwanaku *allyu* kin groups deciding more efficient ways to maximize agricultural production by collective labor enlistment that raised the economic status of all participating members of different clan groups – this procedure led to natural leaders that ultimately formalized their evolution to elite status that magnified in importance as continued success resulted from the labor intensive infrastructure projects they conceived and managed with successful outcomes. While this path to cultural development has elements of evolution from societal consensus on best ways to proceed, many historical examples demonstrate that a single government body (or individual) dominates all other segments of a society from the outset and secures dominance through claims of alliance with, and selection by, celestial deities to enhance dominance over of all classes of society. Other examples of single party leadership rule proceed from individuals that have guided military dominance over rival groups or promote plans for dominance backed up by the propaganda apparatus to sway unquestioning minds; other less adversarial single party dominance derives from upholding and continuing philosophic and religious principals of a dominant deity (or deified person) universally respected by previous generations. As opposed to complexity models that derive from consensus of early *allyu* group populace segments that initially agreed to develop a top-down management structure likely typical of the Andean world, Asian history, and to some degree early Middle East societies, appear dominated by dynastic rulers that totally controlled and dominated all aspects of large populations to their will. For both cases, only the monumental royal architectural and religious building complexes remain to verify the presence of an elite hierarchical privileged ruling class – the early societal structural development to arrive at this final stage state-level remains conjectural although Complexity Theory provides some insights to the path to social complexity and state level development.

While examples from Andean history describe the path toward societal complexity according to Complexity Theory, it is likely that other formative societies in different parts of the world at different historical times duplicated the same route toward social complexity development. An example from Early Iron Age sites in Jordan (Porter 2013) describes the initial phase of community formation defined by cooperative community labor to construct dwellings, walls, defenses and food storage facilities that required more people than a single household could provide. Such "built spaces" reified community identity and bonds by giving expression to the benefits of pooling community labor. Originating from these origins, a leadership class emerged employing patrimonial and charismatic ideologies together with success in management of labor-intensive resource projects related to agricultural productivity and monumental structures to articulate their authority and status. This early path to collective labor in Porter's views did not necessary lead to state level development but rather self-autonomous groups that formed and dissolved because of local contingences (drought, famine, war, political unrest) that dictated subsistence and social life change. After an "initial growth phase," a "conservation phase" originated where certain households began to express leadership and authority in managing food resource strategies and crisis management roles. While Complexity Theory envisions this path toward societal complexity and emergence a top-down management structure, it also allows for disparate groups (X_β, . . .) with less access to resources with little voice in their future prospects to migrate or join other communities, or in a severe crisis mode experienced by all levels of a society, to simply disappear from the archaeological record as Equations (4), (5) and Figure 1.22 indicates.

Reflections on Complexity Theory applications to determine state evolution

Complexity Theory format prescribes the use of convincing arguments based upon economic advantage for all parties concerned to sway group thinking into a unified collective consensus of cooperation. The social binding and cooperation between different *allyu* groups with different ethnic identities and beliefs was the basis for the societal, economic and personal safety advantages of city life as compared to the challenges of isolated rural existence for separate *allyus* occupying separate ecological locations. Given human nature, collective group economic benefits do not always translate into and are perceived as individual member benefits thus leading to subgroups within a society choosing to retain former group identities and practices. Given counseling methodology, the best promotion of different thoughts to the firmly held thoughts of contrary thinkers is to discuss, not prescribe, new thought options and let the contrary thinker develop thinking to the effect and benefits that new perspectives bring on an issue originate from, and are the creation of, his own mind. This process is aided by observing the tangible advantages of group collective action as compared to separate group existence – particularly in agricultural areas where collective action provides economic benefits to all members in terms of less labor necessary to produce greater agricultural output (as discussed in the in Chapter 2 on Similitude). Many different forms of (α) strategy promotion exist to convert (X_β) thinkers into a belief pattern characteristic of (X_α) beliefs given insight into the mindsets of different groups with different origins and ingrained beliefs. In this regard, the draw of Tiwanaku city life with its economic, societal and survival benefits against exterior threats was confirmed in time by orderly transition from early stage *allyu* groups to later state and empire status over the centuries. The appearance of subgroups within the Tiwanaku society (Kolata 1993) producing craft specialties (ceramic production compounds, farming land division overseers, textile manufacture, for example) may represent specialists from different *allyus* that band together to pursue common interests.

Equation (1) is devised based on only two competing groups (X_α and X_β). Given human nature, new challenges and sociopolitical changes in a society can alter previous norms and produce many additional competing factions (X_α, X_β, X_γ, X_δ, . . .) that vie for dominance by a single group. Equation (1) can be modified (Nicolis and Prigogine 1989) to accommodate additional X factions promoting different strategies whose success in winning over converts relies on the convincing powers of group leaders to achieve consensus thinking within their group. With respect to Tiwanaku, Goldstein's (2004) interpretation of the evolved societal structure is consistent with Complexity Theory as determined from the formative processes that lead to state development, here "the dynamic interplay among myriad counterposed factions in both state and periphery" represents different X factions interacting to produce behavioral procedures that define their society. The Complexity Theory analysis thus far developed for Tiwanaku and its Moquegua Valley colonies thus is a dynamic consensus process and counters state development models based upon neoevolutionary theory that have more deterministic paths toward the evolution of hierarchical governing systems. Here Complexity Theory models allow for competition for power and how when it is achieved, how ruling classes further exercise, negotiate and resist further challenges to their continuance. This is accomplished by the introduction of a multigroup equation set with numerous X competitors to replace Equation (1) for only two (X_α and X_β) competitors. This model is particularly applicable to Goldstein's (2004) observations of Tiwanaku's diaspora Moquegua Valley colonies where X_α represents Tiwanaku city doctrines and X_β represents portions of these doctrines that were acceptable and used by the colonists.

Given that Tiwanaku city was composed of assemblages of many different *allyu* groups each bringing distinct views and ethnic identities particular to their origins from different locales, the presence of uniformity of beliefs toward a common basis for all society members as directed by an elite governance class, was aided by celebratory bonding ceremonies for all city members, worship and sacrifice to deities that oversaw Tiwanaku's welfare, and the belief that city life brought economic, social and personal safety advantages beyond what isolated *allyu* group existence offered well beyond what rural existence provided. Through mutual cooperation between *allyu* groups, a city could arise with much in common for all its inhabitants. As for other models of state expansion and evolution beyond Goldstein's agency model, many aptly apply to different Andean and other New and Old World societies – but for Tiwanaku and its colonies in particular, Complexity Theory models appear to present a path toward the understanding the development of this society.

Historical thoughts on the optimum structure of a society

Discussion of the optimum structure of a society from philosophers of the Greek Golden Age, particularly the discussion between Plato and Aristotle (Hale 1965:331–332), relates to the views of land-apportioned individuals enlisted to and committed to voluntary communal labor to achieve community goals – this being expressed as "friends' property is common property" according to Plato's *Ideal State* from his *Laws*. Aristotle's modification (from his *Politics*) of Plato's propositions involves private ownership of property with the proviso that "friends' goods are common goods" meaning that results of individual labor (primarily agricultural products) are available to neighbors upon request. Imbedded in Aristotle's view is the human nature consideration that all workers will not (or cannot) contribute equal amounts of voluntary labor but yet want to share an equal amount of produce nevertheless. Another view from Epicurus' *Rules for Living* would have a "state run by wise men" for the ideal state. Further research (Romilly 1991; Freeman 1950) traces the conditions of state success and the accompanying organization of power necessary for this to be achieved in Greek city states. A view from the 17th century (Hobbes 1991, 1996 [1658, 1642, 1651]) poses the state and its rule of law as the antidote to "war of all against all" and the "fear of violent death at the hands of another" as the way to peaceful coexistence between enemies. The crucial role the state plays in the domestication of humanity is foremost in his *Leviathan* or *The Matter, Form and Power of a Common-Wealth Ecclesiastical and Civil Society*. Further thoughts on the role of the city as a civilizing necessity expressed in later centuries referring to Roman presence in the British Isles (Collingwood and Meyers 1936) as "to live well instead of merely living, was the membership in an actual physical city." From such thoughts and writings about how societies could be ideally composed and the leadership necessary to achieve this end, history has demonstrated much experimentation along these lines with various degrees of success for equality and individual freedoms integral to state formation. Later scholars, among them Childe (1950) laid down ten traits (presence of cities, labor specialization, agricultural surplus, monumental architecture, a ruling elite, science, writing, standardized artwork, long distance trade, solidarity based on residence rather than kinship) that define state formation – although all elements need not be present for transition from farmer collectives to chiefdom level societies. Liverani (2006:6) concluded that state formation was not inevitable but relied on agricultural surplus that was "not for consumption within the family but for the construction of infrastructures and for the support of specialists and administrators – the very authors of the revolution itself." The difference is that the transition is based on communal use of agricultural surplus and an entirely new administrative class is supported dedicated to the state's infrastructure.

Examination of the archaeological record of the societal structure of the Bolivian Tiwanaku society (~300 BC–AD 1100) from its elementary origin to state level appears to follow several elements of Childe's and particularly Liverani's precepts regarding the agricultural surplus provided by the hydrological science supporting raised field agriculture and the later embellishments of an elite class harbored within their own moat-bound compound as well as religious based architectural structures (Chapter 5). Other authors (Sarton 1931; Bartlett 1958; Koestler 1964) emphasize the role of relating factual scientific and engineering material observations obtained by experiment and observation over time, which were previously kept as seemingly unrelated and independent bits of information, to be later shown to relate to each other through an encompassing theory that tied together the previously scattered information bits into "a close and effective relationship" (Bartlett 1958). To a degree, this path is that of the evolution of modern western science (Koestler 1964:224–254) as many examples indicate from early 18th-century Enlightenment Period origins (Sloan 2003). The presence of highly sophisticated hydraulic technologies exhibited by early and late period Andean societies most likely followed this evolutionary path over the centuries of their existence as engineering technology proceeds through cumulative incremental verification steps before culmination in theoretical form. While western science essentially encodes all hydraulic phenomena in solutions of the Navier-Stokes equations (now solvable to a large degree by finite element CFD methods), hydraulic laboratory test and experimentation still plays a major role in discovering hydraulic phenomena. In this sense, early Andean hydraulic water engineers become contemporary with their later western counterparts.

Essentially in ancient and even in modern times, there is still experimentation regarding best political/economic state models given that new considerations of vastly increased population size, new technologies, new societal behavior norms, personal and societal security concerns, personal welfare demands as well as climate change effects that affect societal sustainability and thus modify or erase state models that have been tried and failed. A prime recent history experimental example model of state control illustrating effects of single party control ($X_\beta = 0$, $A_\beta = 0$) exists from Chinese history in recent times regarding large population growth demands on living space land use competing with limited agricultural land resources (see also Chapter 3). Early Chinese dynasties enlisted vast labor resources for major civil engineering infrastructure projects to extract value from surplus labor – this in the form of projects for agriculture, communication, defense purposes (the Great Wall, the Grand Canal (Coats 1984), road building and irrigation agriculture and flood control projects) as well as state power demonstration projects in the form of early dynasty royal palace compound construction.

Later 20th-century phases of Chinese society faced similar problems related to productive utilization of unconstrained population growth. To provide a modern society framework to effectively utilize vast amounts of labor incorporating different useful skill sets productively, first different classes of society were defined as to their potential to accomplish state goals. Here different societal classes with different importance to maintain state-sponsored societal goals were given rules provided by the ruling elite. For example, in the Chinese Great Cultural Revolution in the mid-20th century, four societal classes were defined by the ruling elite: (1) the working class, (2) the peasantry, (3) the urban petit bourgeoisie and (4) the national bourgeoisie represented by the elite ruling party class. Each class had government rules to live by commensurate with their potential to contribute to the party's goals consistent with a common ideology for all societal classes with no constraints on population growth – essentially all classes of society were subject to a common "many people, one mind" ideology defined by the elite leadership (4) class. In the later post-revolutionary society, revision corrections of former practices were implemented acknowledging that extremity of state-controlled thinking produced a society not compatible or competitive with the rest of the global economic structure. The societal isolation model consistent with this model clearly isolated China from the outside world's path toward greater economic benefits consistent with capitalistic models. Practical considerations to relax the economic model toward a mixture of the best of capitalism and socialism under state control of personal life behavior returned as unconstrained population growth with no apparent economic contributions to the central government to carry forward their programs strained economic resources available to the central government to carry forward their thoughts that a moderately affluent, educated middle class works best to reinforce the economic model created by state control. To solve the unconstrained population growth problem, administered reproductive constraints limiting family size were next put into place in post–Cultural Revolution government practice to counter increasing urban housing demands that absorbed agricultural land resources. A further measure of population control mandated by the state control model to limit elderly population from becoming wards of the state was provided by laws that allowed parents the right to sue their eldest male progeny for support. As all land and business enterprise was owned and controlled by the state, central state management directed portions of domestically occupied and agricultural rural lands to be used for state infrastructure projects (roads, bridges, tunnels, dams, canals, rail lines, airports) that by edict promoted economic development through increased communication and accessibility between formerly semi-independent zones of the country. These projects utilized vast labor resources productively which accorded to state directives of using later versions of class (2) and (3) labor to accomplish state projects that demonstrated the managerial competence of state-controlled agencies. To conserve urban land better used for factories, business enterprises and industrial agriculture designed for export, concentrated urban multistory apartment complexes for more efficient storage of city populations around city centers became the norm together with government projects that emphasized state power to manage the large population effectively and productively. Many of these infrastructure projects were designed to integrate formerly quasi-independent regions (Tibet, Hong Kong, far eastern Chinese provinces) into central state control by facilitating communication means designed for rapid transfer of indoctrinated mainland "loyal" population to these areas to dilute traditional cultural norms and independence from state control well established in prior times. State sponsored infrastructure projects also served as economic stimulus to employ labor productively but at the cost of increased national debt to support massive and costly projects. Here the importance of money accumulated by international trade imbalances was vital to demonstrate the power of the state to provide visible signs of state competence and progress toward improved living conditions. Further government imposed programs limiting migration to cities for unskilled workers seeking better economic conditions than those available in the rural farming environment were in place to maintain the economic foundation of cities only made possible through educated, higher class society members participating in productive labor activities conducive to ruling party goals. This class division was achieved by prohibitive inner-city real estate rental and ownership costs that concentrated poor migrant populations into subclass housing districts distant from city centers. As population control limiting family size and migration was applied selectively to different societal classes, the overall government plan dictated that the more economically productive, segments of the population were to be concentrated in city areas that promoted wealth through interactive internal and external country business activity. While rural populations remained largely self-sustaining through individual farming areas and collective organization of larger government-controlled areas, these farming areas supplied much of the food basics to the wealth producing urban population concentrations. A coexisting urban and rural society with advantages given to the wealth producing segments of the population appears to be a modern solution to manage largely unconstrained population growth effectively under elite state control. The dictates from the central government appear to rule many (but not all) of the compliant population that sees economic advantages – particularly those in compliance with ruling party dictates through party membership. Although Complexity Theory can formulate paths toward statehood and elite control in early societies, the preceding example reinforces that once tenuous elite control dominates all economic aspects of a society, the next step is mind control to assure compliance to stare objectives. As Durant and Durant (1968) observe from histories' lessons, excesses of this type foment revolution with restarts toward control by popular democratic demands rather than an elite class that assumes superiority and the right to rule lower elements of society.

From this discussion, although the route to elite control may first arrive by consensus (voluntary or not), once in place, it can have a mind of its own to ensure its continuance by enforcing its doctrinal message to a compliant population oblivious to later consequences of state rule. In an extreme case, a "social credit" score is maintained by the ruling elite in present-day China to judge individual citizens' loyalty to governance dictates. Once control is established over the population the ruling elite seeks expansion of its domain to increase access to new resources vital for its expansion – a tradition from early dynastic rule extended to the present day. Since opposition exists from modern world powers with greater military resources to challenge direct expansionist policies, more subtle modern economy means to ensure access to vital resources with little threat and transparency to other world powers is used. This may take the form of infrastructure loans to desperate countries that cannot logically be repaid leading to seizure of vital collateral ports that guard access shipping routes to vital resources as well as elements of control of leadership elites of those countries. Thus while methodology of extending control over native populations is well advanced in some modern socialist dominated economies controlled by an elite ruling class, the next step in ambitions is to expand outward to dominate on a world scale and propagate their governance principles to create an orderly society under their control – this being a key objective continuing from ancient dynastic times with only the means to achieve these ends varying from means available from modern political practice.

Outside of China in Southeast Asia (primarily Vietnam, Laos and Cambodia) government control and ownership of land exists; in Vietnam in particular renewable 50-year land plot leases to individual families and businesses are available to promote small business activities continuation over generations. Major Vietnamese cities show remarkable growth on the order of one million population increase every five years with commensurate apartment complex and commercial construction activity to accommodate population growth and new business activity. Vitally important to all Southeast Asian societies is preservation of historic architectural sites within cities related to Buddhist historic monuments that show the unity of the current generations to preserve elements of ancestral traditions and past customs and traditions. Elsewhere in Laos and Cambodia, government control of land exists but land can be bought as private property. While Laos appears largely influenced by Buddhist religious tradition and rituals evidenced by large numbers of well populated monasteries, less apparent government behavioral control is apparent. Cambodia, on the other hand, is ruled by a central government encouraging economic growth through investment in infrastructure by foreign and domestic sources to promote sources of wealth generation through tourism of its historic sites – primarily Angkor Wat which draws ~1.5 million tourists each year. These countries, as observed from a recent visit, demonstrate a version of small-scale capitalism under socialist governmental control that limit state mandates on correct thinking and behavior but never-the-less oppose nonconformist attitudes to government practices. Here elementary capitalism takes the form manufacture and selling of consumer level products by successive generations of city and rural shop keepers that maintain the same business through generations by renewal of 50-year property leases. Apparently the further distance from the Chinese mainland and its central government ideology, the less influence from Chinese economic and social behavioral control models exist in Southeast Asian countries with their emphasis on economic growth allowing individual enterprise and foreign investment to promote construction and development of new businesses. The western economic model appears to have appeal to promote economic success and raise living standards of the population but with the difference of one party rule that controls the state financial distribution structure along socialist principles lines. The central governments of these countries realize that vast infrastructure projects, many of which are designed to support tourism activities and integrate communication connections to formerly isolated parts of their countries are first required to support vital economic growth and this is best accomplished under one party rule to avoid conflict between parties and individuals with different views on paths toward societal progress. While consumer level capitalism appears dominant in these countries to provide the economic survival base for most of the population of these countries, the educational infrastructure to provide advances to higher levels of attainment for the population appears not a government priority to achieve given the lack of monetary resources to implement. On this basis, these societies remain noncompetitive to western economic societies and unable to elevate living standards substantially for their populations for times to come. In western societies advanced governmental support features are prominent that involve distribution of monetary resources through taxation available to government agencies to develop and advance to higher levels of societal attainment; The lack of monetary resources and educational advancements available to Southeast Asian governments sustaining low levels of economic activity prevent change in their competiveness levels compared to western societies. From Equation (1) considerations, it appears that Southeast Asian socialist governments do not support contestive views of government programs by their populations and given the low level of government monetary resources available from taxation of minor business enterprises, will remain at their current economic status with few options of change available for the immediate future.

Conclusions

Surface and groundwater-based irrigation systems formed the basis for sustainable agriculture in the Andean world. Vital to the expansion of agriculture was the co-development of a technical base in the hydraulic and hydrological sciences that guided the efficient use of water to maximize food production. Given that Peruvian coastal sites involved many different ecological and water availability conditions ranging from coastal deserts intersected by intermittent flow rivers to mountainous areas dependent upon rainfall for agriculture, different water usage technologies needed to be developed for different agricultural systems. The continuous preoccupation over ~4000 years from the Early Preceramic to Late Horizon times to develop sustainable agriculture and its associated water control technologies to provide the foundation for self organization into hierarchical management polities noted in the archaeological record. Accompanying the development of efficient, sustainable agriculture were rituals, ceremonies and communal celebratory events observed by all Andean societies to acknowledge the beneficence of nature deities that oversaw their creation and, through sacrifice and offerings, maintained their connection and devotion to the nature deities. Through ritual and ceremonial use of *pacchas* of different types combined with nature observations, water revealed its secrets to guide sustainable agriculture and, once in place, societal advancement to higher organization levels progressed as the archaeological record demonstrates. From Collier's (1982:17) writings, Wittfogel (1955) concludes that "in contrast to the stratified societies of Medieval Europe, Andean societies failed their own inner forces to evolve beyond their general pattern of social stratification." From the discussion thus far presented, plus chapters to follow, it is clear that the hydraulic/hydrological technical base underlying Andean irrigation societies was in an advanced state of development that in many cases, anticipated many discoveries made and formalized by modern hydraulic science many centuries (or millennia) later. This, despite stratification into different societal classes characteristic of major Andean societies, provided benefits for all classes in terms of societal continuity, sustenance through productive economies and a deep pattern of beliefs and customs shared by all classes of society. Many of the Andean societies had advanced water control technology coexisting with elaborate rituals and ceremonies acknowledging the source of knowledge from nature's instruction and the beneficence of deities that controlled their fate all of which bound them together in common beliefs and societal practices. Unlike western, post-Enlightenment European societies that ultimately separated and decoupled religion from science after Medieval Period episodes of papal overreach viewing scientific explanations contrary to their version of a Supreme Being controlling all aspects of the physical world, Andean societies lived without this episode in western history. It may be that "not evolving beyond social stratification" was purposeful and involved a cooperative, collective type of societal organization that provided relevance to all levels of society and universally accepted an elite, hierarchical form of leadership that brought a degree of order, predictability and security to their uncertain existence and, as such, was an evolved form of social stratification meaningful to the challenges of their world. In this case, even with different elements of a stratified society with different capabilities, ambitions and viewpoints, effective elite management could enlist labor from all stratified societal classes to cooperate in large scale projects that provide economic benefits for all societal classes through industrial scale agriculture projects as well as for projects that enhance the stability and sustainability of all society members to enhance their well being and self importance. Here the complacency that western societies have achieved the best possible societal outcomes may be questioned by the longevity of Andean societies and the continuity in form and stability of many of their societies over many centuries.

2 Similitude in archaeology

Examining agricultural system science in pre-Columbian Peru and Bolivia

Introduction

The cognitive ability of ancient civilizations to conceptualize, design and build water supply systems for agricultural use is examined through mathematical models that predict the optimum use of land, water, labor and technology resources to maximize food production. From the archaeological record of agricultural systems used by several pre-Columbian societies of ancient Peru and Bolivia, knowledge of agricultural system configurations permits comparison of actual to theoretically optimum agricultural systems. This comparison permits evaluation of the agro-engineering knowledge achieved by Andean societies subject to different ecological and geographic location conditions and provides insight into their technical achievements produced by evolutionary trial-and-error empirical observation of system improvements and/or engineering foresight to conceptualize an optimum design and put it into use. Use of a basic equation derived from similitude methods provides the basis to replicate the thought process and logical decision making of ancient agricultural engineers albeit in a format different from western science notational conventions. Examples of agricultural system designs from coastal Peru canal-supplied (AD 900–1450) Chimu irrigation systems, groundwater based raised field agricultural systems of the (AD 300–1100) Tiwanaku society of Bolivia and later (AD 1400–1532) Inka terrace systems are used to illustrate conclusions derived from a first application of similitude methods to archaeological analysis.

Investigation of economic progress and intensification (defined as increased agricultural productivity from land, water, labor and technology resources) in the pre-Columbian Andean world is examined from the vantage point of technology advances, governmental bottom-up or top-down resource management, environmental and climate factors and societal structure. A summary of progress in the field of agricultural field system designs, strategies and intensification for societies in the Lake Titicaca basin area is available (Janusek 2004; Kolata 1991, 1993, 1996a; Kolata and Ortloff 1989a; Ortloff and Kolata 1993; Ortloff and Moseley 2009; Ortloff and Janusek 2015; Scarborough 2004; Scarborough 2015; Tung 2012) as well as for Peruvian north coast and highland societies (Lane 2009, 2017; Moseley 2008; Ortloff 1993, 2014a, 2009, 2014b) and Inka society (Doig 1973:529–531).

For Lake Titicaca basin studies, Scarborough's (2004) reference to "wide interpretive distance" between several authors' conclusions on aspects of Lake Titicaca basin agricultural strategies and intensification conclusions point to different views of the intensification processes from the same data set. To help quantify different interpretations of archaeological data, an alternative analytic approach is proposed using similitude analysis to produce an understanding of the generation and maintenance of intensification as related to agricultural productivity given land, water, labor and technology resources.

The similitude approach offers a path to relate these resource factors to agricultural production incorporating the many complications related to different societies with different political economies operating in different environments subject to different climate and weather variables. When observed archaeological field data coincides with similitude predictions of the optimum use of land, labor, water and technology resources to provide economic stability benefits to the populace, advances in intensification signal the management of resources directed by an intelligent management elite whose continued rule is assured by populace consent continued in time. Similitude methodology provides insight for discovery of the science base used for agricultural intensification by several Andean societies for which extensive archaeological field data exists on their agricultural systems. Additionally, similitude methodology predicts the economic benefits of one agricultural strategy over others; such benefits may lie behind decisions made to implement or alter agricultural strategies by the ruling elites signaling a form of intelligent decision making based upon a rational analytic basis.

In summary, the use of similitude methodology examines use of land, water, labor and technology for several Andean societies to determine if observed progress toward intensification noted in the archaeological record follows optimal use of resources predicted by mathematical models based on similitude methods. Provided observed field system design and use strategies from the archaeological record follow predicted optimum agricultural usage patterns, then an Andean version of science/engineering underlying agricultural strategies exists – albeit in different formats from western science. While closeness to predicted optimum conditions may exist through repeated trial-and-error corrections, an alternate but complementary interpretation involves cognitive intelligence proceeding from interpretation as to how nature works in an optimal manner. As western science has a similar historical, multi-century experimental path toward discovery of basic scientific principles (Bartlett 1958), this same observational, recording and learning path derived from successful (or failed) agricultural experiments may also be attributed to Andean societies when their agricultural constructions ultimately match science-based optimal use predictions. Provided correlation exists between observed and theoretical predictions, the net result points to an active, societal intelligence at work by ruling elites to provide economic improvements for their populations. This posits a dynamic nature for Andean societies based upon technical achievements similar to the basis of modern societies. The first task then is to determine the optimum use of land, water, labor and technology resources and then determine how closely select Andean societies duplicated this path to maximize agricultural production. Again, if the archaeological record demonstrates closeness to optimum use of resources, then the basis for the intensification process is founded upon some as yet codified form of scientific principles applied to agricultural productivity.

The question of the development societal structure as dependent upon water management techniques has a long history. Clark identified the pivotal role of water in his publication *Water in Antiquity* which reviewed available archaeological evidence. This treatise on water management and its relationship to societal structure became the foundation for a theoretical hydraulic hypothesis of the composition and structure of an oriental society in Wittfogel's 1957 publication *Oriental Despotism: A Comparative Study of Total Power* (Wittfogel 1956, 1957a, 1957b) that states that societies in Asia depended upon the building of large-scale irrigation works as the source of their dominance and bureaucratic power.

Here major irrigation and flood control projects required organized, forced labor and a large and complex bureaucracy to manage projects that enhanced the basis for despotic rule. Steward followed this hypothesis in his irrigation civilization study, claiming that irrigation was the catalyst for state formation. Adams attempted to test the hydraulic state hypothesis with regard to the rise of Mesopotamian civilization and found that complex systems of canals used for agricultural irrigation projects came *after* the appearance of cites and the indicators of bureaucratic statehood rather than before as Wittfogel had proposed; the same conclusion was found with regard to the emergence of the Mesoamerican archaic state (Scarborough 2004, 2015). With further research based upon socioeconomic development in different parts of the world (Wittfogel 1956, 1957a, 1957b; Wright 1977; Lamberg-Karlovsky 2016; Marcus and Stanish 2006; Romer 1986) as well as climate change effects on ancient world societies (Kerner et al. 2015) it was evident that societies throughout history had developed sophisticated techniques of water management but all did not evolve into states and that many water management systems utilized reservoir management rather than irrigation (Lane 2009).

Thus a far more complex and diverse association between water and society than Wittfogel, Steward and Clark had envisioned now prevails as exemplified by cross-cultural studies (Scarborough 2004; Routledge 2013; Smith 2003; Wright 1977; Marcus and Stanish 2006; Coward 1979; Downing and Gibson 1974; Hunt 1988; Mitchell 1973; Sanders and Price 1968; Steward 1955; Albarracin-Jordan 1996) that illustrate a large variety of methods of water control and their relationship to land, water, labor, between technology and societal structure. Here use of similitude methods provide new insights to the connection optimum uses of land, water, labor and technology and the societal structure that attains these ends through utilization of a form of agro-science not previously considered in the literature of Andean states. In later chapters, a new perspective on the relationship between technical mastery of water control principles and the societal structure that originates and cultivates this knowledge base is offered to clarify the relation and interdependence of these two factors.

To implement the search for strategies that produce optimal use of resources and maximize agricultural productivity, the use of similitude methods (Buckingham

1914, 1915; Murphy 1950) is made to derive a basic equation governing optimum ways to maximize food production taking into account the widely varying ecological conditions at different geographic locations experienced by different Andean societies. The basic equation subsequently derived from similitude methods is then used to determine the degree to which major pre-Columbian societies of Peru and Bolivia originated technology that conformed to predicted optimum productivity methods; this is achieved through examination of field system designs and strategies from the archaeological record. Where conformance to basic equation predictions exist from the archaeological record, insight into the creativity and engineering science underlying Andean societies' agricultural strategies and field system designs offer perspectives on their innovation, creativity and path toward agricultural intensification and progress. Theories exist to explain the connection between societal structures as influenced by their city and agricultural water supply and distribution systems (Renfrew 1979; Renfrew and Bahn 2000; Routledge 2013; Hastorf 1993:1–8); alternatively, city housing, administrative and public use structure architectural plans according to city planners can specify water supply system designs. The present analysis based on similitude methods provides an alternative quantitative methodology to determine the basis that underlies the decision making related to agricultural designs of field systems used by several Andean societies. Again, this course involves determination of optimum use of land, water, labor and technology that provide maximum economic advantage considerations; here the optimum path is provided from Similitude Theory considerations. Implementing plans to achieve this end challenges the elite management structure of a society to effectively carry forward such plans much in the same way that modern technological developments challenge and influence the future of present-day societies.

Basic equation development

Archaeology examines the cultural remains of a society and through anthropologic models, attempts to interpret the thinking and societal/political/economic structure underlying the creation of ceremonial and secular architectural remains and iconographic material objects – this abetted by historical reports and surviving literature from ancient sources. These cultural and architectural remains inform how a society prioritized resources to support its institutions and how it was organized to achieve higher levels of sustainability. Further insight derives from examination of the use of land, labor, water and technology to optimize agricultural production to maintain societal continuity and sustainability through natural weather/climate/geophysical landscape variations and/or man-generated disasters (Moseley 1983; Renfrew 1979). The development of technologies vital for agricultural yield increase follow from observation, recording and encoding of "what works best" to increase agricultural production; these heuristic principles refined over time by agricultural experiments, chance discoveries and intuitive insights ultimately improve field system designs toward optimum designs and irrigation strategies. While the archaeological record demonstrates the progression of engineering skill applied to agricultural systems as they evolved over time, the science underlying changes in agricultural system evolution remains elusive as the archaeological record shows evolutionary changes but not the development and codification of the science underlying these changes.

To investigate the science underlying evolutionary changes in agricultural systems, one strategy is to determine the theoretical optimum use of land, water, labor and technology given a society's resource base and then determine, from the archaeological record, how close this optimum had been achieved. Given the wide variety of agricultural methods applied by Andean civilizations adapted to different environmental conditions (Figure 2.1) and archaeological knowledge of societies that occupied pre-Columbian Peru and Bolivia (Moseley 2008), examples illustrating agricultural strategy changes due to climate and geophysical landscape change provide illustrative cases of how these societies modified land, labor, water and technology to maintain agricultural productivity.

Given that different productivity optimums exist for different environmental resource conditions in different areas occupied by different Andean societies,

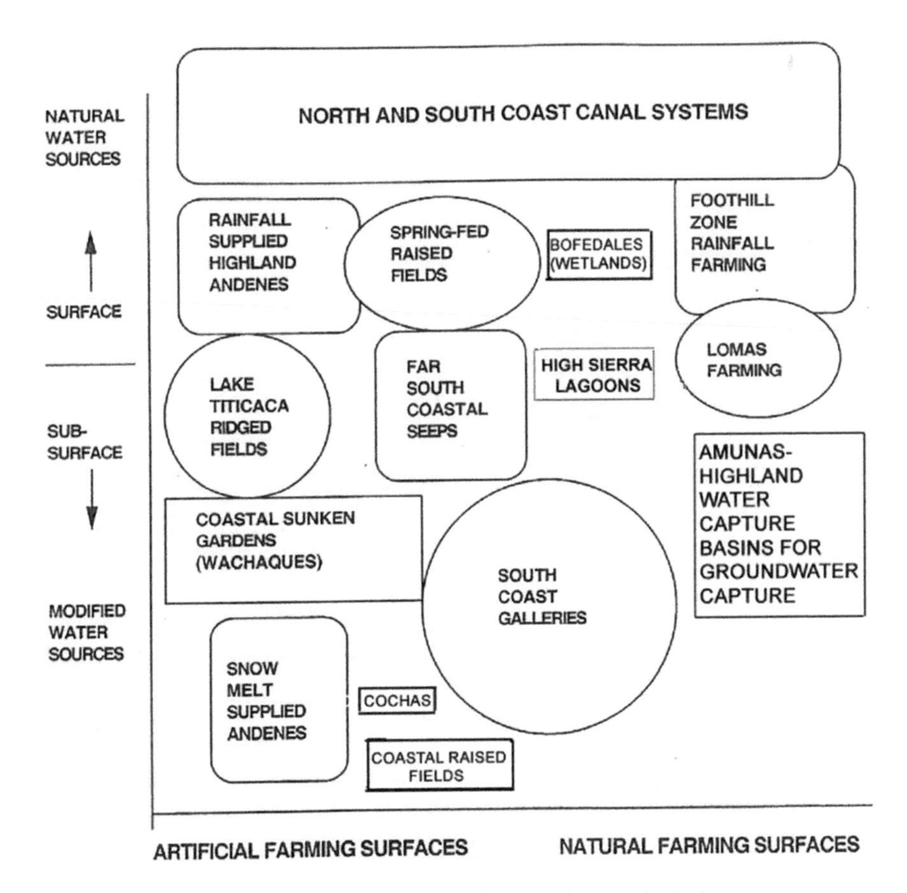

Figure 2.1 Agricultural techniques used in different Andean ecological zones.

different optimum conditions exist. A basic equation derived from similitude methods that incorporate land, water, labor and technology parameters as independent variables and the rate of food production as the dependent variable are required to evaluate agricultural strategy evolution and the optimum use of resources. When coincidence exists between basic equation predictions and archaeological record observations of agricultural field evolution technology to achieve optimum food production, then insight into the thinking of ancient Andean societies follows.

The derivation of a basic equation that relates optimum use of land, water, labor and technology variables to the rate of food production is achieved by use Buckingham Pi-Theorem similitude methods (Buckingham 1914, 1915; Murphy 1950) – the Appendix provides explanatory details of this methodology. The first step is to express the units of key variables in terms of length (**L**), mass (**M**) and time (**t**) as follows:

Variable	*Description*	*Units*
F	food supply mass	**M**
dF/dt	rate of food production over a time interval (time derivative)	**M/t**
r	average consumed food mass/perso **M/r** = total population over a time interval (r = constant) (number of persons)	**P**
$\mathbf{L_c}$	total canal or water trough lengths to supply land area A	**L**
A	area under cultivation	$\mathbf{L^2}$
V	volume of terrain removed/altered to produce terraces or canals; this parameter relates to the labor force size and soil volume transfer capacity; in terms of units, k_1 **M** = labor force = k_1 r **P**, k_2 $\mathbf{L^3}$ = total soil volume removed and/or altered on a project by the k_1**M** labor force; $k_1 < 1$.	$\mathbf{L^3}$
Q	water volumetric supply rate	$\mathbf{L^3/t}$
t	time	**t**
Φ/Φ*	technology level (low angle surveying (low) 0 < **Φ/Φ*** < 1 (high)	**nondimensional**

The Φ/Φ* term, as defined earlier, holds for coastal based irrigation systems where surveying technology is the key technology vital to expand arable areas. While other technologies such as soil fertilization and special water supply systems for specialty crops as well as seed generation areas exist (Eling 1986) to add to the technology base, little research in these technologies exist to amplify the Φ/Φ* definition used in the present analysis for coastal irrigation systems. For other types of farming systems such as raised fields, technology is defined more broadly in terms of groundwater height control for specialty crops as described in Ortloff and Kolata (1993).

Next, the equation for optimum use of land, water, labor and technological resources is expressed in general functional form as:

$$dF / dt = f\left(L_c, A, V, Q, t, \Phi / \Phi^*\right)$$

Here the rate of food production derivative (dF/dt) is specified as a general function of the extensiveness of the canal irrigation system (L_c), the area under cultivation (A), the labor involved to implement projects (V), the water flow rate to field systems (Q) and the technology base available (Φ/Φ^*) to implement projects. Applying the Pi-Theorem (Buckingham 1915), the preceding equation in nondimensional group form (i.e., all terms have no dimensions but are just numbers) is (Ortloff 2009:156–157):

$$\left(A^{3/2} / Q\ F\right)\left(dF / dt\right) = f\left(L_c / A^{1/2},\ V / A^{3/2},\ Q\ t / A^{3/2}, \Phi / \Phi^*\right)$$

Rearranging and selecting a functional form with population P related to food supply F and the food mass consumption per person r (in a time interval) by P = F/r, then the basic equation becomes:

$$\begin{aligned}&\left[\left(A^* - A - \Delta A^*\right)^{3/2}\right) / \left(Q^* - Q - \Delta Q^*\right) F\Big]\left(dF / dt\right) = \\ &\left[\left(L_c{}^* - L_c\right) / \left(A^* - A - \Delta A^*\right)^{1/2} + \left(V^* - V\right) / \left|A^* - A - \Delta A^*\right|^{3/2}\right] \qquad (1)\\ &\bullet \left| Q - Q^* - \Delta Q^* \right| t\left(\Phi / \Phi^*\right) / \left(A^* - A - \Delta A^*\right)^{3/2}\ ]\end{aligned}$$

where ΔA^* = fraction of maximum land area A^* unusable for cultivation due to non-accessibility by an irrigation canal or landscape contours and ΔQ^* is unusable water due to limited irrigation canal inlet access. Equation (1) incorporates a constraint condition guaranteeing maximum food production (dF/dt = 0) when all land (L_c), water (Q), labor (V) and technology resources (Φ) are maximally exploited (* notation) and ΔA^* and ΔQ^* are small. For this condition, labor is in balance with land, water and technology resources to maximize food production in an economic equilibrium sense. For cases for which $V > V^*$, excess labor is being applied to no productive advantage as a subsequent example case explains. Similarly, when $L_c > L_c{}^*$ and $Q > Q^*$, canal length extension and excess water over that required for maximum food production with balanced labor, water supply and canal lengths results. Thus the * conditions represent the optimum employment of L_c, A, V, Q and Φ resources for maximum food production when dF/dt = 0 – a result from elementary calculus. While different functional arrangements are possible, the selection given reproduces known economic laws and provides interpretation of archaeological data related to agricultural field systems and irrigation networks. When L_c, V, A and Q equal their superscript * counterparts, dF/dt = 0 given that $0 < \Delta A^* << A^*$ and ΔQ is small. In an F vs. t (time) plot (Figure 2.2), the zero derivative coincides with zero slope of the maximum of the F vs. t curve.

A positive slope of the F vs. t curve, dF/dt > 0, indicates potential growth of a society's food supply due to large exploitable resources of land, water, labor and technology ($L_c < L_c{}^*$, $V < V^*$ and $Q < Q^*$). For a negative slope, dF/dt < 0, land, water, labor and technology resources are diminishing from flood deposition and erosion events, farming land area contraction due to drought, geophysical land wasting effects, water supply decline and/or labor decline due to plagues, migrations and political turmoil, among other degrading causes, affecting agricultural

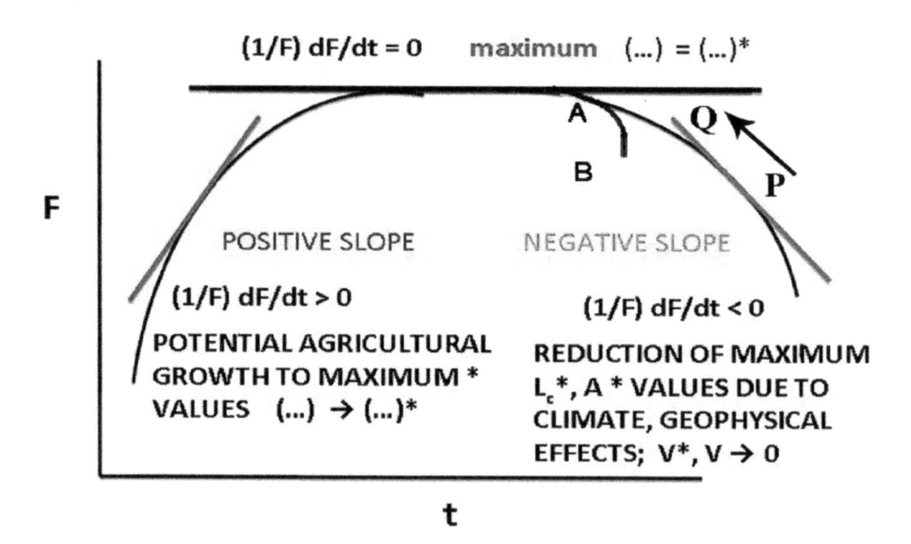

Figure 2.2 Food supply (F) as a function of time (t). Different dF/dt tangent slopes to F vs. t curve (left to right) indicative of increasing, maximum and diminishing food supply with time.

production. When L_c* decreases rapidly, indicating rapid contraction of the irrigation canal network due to drought or geophysical land wasting inflation/deflation effects, and V > V* represents a larger than required excess labor force working less available land, no options remain to sustain food production as A* contracts and Q can exceed Q* indicative of available, but non-usable water due to contraction of arable land and the canals necessary to support irrigation. In Figure 2.2, line A-B represents a rapidly developing negative climate/weather, geophysical or crop sustainability event (tsunami, massive flood, earthquake, crop pest infestation, sand inundation, invasion, epidemic, etc.) with no anticipated or rapid defensive response to sustain agriculture; such events are in the realm of catastrophe theory (Renfrew 1979; Poston and Stewart 1978) where dF/dt becomes highly negative over a short time period with a transition to a collapsed state.

To illustrate application of the basic equation, several scenarios are examined. The first considers too large a labor force (V > V*) exploiting a small farming area (A <<A*) with low water resources (Q << Q*) using a low level of technology in the form of short irrigation canals (L_c small) emanating from a river source. Here

$$\left(A^{*3/2}/Q^{*}F\right)\left(dF/dt\right)=\left[\left(L_c{}^{*}\right)/\left(A^{*}-\Delta A^{*}\right)^{1/2}-V/\left(A^{*}-\Delta A^{*}\right)^{3/2}\right]$$
$$\left(Q^{*}\right)t/A^{*3/2})\left(\Phi/\Phi^{*}\right)<0$$

where the dominant (V >V*) term from Equation (1) is negative and A*- ΔA* > 0 leading to (dF/dt) < 0. The preceding expression is a version of the economic *Law* *of Diminishing Returns* in that excess labor ($V > V^*$) added to balanced, equilibrium resource assets (V, L_c, A, Q) does not increase food output without a corresponding increase in the resource assets that utilize the additional labor resources. This law applies to the rightmost branch of the F vs. t curve in Figure 2.3 as excess labor only is not sufficient to exploit limited land, water and technology resources.

A second example involves a small labor force exploiting large farming areas with large water resources using a high level of technology. Here $V << V^*$, $A < A^*$, $Q < Q^*$ and $\Delta A^* << A^*$ and the basic equation becomes:

$$\left(A^*\right)^{3/2} / Q^{*2} F)\ \left(dF / dt\right) = \left\{ \begin{array}{l} \left[\left(L_c^* - L_c\right) / \left(\Delta A^*\right)^{1/2} + V^* / \left(\Delta A^*\right)^{3/2}\right] \\ t\ Q^*\ \left(\Phi / \Phi^*\right) / \left(A^*\right)^{3/2} \end{array} \right\} > 0$$

where $(1/F)(dF/dt) > 0$ indicates the potential to increase food production. This result is typical of the leftmost branch of the F vs. t curve in Figure 2.3. Here the *Law of Diminishing Returns* works to increase food production as excess labor is available to exploit excess land, water and technology resources to * levels. For both these cases the *Law of Diminishing Returns* is imbedded within the formalism of the basic equation. A further example originates from extreme drought conditions imposed upon a canal irrigation system. Here $Q \rightarrow 0$ with a large population reacting to incipient drought resulting in $Q << Q^*$ and $A << A^*$ values; the basic equation reduces to a more simplified form with the introduction of variables $F' = F/F^*$, $A' = A/A^*$, $L_c' = L_c/L_c^*$, $Q' = Q/Q^*$ and $t' = t/t^*$, where t^* has dimensions of (one year)$^{-1}$. To render variables nondimensional, the (') notation has been dropped in the development to follow. For severe drought conditions for agricultural technology at an elementary level ($\Phi/\Phi^* \sim 0.5$) with surveying

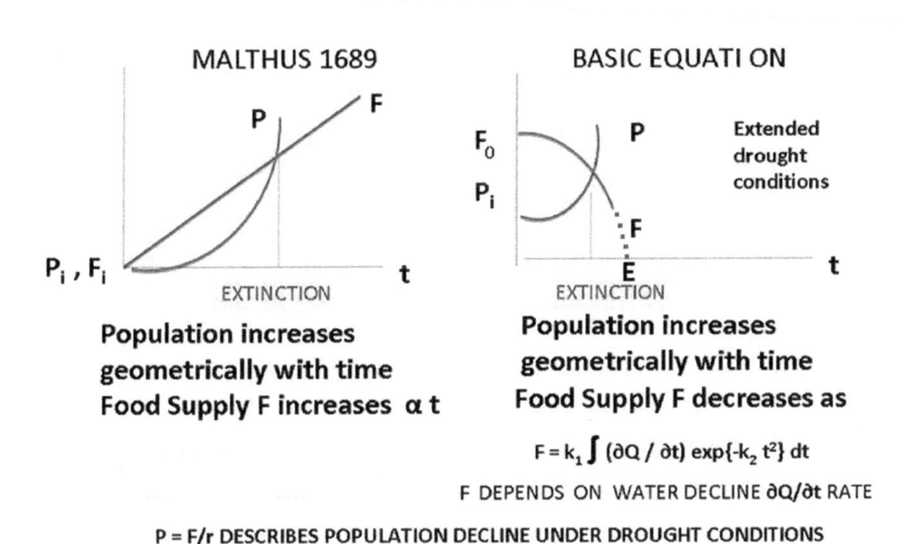

Figure 2.3 Extinction theories: Malthus (1689) and the Basic Equation.

accuracy for canal development limiting exploitable land area using marginal labor and water supplies for agriculture well below that available from a river source, the $L_c << L_c^*$, $A << A^*$, $Q << Q^*$ and $V << V^*$. From the general case, after integration of the basic equation with respect to time,

$$F = F^{*-1}\, exp\left\{\frac{\left[L_c^*(1-L_c)/A^{*1/2}(1-A)^{1/2} + V^*(1-V)/A^{*3/2}(1-A)^{3/2}\right]}{Q^{*2}\, Q^2 t^2 \Phi / 2\Phi^*(1-A)^3}\right\}$$

with

$$c_d = \frac{\left[L_c^*(1-L_c)/A^{*1/2}(1-A)^{1/2}\right] + V^*(1-V)/A^{*3/2}(1-A)^{3/2}](\Phi/\Phi^*)}{\left[Q^{*2}/2A^{*3}(1-A)^3\right]}$$

Under drought conditions, c_d simplifies to:

$$c_d \approx \left[-|L_c^*/A^{*1/2}| - |V^*/A^{*3/2}|\right](\Phi/\Phi^*)\left(Q^{*2}/2A^{*3}\right)$$

where c_d can be negative indicating a canal can be of great length to a distant water source, V can be large indicating a large labor force to maintain a long canal and A small indicating less land available for agriculture due to drought. Here t denotes the drought duration in years. For this condition,

$$F = (1/F^*)\, exp\{-|c_d|\, Q^2 t^2\} \tag{2}$$

so that

$$\partial F/\partial Q = -(2|c_d|/F^*)\, Q\, t^2 exp\{-|c_d|\, Q^2 t^2\} \tag{3}$$

Equation (2) indicates that the food supply can decrease exponentially as negative $(\text{time})^2$ under severe and sudden drought conditions in a closed system where food resources are only available from land, water and labor within a closed geographical boundary with no outside import resources available. Recent studies (Tung et al. 2017; Tung 2012; Kurin 2016) detail the societal collapse of a contemporary Middle Horizon Wari site due to food shortages; here dietary shifts associated with extended 10th- to 11th-century AD drought assign limited food resources to productive society members and leadership individuals capable of sustaining and guiding the society into a climate recovery period (which apparently never happens as Figure 2.4 indicates as the drought period appears to extend for a long time period).

As P = F/r, Equation (2) indicates population also exponentially declines as negative time squared as drought deepens consistent with Tung's (Tung et al. 2017; Tung 2012) results for an examined Wari population during 10th- to 11th-century AD drought. These results are consistent with dual competing prey-predator species

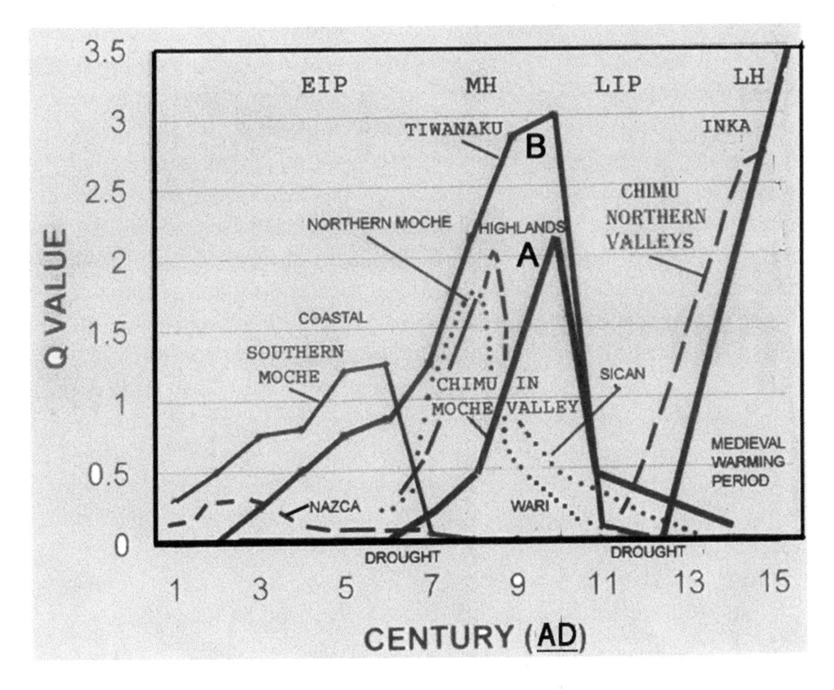

Figure 2.4 Time spans of major Andean societies based on their sustainability (Q) subject to climate change duress – from Part 1(Ortloff and Moseley 2009).

extinction theories that predict the decline rate of one species after their food supply (the other species) is exhausted (Cooke 1979; Kurin 2016). While these closed society, limiting relations apply for rapid onset of drought characteristic of Figure 2.4, A-B, societal extinction does not necessarily follow if population can somehow adjust to diminishing food supply by dietary restrictions given intermittent periods of rainfall or promote migration to balance food supply with reduced population if drought slowly develops. Under conditions of variable drought duration and intensity, population levels may adjust correspondingly but this may lead to selective "pruning" of key management and labor force personnel in a society that decrease the coherence of a society to function efficiently and collectively.

While some scholars (Janusek 2004; Goldstein 2005) assign Tiwanaku collapse to internal social unrest, clearly a catalyst for societal turmoil would be food shortages as Tung's recent research dramatically illustrates (Tung et al. 2017; Tung 2012). A further example from the archaeological record resides in 6th-century AD Moche settlements in the Jequetepeque Valley (Bawden 1999; Dillehay 2001); here a large Moche population exploiting a shrinking agricultural resource base depleted by climate extremes (Shimada et al. 1991; Thompson et al. 1994; Thompson and Moseley-Thompson 1989; Chepstow-Lusty et al. 1996) experienced conflict between groups occupying multiple walled settlements competing for water and agricultural land resources. The former coherence of Moche society exhibited through common religious practices and rituals at major urban centers appeared to be fragmented in the post-6th-century drought environment (Mauricio 2018; Dillehay 2001) as scattered settlements replaced several of the major urban centers on the north Peruvian coast. This example of conflict represents an extreme, but perhaps common, solution resulting from a population exceeding its resource base resulting from geophysical climate change effects.

As a further example of the descriptive content of the basic equation, an increase in low angle surveying technology (Φ/Φ* increases toward unity) implies less labor is required to increase food production (dF/dt) for the same amount of land A, water Q and canal lengths L_c. For V < V* and Φ/Φ* increasing, then dF/dt increases consistent with the results of previous scholars (Boserup 1965; Morrison 1994). Of the many theories of state formation as dependent upon population growth, resource availability, warfare, leadership capability, trade and exchange and environmental factors (Boserup 1965; Carneiro 1970, 1981; Wittfogel 1956, 1957a, 1957b), the present analysis supports technological advancement as one defining source a hydraulic society. From similitude analysis, technical progress increases the rate of food production (dF/dt) given fixed land, water and labor resources; conversely, an increase in food production leads to an increase in population. The well-known Boserup hypothesis (1965) can be stated as:

(population growth (V*) → new farming (Φ/Φ*) → increase in agricultural (dF/dt)
technologies production

based on introduction of new farming methods and supportive technologies to produce increases in agricultural production, as shown in (Renfrew and Bahn 2000:477), this statement apparently works in both directions according to similitude analysis with technological advances key to this process. Here, for L_c << Lc*, A << A* and V < V*, the basic equation yields

$$\left(A^{3/2}/QF\right)\,(\mathbf{dF/dt}) = \frac{\left[L_c*/\left(A*-\Delta A*\right)^{1/2} + (\mathbf{V}*-\mathbf{V})\right]/\left(A*-\Delta A*\right)^{3/2}}{Q*t(\Phi/\Phi*)}$$

confirming the Boserup hypothesis. Alternate hypotheses presented by Netherly's studies (Netherly 1984) of the Chimu's (AD ~800–1400) Late Intermediate Period occupation of the Chicama Valley appears to indicate that there is no direct association between the Chimu political center at Chan Chan (Figure 1.23) and agrarian expansion in the northern Chicama Valley. In this valley, Chimu occupation was limited thus permitting more opportunistic farming regions controlled by kin groups outside of the direct control and interest of Chimu hierarchy which was more concerned with Moche Valley agriculture lands adjacent to Chan Chan and its supporting water supply networks providing the sustenance for the Chimu capital of Chan Chan. The concentration of canal systems in the Moche Valley area necessary to irrigate vast north and south side field systems (Figure 2.9) vital for the sustainability of Chan Chan's elite compounds and agricultural base required concentration on water, land, labor and logistics management skills combined with hydraulic technology skills (Moseley and Deeds 1982) to maintain this vital irrigation system.

The adjacent valley rural areas could function under local farmer's control (Netherly 1984) as the Moche Valley system was self-sufficient to provide agricultural resources without imports from adjacent valleys during normal climate periods where rainfall runoff into the Moche River was sufficient to provide for the city. The later development of the Chicama-Moche Intervalley Canal (Ortloff 2009) changed this situation as 10th- to 11th-century drought intensified and supplemental water resources were necessary to import from the Chicama River to sustain the city of Chan Chan's intravalley farm systems.

From the basic equation, a new form of civilization extinction based on declining water supply for agriculture can be derived and compared to Malthusian predictions. Figure 2.3 shows Malthusian theory where population increases geometrically with time while food resources increase linearly with time. Past the time when food supply and population curves cross, civilization extinction prevails. Although Malthusian theory (*Essay on Population* 1798) likely had relevance given limited European 17th- to 18th-century agricultural technology and population growth trends, the development and growth of the green revolution in later centuries together with more efficient industrial agricultural practices and available contraception methodologies in later centuries could then match geometric population growth to food supplies in advanced countries thus relegating Malthusian theory to historical insignificance in later times. As Durant and Durant (1968:21–22) would phrase it "the multiplication of consumers was the multiplication of producers" in later centuries in advanced civilizations to limit the relevance of Malthusian theory in modern times. Malthus viewed nature's three agents to restore the balance: famine, pestilence and war – without these, population would soon exceed its ability to feed itself. Not that these factors lack examples in later centuries but recovery to affected societies is apparently possible through generosity and giving of prosperous nations, medical advances and international agencies tasked with maintaining peace.

To implement the basic Equation (1) extinction relation, the food supply change dependent upon water supply change given land area and canal lengths is given by $\partial F/\partial t = (\partial F/\partial Q)(\partial Q/\partial t)$ with A, A*, L_c, L_c* constant. A relation developed from Equation (2) for extended drought conditions between t_0 and t_1 time limits (in years) describes the decline in the food supply F as:

$$F = -2\left|c_d\right| Q \int_{t_0}^{t_1} t^2 exp\left\{-\left|c_d\right| Q^2 t^2\right\} \bullet t^{-2} exp\left\{-kt^2\right\} dt \tag{4}$$

where the water decline rate ($\partial Q/\partial t$) between time limits t_0 and t_1 (in number of years) due to drought may take different forms other than that shown in Equation (4) depending on the average water supply decline time history during extended drought. Specifications of Andean water decrease for agricultural use over times of extended drought are available (Thompson et al. 1994; Thompson and Moseley-Thompson 1989; Chepstow-Lusty et al. 1996) but require formatting into a mathematical format appropriate for use in Equation (4). As an example case, for rapidly occurring drought, a hypothetical rapid water supply decline relation is given in terms of nondimensional variables as $\partial Q/\partial t = -t^{-2} \exp\{-k\ t^2\}$ which describes a severe drought water supply decline decreasing even more rapidly than exponential minus $(\text{time})^2$. This rapid water decline case seriously affects the food supply F. Here Q* is the initial water supply at t = 0.

Substituting and integrating Equation (4),

$$F = F_0 - 2\ |c_d|\ Q^*\ F^*\left(\pi^{1/2}/2a\right) erf\left(a\ t\right)\Big|_{t_0}^{t_1} \tag{5}$$

where $a = (|c_d|\ Q^2 + k)^{1/2}$. Here K and k are constants used in the drought specification term. In Equation (5), the error function

$$erf\left(\mathrm{x}\right) = \left(4/\pi\right)^{1/2} \int_0^x exp\left(-s^2\right) ds \tag{6}$$

is numerically tabulated (Abramowitz and Stegun 1964). Equation (5) then provides the F vs. time curve for Figure 2.3. From Figure 2.3, when the P (population) and F (food mass decline) curves cross, insufficient water is available to support food supplies necessary for a geometrically growing population under extended drought conditions causing societal extinction to occur. As the basic equation applies to population decrease commensurate with food supply (P = F/r) decrease from either natural or man generated disasters, societal extinction occurs when population approaches Point E, Figure 2.3. Well before that point, however, questions arise as to the coherence of a society as key leadership, managerial and labor force personnel are randomly subtracted from the population upsetting normal societal task assignments and functions leading to secondary collapse mechanisms through political instability. Such a path has been suggested (Janusek 2004; Janusek and Kolata 2004; Ortloff and Kolata 1993) to describe the collapse of the Tiwanaku society under post-11th-century drought conditions as well as for other Old and New World societies subject to climate variations (Fagan 2008; Manzanilla 1997; Chepstow-Lusty and Bennett 1996).

To illustrate application of the preceding result to archaeological data, the following illustrative example is given using available or surmised field data from an early (5400 BP) coastal Zaña Valley preceramic Peruvian canal system (Dillehay et al. 2005). For a group of 100 people using elementary technology ($\Phi/\Phi^* \sim 0.5$), and for ~1 kg of food per person per day, an average of ~1.0e-1 m^3 per person-day of irrigation water is necessary to sustain the food supply given evaporation and seepage losses that leave only ~10–15% of available water available for crop root systems (Gleik 1996). Given that pre-drought conditions were optimal (*) conditions, over a year, $F_0 = 3.65e4$ kg of food/year is required from $Q_0 = 3.65e3$ m^3/year of irrigation water under normal, non-drought water supply conditions at $t_0 = 0$ before drought begins. For shallow, short $L_c \sim 100$ m canals to a field area of ~ 1000 m^2, some simplifications can be assumed for the example case. Here $L_c << L_c^*$ as the full potential of arable land is not utilized, $V << V^*$ as only a fraction of the population is used for agricultural labor purposes and $Q << Q^*$ as water for a small agriculture area is only a fraction of that available from a river source.

For drought conditions,

$$c_d = \left\{-|L_c^* / A^{*1/2}| - |V^* / A^{*3/2}|\right\}\left(\Phi / \Phi^*\right)\left(Q^{*2} / 2A^*\right)$$
$$= (3.16 - 0.06)(0.5)(1)(0.01) \approx 0.2.$$

After $t_1 = 1.0$ year, using $\partial Q/\partial t = -t^{-2} \exp\{-k\, t^2\}$ to characterize the rapid decline in water supply, only ~ 25% of the initial food supply at t_0 is available for 0.35 of the pre-drought water supply. Further years of decreasing water supply show lower values of yearly food supply as expected. Application of Equation (5) for times $t > t_0$ then would provide the food supply curve E shown in Figure 2.3. At times when the available food supply becomes less than that to sustain the population, population decreases accordingly; if drought continues, then extinction results. Here a closed system is assumed with a stationary population fixed within a closed area with no exterior food sources available for import. Application of Equation (4) to other societies using data taken from their agricultural field area and irrigation canal systems can be used to compute the decline of agricultural output under drought stress. Such calculations aid to understand the canal flow rate contraction drought response patterns mirrored in Figure 2.8. Here the Peruvian north coast Chimu society experienced continually decreased canal system flow capacity as indicated by time-contracted canal cross-section profiles (Figure 2.8) to accommodate decreased water supplies in response to continuing 10th- to 11th-century AD severe drought (Thompson et al. 1994; Thompson and Moseley-Thompson 1989; Chepstow-Lusty and Bennett 1996) experienced in both Peruvian and highland as well as the Bolivian altiplano. Here the food production derived from the irrigation water of the main N1 canal (Figure 2.9) declined substantially affecting the continuity of the Moche Valley's food supply. Given that other valleys than the Moche Valley with greater water and land resources were available to the Chimu, dispersal of population to satellite valleys was inevitable (Moseley and Deeds 1982; Ortloff et al. 1985). The supplemental water source from the Chicama-Moche Intervalley Canal (Thompson and Moseley-Thompson 1989; Chepstow-Lusty and Bennett 1996) experienced in both Peruvian coastal and highland may not) have functioned to reinvigorate the Moche Valley agricultural system as only further research can establish. Earlier 6th-century AD drought (Shimada et al. 1991; Thompson and Moseley-Thompson 1989; Thompson et al. 1994; Thompson, Moseley-Thompson and Davis 1995) had similar agricultural contraction effects on the Moche Valley heartland with the collapse of major ceremonial center and agricultural systems (Bawden 1999; Moseley 2008). Thus from knowledge of drought duration and severity (characterized by a $\partial Q/\partial t$ term) and knowledge of the field system and supply canal design and drought response data shown in Figure 2.8, use of Equation (5) can be used to calculate the decline in food availability in the Moche Valley over time as a source of political change and resettlement patterns characteristic of the close of Moche V settlement (Moseley 1983, 2008).

That major mass wasting and drought events influenced Andean history and determined the time dependence and magnitude of the $\partial Q/\partial t$ term is apparent from the archaeological record. Figure 2.4 shows the sustainability of major ancient South American civilizations (Ortloff and Moseley 2009) occupying coastal and highland areas within present-day Peru and Bolivia. Major ~30-year 7th century and ~200-year 10th- to 11th-century AD drought (Shimada et al. 1991; Thompson et al. 1994; Thompson and Moseley-Thompson 1989; Chepstow-Lusty et al. 1996) caused cultural decline and societal transformation of several major Andean societies in their homelands while other societies were able to cope with changes in irrigation water supply through modification of their canal systems and/or massive construction projects transferring water from distant sources. An example of a coping strategy originates from the Late Intermediate Period (AD ~1000–1400) Chimu society of north-coast Peru involved construction of the ~75-km-long Intervalley Canal in the 10th to 11th centuries AD that was intended to transfer water from the Chicama River to reactivate north side Moche Valley desiccated fields (Ortloff 1993, 2014a, 2009; Ortloff et al. 1982; Ortloff, Feldman et al. 1985; Ortloff and Moseley 2009) shown in Figure 2.9 through reactivation of the main Vinchansao supply canal. In other cases, migration to zones with larger water and land resources occurred as exemplified by 7th-century AD Moche presence and settlement in the northern Jequetepeque and Lambeyeque valleys (Bawden 1999; Dillehay 2001) as well as LIP Chimu colonies north and south emanating from the Moche Valley base during 10th- to 11th-century drought. Other examples from Andean history demonstrate conquest of weaker states and population transfer (14th- to 15th-century AD Inka expansion and conquests (D'Altroy 2003), Chimu military/cultural expansion into north coast Peru valleys in the 10th century AD (Moseley 2008) that were likely influenced to some degree by climate change effects. Examples of societies dispersing and/or disappearing from the archaeological record unable to cope with ecological disasters, geophysical landscape change, drought and flood events altering field system's productivity are indicated from the archaeological record. These events include 6th- to 7th-century AD drought and sand encroachment into the Moche Valley causing decline of the classic southern Moche V society (Moseley 1983); the Late Archaic Period Caral society of the Supe Valley (~2500–1600 BC) experiencing contraction of coastal agricultural lands due to flood erosion, beach ridge formation, bay infilling and aeolian sand transfer compromising their agricultural and marine resource base (Murphy 1950; Ortloff 2009; Sandweiss et al. 2009) as described in Chapter 7. Further examples include 10th- to 11th-century AD contraction of Chimu Moche Valley agricultural lands and irrigation canals from drought and tectonically induced river downcutting (Ortloff 1993, 2009; Ortloff et al. 1985) and the 10th- to 11th-century AD collapse of Tiwanaku raised fields and Wari *andenes* farming sites due to drought (Ortloff 2009; Ortloff and Kolata 1993; Chapter 5) as well as early coastal societies in far south Peru that experienced El Niño mass wasting effects that caused societal collapse (Kolata 1991; Moseley et al. 1998; Ortloff 2009; Niles et al. 1979). Figures 2.5, 2.6 and 2.7 illustrate geomorphic A* landscape changes in the north central region of Peru that started in the Late Archaic Period that marginalized former agricultural and marine resource areas. These landscape changes are still evident and ongoing from continued El Niño and other mass wasting effects observed from satellite imagery taken over time as well as archaeological excavation data (Niles et al. 1979; Moseley and Deeds 1982; Moseley et al. 1992, 1992) on the Peruvian north-central coast.

All of these mass wasting geomorphic changes resulting from large El Niño erosion and deposition events over ancient society occupation periods reduced A and A* agricultural lands (and increased ΔA*) that altered the agricultural landscape and induced some form of societal structural change. Figure 2.8 illustrate results from archaeological excavations of canal profiles taken from the Huanchaco canal N1, N2 and N3 profiles. Successive contractions of canal profiles mirrored the decline in available water supply from the Moche River through canal systems during the 10th- to 11th-century AD drought (Ortloff 1993, 2009;

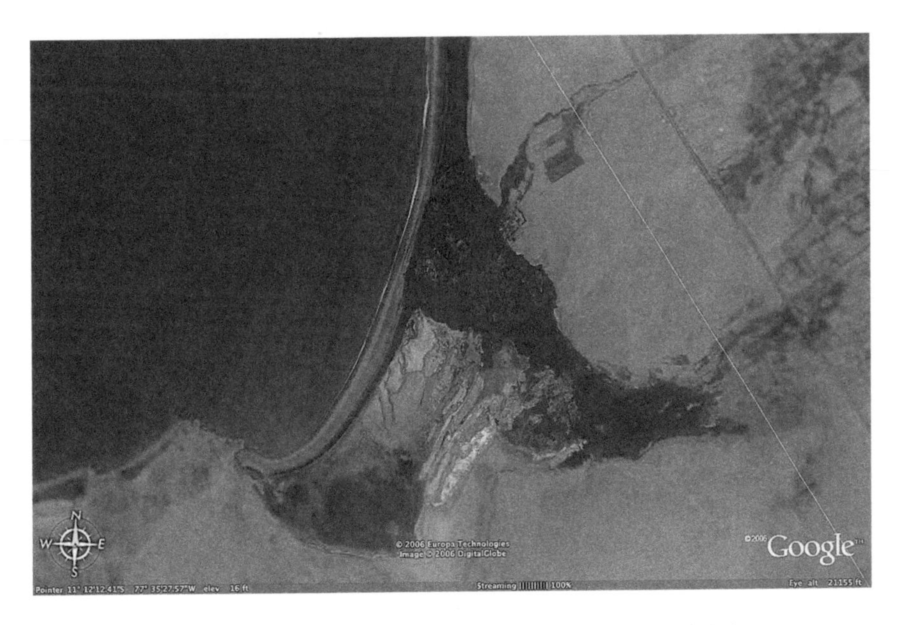

Figure 2.5 Marsh-filled coastal bay resulting from blocked flood debris by an accumulating beach ridge; area shown located is on the Peruvian north-central coast area south of Aspero demonstrating an A* loss of arable land and loss of the marine resource base. Loss event occurred in the Late Preceramic Period 3000–1600 BC.

Figure 2.6 Infilled bay behind a beach ridge in the Peruvian Supe-Huanchaco Valley coastal zone formerly supporting agricultural land in the Late Preceramic Period.

Ortloff et al. 1982; Ortloff, Feldman et al. 1985; Thompson et al. 1994; Thompson and Moseley Thompson 1989). These changes reduced Q and Q* in the 10th- to 11th-century AD time period that limited water available for agriculture for field systems adjacent to the Chimu capital of Chan Chan. These examples document A, A* land loss and Q, Q* water loss events originating from natural climate change events that later influenced intensification progress and defensive measures against flood damage and drought as detailed in Chapters 3 and 4. Decreased food production rates over time, as can be calculated by Equation (4), also affected the marine resource base of coastal societies as Chapters 3, 4 and 7 illustrate. Equation (4) provides the mechanism to estimate changes in local population size responding to local changes in irrigation water and/or groundwater supply used for agriculture as dependent upon the severity and duration of a drought. Water availability changes over time likely motivated Andean societies to develop maximum productivity strategies through technology advances to get the highest crop yields per unit amounts of water and land to meet population demands together with developing crop storage housing (Bawden 1999) to sustain

Figure 2.7 Aeolian sand infiltration – Supe Valley agriculture moves from wide, down-valley to narrow, up-valley field areas in the Late Preceramic Period due to sand covering field systems.

population through drought periods. This resource maximization strategy likely drove the pursuit for optimum field system design and use designs; key to this were advances in technology as the means to increase food productivity given fixed amounts or arable land, water and labor; details of the optimization procedure used are subsequently detailed.

An additional exposition of the use of Equation (1) derives from theories related to the development of societal complexity and political development (Hastorf 1993:16–17) based upon population size and density (Spooner 1972; Cohen M. 1977; Cohen, R. 1978). From Equation (1), as V population exceeds the equilibrium V* value representing food supply balanced with a population needs,

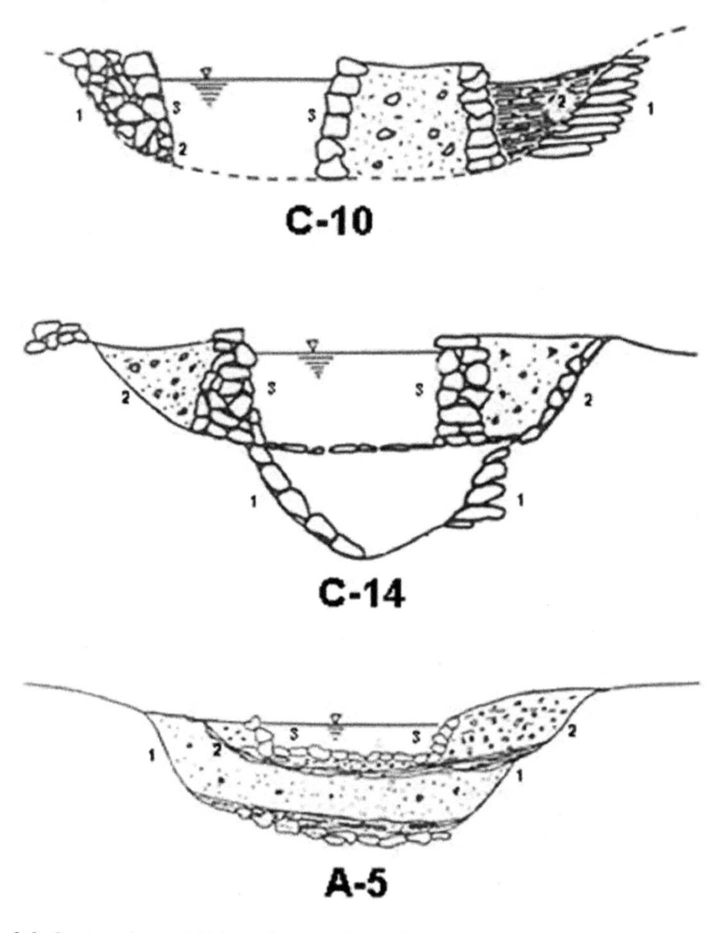

Figure 2.8 Contraction of Chimu Pampa Huanchaco canal cross-sections due to 9th- to 10th-century AD drought. Early (1) to late (3) canal profiles.

then (1/F) (dF/dt) becomes negative. This condition is represented by point P in Figure 2.2. Clearly as this condition develops, it is apparent to the ruling elites that remedial action is necessary to reverse this trend before privation and disassembly of previous societal norms occur to alter the earlier established composition of a society based on an adequate food supply. From Equation (1), a remedial action is to increase A*, the amount of available arable land that can be put into production with extended use of low slope canal systems (L_c*) provided by an increase in Φ/Φ* and an increase in Q* – provided extended arable land, water and technology resources are available. Note that from Equation (1) that (1/F) (dF/dt) has an area dependence of $(A^* - A - \Delta A^*)^{9/2}$ so that (1/F) (dF/dt) rapidly

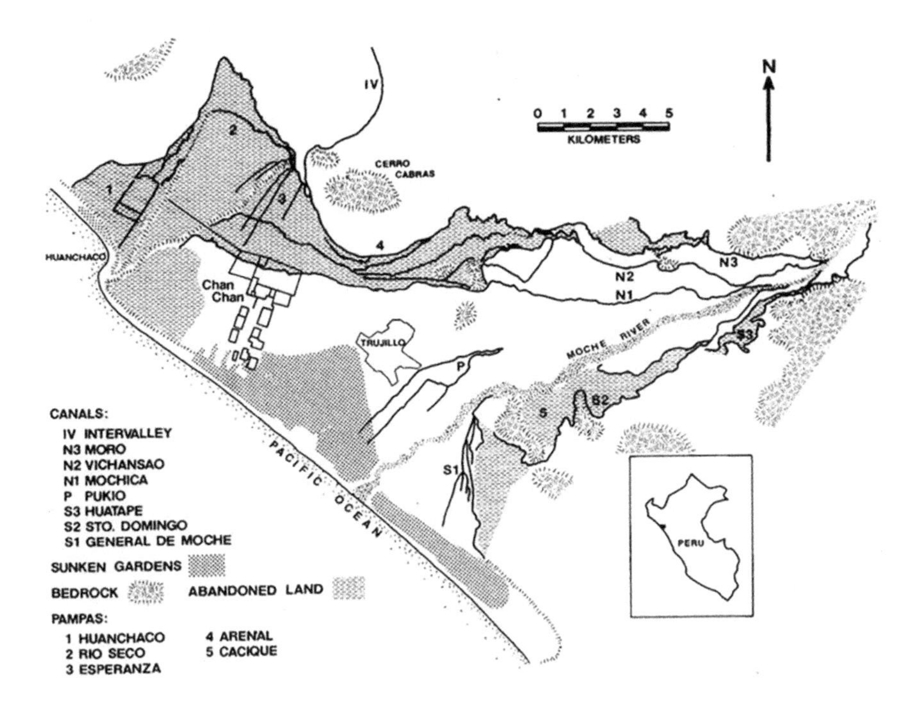

Figure 2.9 The Pampa Huanchaco canal system with notation of canal cross-section profile locations.

becomes a smaller negative value as A* increases. This leads to surplus labor being employed on surplus arable land to promote a near zero slope in Figure 2.2 for (1/F) (dF/dt) which then approaches a new maximum in food supply to feed the increasing population. The expansion of canals to irrigate previously unused lands is well represented by intravalley canals in Peruvian north coast valleys as well as use of intervalley canals to bring water over long distances from large flow rate rivers to valleys with fewer water resources to address concerns related to increasing population in areas with limited arable land and water resources (Kosok 1965). From Figure 2.2, point P then transforms to point Q consistent with Equation (1) formalism. This step is consistent with elite management supervision to direct changes in the social structure to implement changes in agricultural strategy proceeding from over-population effects.

Advances in technology (Φ/Φ*)

The benefit of higher agricultural technology levels is next considered. The importance of technology as it relates to the enhancement of the political economy of a society as well as its path toward complexity has been noted by key authors involved in modern economic theory (Romer 1986; Hidalgo and Hansmann 2009). For the present analysis, the ability to measure canal slopes with greater precision relates directly to the amount of land that can be productively farmed downslope of a canal. Larger coastal field areas can be used when lower canal slopes are achievable as low slope canals placed high on hillsides include more downslope farming area. Assuming additional farming land is available for canal placement, then given surveying accuracies sufficient to measure small canal declination slopes, A/A* is assumed proportional to the technology ratio Φ/Φ*. Here Φ* represents the minimum canal slope possible given the accuracy achievable by contemporary surveying instruments used in conjunction with known *yupana* calculation and *quipu* and *chimpu* data storage methods (Ortloff 1995, 2009; Menninger 1970) to process data. Substituting A/A* proportional to Φ/Φ* into the basic equation and integrating over time, the food supply F becomes:

$$F = F^* exp\{(L_c^* - L_c)/A^{*1/2}\,'\Omega^{1/2} + (V^* - V)/A^{*3/2}\,'\Omega^{3/2}]$$
$$(Q^* - Q)|Q - Q^*| \bullet (\Omega + 1)\, t^2 / 2A^{*3}\,'\Omega^3\} \qquad (7)$$

where $'\Omega = (\Phi/\Phi^*) - 1$. Note that when land, labor and water are optimally employed (i.e., these quantities are equal to their * values) then F = F* as expected. As $\Phi/\Phi^* \rightarrow 1$, $'\Omega \rightarrow 0$ and F increases. For a fixed water resource usage $Q < Q^*$ implying that $\Delta Q^* > 0$, or equivalently, not all water available can be used due to topographic landscape accessibility conditions. Given canal lengths $L_c < L_c^*$, labor resources $V < V^*$ and $A < A^*$, then as $'\Omega$ decreases toward zero, F increases due to small denominator $'\Omega^{1/2}$, $'\Omega^{3/2}$ and Ω^3 terms. This indicates the importance of surveying technology for coastal societies to build low slope canals that increase downslope farming areas that increase food output when land, water and labor resources are utilized at their maximum levels. Figure 2.10 shows surveying accuracies (in degrees) for several major Andean societies in different time periods and illustrates that from early surveying accuracies available at ~2500 BC at the preceramic site of Caral, refinement in surveying accuracy occurred over time to accurately measure lower canal slopes. Included in this figure at a comparable time period is a band representing the minimum slope capability of Roman surveying engineers to illustrate the importance of this technology to both Old and New World societies. While less steep slopes are commonplace along the length of many Roman aqueduct constructions (Hodge 2011:347), one major difference exists between Roman and later Late Intermediate Period (AD 1000–1430) Chimu canal designs: Roman water engineers kept the canal cross-sectional width constant and varied the sidewall height according to water depth changes brought about by slope decrease sections. Some exceptions to this rule are noted (Gallo 2014:122) where canal sections are locally widened to reduce flow velocity to precipitate settling of entrained silt and soil particles to facilitate, by decantation, cleaner water to potable status. The Chimu in certain specific cases utilized a different canal design strategy that varied canal cross-sectional shape, wall roughness and slope in a manner cognizant of sub- and supercritical Froude number

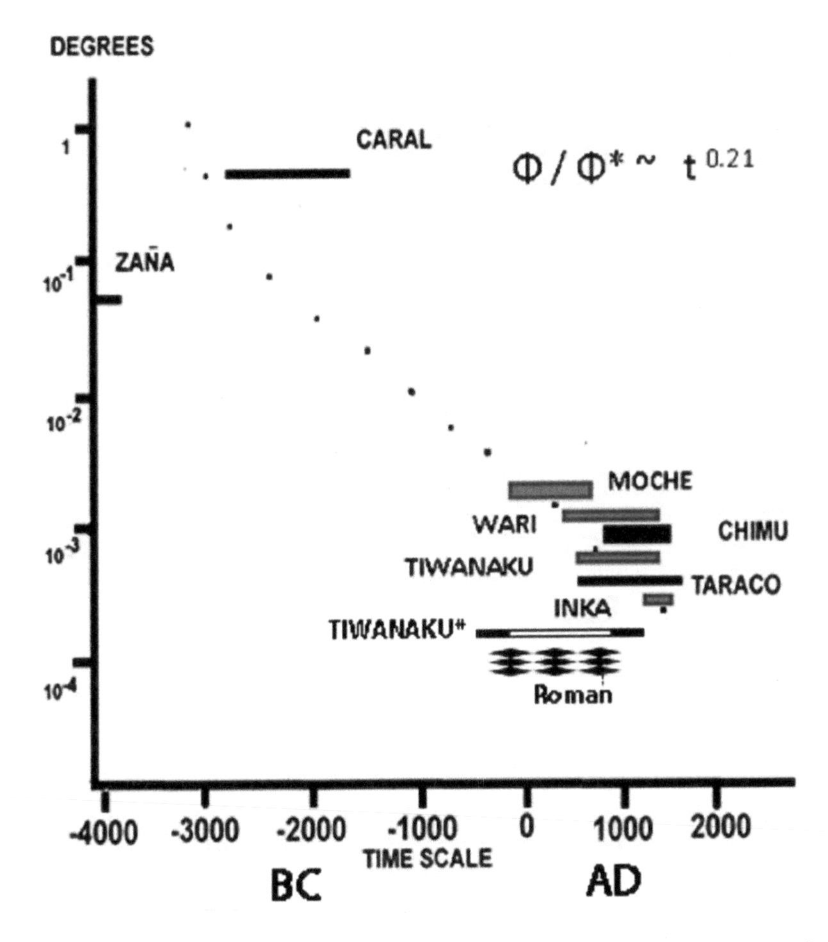

Figure 2.10 Surveying accuracy improvements over time for major ancient Peruvian societies.

effects on cross-sectional shape changes, particularly seen in canal expansion and contraction sections (Ortloff 2009; Ortloff et al. 1985) in their Chicama-Moche Intervalley Canal. The Chimu canal designs thus anticipate formal discovery of open channel hydraulics technologies in western science by ~1000 years. One canal (now destroyed), labeled Tiwanaku* in Figure 2.10 emanating from the Tiwanaku River about 1 km distant from the city of Tiwanaku (the Waña Jawira canal, Kolata 1993:226) had the lowest slope (0.0001 degrees) thus far encountered in the ancient Andean world; adjacent to this canal was a drainage canal flowing in the *opposite direction* to the Tiwanaku River. From Figure 2.10, surveying technology apparently advanced by technical transfer and/or diffusion between societies or from a creative invention within a society with little contact with more advanced societies in the post-Middle Horizon time period.

Although disruptions in societal continuity, as indicated in Figure 2.4, occurred from extended drought in the 7th and 10th to 11th centuries, some technical transfer likely occurred between successive LIP societies sharing the same coastal areas and sharing common interest in canal based irrigation systems as indicated in Figure 2.10. Elsewhere, for societies in different ecological zones (the Tiwanaku in particular using raised field agriculture), surveying accuracy appears to be independently developed to a high degree as evidenced by one canal leading water from the Tiwanaku River to a field system showing a slope of 10^{-4} degrees. Figure 2.10 indicates a steady progress in surveying accuracy refinement supporting the need for bringing into production more land area to support growing populations.

For any of these paths of technical advancement, guidance as to how Andean technology developed can be obtained from basic equation insights based upon the agricultural system optimization goal. If the rate of technology growth $\partial(\Phi/\Phi^*)/\partial t$ of an Andean society originates from an existing, highly developed body of surveying accuracy knowledge ($\Phi/\Phi^* \rightarrow 1$) then the rate of technical advancement derives from an extension from the existing technology level. In mathematical terms, $\partial\Phi/\partial t = k\Phi$ or $\Phi = C\ e^{kt}$ where k and C are constants. The preceding result indicates exponential growth in technology. Substituting into the basic equation, integrating, and examining time dependence, the food supply $F \sim (e^{kt})\ (k\ t - 1)$ results from which $\partial F/\partial t \sim k^2\ t\ e^{kt}$ governs the increased rate in food production with time as a result of exponential technology increases. This indicates that high levels of technology growth proceed from an earlier base of technology similar in form to that observed in modern societies where knowledge and invention multiply rapidly over short time periods. More typical of modern societies, exponential growth in knowledge is in the form of a branch-like structure where new fields of science originate as new discoveries are made. Although high levels of technology growth are generally beneficial to society this trend also implies concentration of wealth and power into the creator class responsible for the technology creativity process. Thus societies may fragment into privileged classes based upon uneven distribution of wealth and knowledge as the basis for a top-down societal control system ruled by privileged classes. Of course, once a privilege class has been established, there is no guarantee that its rule will be beneficent with regard to the public interest while garnering yet more power and social control – history provides many examples in this regard (Durant and Durant 1968). While Equation (7) is based on mutual benefits of agricultural technology to all levels of a society, there is a point in time where the creativity of individual farmer collectives is superseded by large scale projects that require top-down organization of logistics, labor and higher levels of technology Φ/Φ^* to implement. Project implementation in this case requires a labor tax obligation extracted from a population under state control – a practice familiar to Inka state governance requiring agricultural land management and division for individual societal groups as well as partitioning selected lands used to support royalty and religious branches of government. Two versions of Equation (7) then arise – one version for the general population where a lower agricultural technology Φ/Φ^* level is required for basic crop types that benefits the general population's food requirements and another Equation (7) version requiring a higher level of Φ/Φ^* used for specialty crops on more fertile lands requiring specialty trained labor to farm for Inka royalty's specialty food needs. For this case $V << V^*$, $A << A^*$ and a higher level of Φ/Φ^* is required to farm specialty crops in smaller royal farming areas to provide royalty's food requirements (Chapter 6) describing the Tipon water system is a prime example of specialty crops for Inka royalty. Based on this observation, highest levels of hydraulic technology were not only used for specialty crops but employed for aesthetic water display purposes for royal compounds as such displays required technologies well beyond those required for agricultural purposes.

Another path for Andean technology improvement over time based on surveying accuracy of low canal slopes involves technology advancing by a power law $\Phi = K\,t^n$. From Figure 2.10, the time exponent n is small ($n < 0.2$) over time between different Andean societies indicating somewhat limited inter-societal technology exchange. This follows from Andean societies being widely dispersed geographically over different ecological zones with widely varying water supply conditions (Figure 2.2) which require agro-technologies only useful for their ecological conditions. A further impediment to technical transfer originates from societal collapse/modification periods (Figure 2.4) originating from climate change episodes which interrupt the continuity and sustainability of societies to limit inter-societal technical transfer; this effect would naturally stimulate intra-societal innovation development. For societies with satellite sites operational in different ecological zones to exchange and trade resources (according to Murra's Vertical Archipelago observations, Murra 1975), multiple technological exchanges between sites and a central governing city occurred to broaden the agricultural technology base available to a society. An example (Bandy 2013; Ortloff and Kolata 1993) of innovative technology specific to one society based on its unique ecological conditions (groundwater raised field water supply) underwrote rapid Tiwanaku development from AD 300–1100 and underscores the Ω^2, $\Omega^3 << 1$ dependencies shown in Equation (7). Extension of groundwater control technology to Tiwanaku's city areas through use of a perimeter canal (Chapter 5) encircling the ceremonial and monumental structures of the city to promote rapid post-rainy season ground drying, foundation stability of major monumental structures and draining of the Semi-submersible Temple by groundwater height control over Tiwanaku seasonal rainfall changes is a further example of advanced hydrological technology applied to the Tiwanaku intra-city environment (Ortloff 2014b; Ortloff and Janusek 2015); details of this technology are given in Chapter 5. The Chimu development of open channel flow technology to support Peru coastal valley irrigation systems and intervalley water transport by mega-canal construction (Ortloff 1993, 2009, 2014a; Ortloff et al. 1985) attest to a specific hydraulic technology development attuned to arid ecological conditions and drought remediation. These examples illustrate that different societies develop different technologies specific to their ecological landscape conditions with limited technical transfer from older and distant societies as Figure 2.10 implies. For different societies occupying the north coast Peru area, technical transfer appears to be a stronger possibility. For stable climate periods between major 7th and 10th- to 11th-century drought events, a high rate of internally developed agro-technology apparently existed driven by population increase demands (or conversely, technical development allowed for population increase). In summary, surveying accuracy improvements combined with advanced hydraulic/hydrological technology appears in individual Late Intermediate Period Andean societies sharing common landscape/water resource features with limited technical transfer from earlier sources under different landscape/water resource conditions; thus Φ/Φ^* increase is a main contributor to intensification progress according to the large effect of the small denominator Ω^2 and Ω^3 terms in Equation (7).

Optimal agricultural strategies: coastal Peru, highland Bolivian and Inka field system designs

Questions arise as to the optimum agricultural strategy given land, water, labor and technology resources to exploit the lands' agricultural potential. As initial field and irrigation system designs improve as new strategies and technologies were implemented that contributed to increased productivity, agricultural systems evolved toward an optimum design. The question arises as to what is the optimum field system design for different site ecologies and are these designs achieved initially by thoughtful insight guided by a form of scientific and economic principles known to these societies – or in an evolutionary manner by trial and error observations of improvements in agricultural productivity by different societies. Initially, it may be posed that early societies experimenting with agriculture developed individual field plots (Figure 2.11, Strategy 1) that had individual water sourcing and represented lands farmed by individual kin *allyu* groups. New population transfers from outside locations migrating into the same area and/or additional people from the expanding population base desiring to duplicate similar plots to those already in existence may have had access only to marginal lands sourced by long canals to these regions as the best watered lands had been settled earlier by first arrivals. Here the assumption is that the more easily productive lands requiring the least amount of labor were initially occupied by groups exploiting readily available field areas conveniently close to the water sources (Figure 2.11,

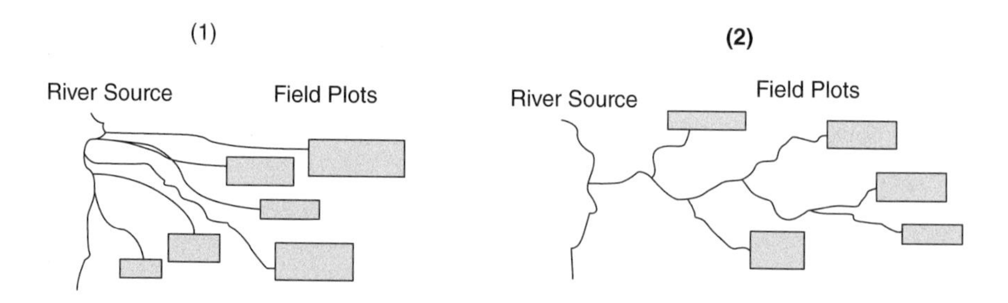

Figure 2.11 Evolution of bottom-up to top-down agricultural strategy – Strategies 1 and 2.

Strategy 1). Subsequent groups arriving at later times thus only had land areas requiring longer canal paths to distant fields (Figure 2.11, Strategy 2) requiring increased labor demand to farm. If realization that voluntary cooperation rather than competition between resource competing groups yielded higher productive use of land/water resources for all involved through canal path redesign and cooperative use of labor under a management consensus structure (Strategy 2), then societal integration proceeds to the first step in a bottom-up manner.

While such Strategy 1 → 2 developments appear a logical step in the path toward social complexity, the basic equation provides a quantitative basis to illustrate that cooperation rather than competition leads to economic advantages for all and provides a logical path with economic advantages to more intricate social structures preceding state development. The combined basic equation is given as:

$$(1/F)(dF/dt)=\left\{\begin{matrix}\left[\begin{matrix}(L_c*-L_c)/(A*-A-\Delta A*)^5\\+(V*-V)/(A*-A-\Delta A*)^6\end{matrix}\right]\\ \bullet(Q*-Q)|Q-Q*|\,t(\Phi/\Phi*)\end{matrix}\right\} \quad (8)$$

Non-cooperative Strategy 1 requires $L_c > L_c*$ and $V > V*$ so $dF/dt < 0$ meaning that Strategy 1 is inefficient compared to Strategy 2 for which $L_c \leq L_c*$ and $V \leq V*$. For Strategy 1 to be sustained for new, incoming migrant groups and/or increased population derived from original kin groups, long canals to outlying land areas require increased labor to manage leading to $(1/F)\ dF/dt < 0$. Here an inefficient bottom up managed system by a core of kin-based farmers (Strategy 1) can evolve into a cooperatively managed top-down Strategy 2 provided voluntary concession to an oversight management body can provide tangible productivity benefits to all through increased field system yields. This process involves original and newcomer groups exploiting agricultural resources collectively and sharing common societal objectives. Alternatively, for a combined field system operated in cooperation mode where L_c is smaller representing a more efficient canal water distribution system requiring less V labor to construct and maintain, a larger A area leads to a smaller $(A* - A - \Delta A*)$ in Equation (8). This results in a rate of food production dF/dt increase. A further interpretation of Equation (8) involves L_c approaching its more efficient L_c* and labor approaching its more efficient $V*$ so that $dF/dt \rightarrow 0$ which represents the maximum food production that the combined Strategy 2 area can produce. These interpretations point to the cooperative Strategy 2 having economic benefits for both groups.

This development process, applied to the Tiwanaku society (Janusek and Kolata 2004; Kolata 1993), can be characterized (Scarborough 2004, 2015) as *labor-tasking* evolving into *techno-tasking* as early, labor intensive development of separate raised field plots yielded to advanced raised field system development with integrated global features such as groundwater height management by large-scale canal bypass water supply and drainage features and mega-construction projects such as the Koani trans-pampa road system (Kolata 1991, 1993) providing a lake rise barrier and transport to across-lake farming and town sites to produce food production benefits obtained through technical advances commensurate with progress from bottom up to top-down elite management structures. Changes to top-down authority usually involve construction of elaborate elite palaces, religious edifices and ceremonial centers only possible from top-down control of the population. An example of this process characterizes management of the Middle Horizon (AD 700–1000) Tiwanaku raised field systems (Janusek and Kolata 2004; Kolata 1993; Bruno 2014) where the technology for surface and groundwater control to support agriculture during altiplano long-term climate and short-term weather fluctuations demanded broad managerial oversight for the collective good of all members of the farming community rather than for individual member groups (Williams 2006). For this example, multiple individual operational center structures occupied by small *ayllu* groups (Goldstein 2005) were successively incorporated into higher-level, larger-scale oversight management structures on the Koani plain with "*ayllu* pre-state social formations that functioned by community consensus" ultimately acceding authority to top-down management oversight for major projects that required enlisted labor from all groups to accomplish agricultural enhancement goals (Kolata 1993). While some kin groups would undoubtedly concede independence to a controlling, oversight group to govern projects based upon best use of labor, other non-cooperative groups wishing to maintain their independence and self-rule identity limit evolution from a bottom-up to a top-down societal structure as evidenced by Late Intermediate Period societies in the Huarochiri area of north central Peru (Lane 2009, 2017). Recent research (Goldstein 2005; Williams 2006; Albarracin-Jordan 1996; Bermann 1994) has indicated that *allyu* groups and immigrant groups establishing residence at Tiwanaku established "nested hierarchies" characterized by groups that specialized in manufacture of ceramics and other specialized crafts while maintaining their ethnic roots thus forming a loose ethnic confederacy of politically autonomous communities who shared only some of the cultural and ceremonial ties to the capital's elite hierarchy. While this separation of groups involved in manufacturing activities may be confirmed by further research, certainly all parties shared interest in cooperating in guaranteeing agricultural productivity through cooperative efforts as this was the security base that permitted separate ethnic groups to pursue their individual pursuits. For the efficiency of agriculture productivity, some concession to elite management with oversight to devise and implement new strategies and collect labor to implement these strategies was vital.

New theories (Borsch and Sanger 2017; Crumley 2017; Dillehay 2017) that argue against the primacy of hierarchy as the desired model of a complex society and that power relations in decentralized form fit into this category. Here these theories question the basic concept that "simplicity" is the starting point and that only "complexity" is achieved; instead equivalent power relations are recognized as requiring large amounts of effort to establish and maintain governance without government. An early example of complexity without a hierarchical leadership structure is found at the Huaca Prieta-Paredones site (Dillehay 2017) in the Chicama Valley coastal zone. Here the emergence of social complexity in Preceramic times (10,000–4000 cal BC) was based upon exchange networks connecting different sites with different products (marine resources, agricultural cultigens, exotic trade goods); no elite architecture or burials with sumptuary goods were found at the Huaca Prieta-Paredones site indicating an early classless, egalitarian social structure. Later Preceramic-Formative Period developments (Dillehay 2017) indicated trends toward monumental architecture and specialty craft production within sites that indicated the emergence of a hierarchical structure to define ritual patterns and ceremonies for societal integration as well as to coordinate labor to build monumental structures, direct organized agriculture, manage trade relations and provide a sense of security under their direction.

An example of kin group control of productive areas using the labor efficiency advantage to rise to political power characterized the Tiwanaku elite political structure (Bandy 2013; Janusek 2004; Janusek and Kolata 2004; Kolata 1993) and the path to social complexity. As locally controlled management is sometimes better than elite corporate level management, bottom up structure may prevail as elite management choose not to disturb a process that functions efficiently without their intervention (Lane 2009; Lamberg-Karlovsky 2016:1–4). A further alternative relies on elite groups exercising control and mandating organization of individual farming groups into a collective. In his case, top-down management ceases to be a cooperative evolutionary end but rather an element of elite group control. An example of this is manifest in Inka societal structure where top-down control was institutionalized into state-controlled lands with elaborate labor taxation imposed upon society members (D'Altroy 2003; Morris 1998; Murra 1982a, 1980a, 1980b; Quilter 2014; Goldstein 2005; Baudin 1961) by the elite governing class. This dominant, top-down, elite controlled social structure prevailed within the Inka domain to assign elements of the population to available land resources (Baudin 1961:62–70) as well as for territories conquered by the Inka where *mit'a* labor service was imposed upon conquered populations to maximize their agricultural productivity to meet state ends. While specific to the Inka society, rule through elite manipulation of culture, traditions, resource management, traditional symbols and rituals applied to underwrite elite control of many New and Old World societies throughout history is manifest (Barwald 2000; Renfrew and Bahn 2000; Routledge 2013; Smith 2003; Goldstein 2005; Durant and Durant 1968).

For the present discussion, agricultural technology is assumed as a catalytic agent in production of a secure agricultural base to support complex societal development. Inherent to this agent and its implementation are additional considerations of alliance relationships between different kin groups, levels of organization and social obligation, sense of community to allow directives from a hierarchical class, competition between elites, control of agricultural technology, cultural identity, demographic pressure, environmental stress – to name a few factors that influence social complexity. While these considerations play an important role in social complexity evolution, without the technical base to support industrial agriculture and the hierarchical leadership to utilize it effectively, societal development is limited. Subsequent chapters describe this technical base and its importance to the development of the agricultural base of major Andean societies.

As climate change in the 13th- to 14th-century AD recovery period resulted in increased rainfall (Fagan 2008; Thompson et al. 1994:66), Inka administrators developed lowland agriculture to redistribute restive populations of conquered territories to farmlands distant from their homeland (D'Altroy 2003; Morris 1998; Murra 1980a, 1980b; Quilter 2014; Renfrew and Bahn 2000). This reorganization process was an integral part of Inka top-down management of resources to promote intensification. Some change from *andenes* terraces to more productive bottomland agriculture occurred due to well-understood canal irrigation practices, predictable and longer duration river water supplies than rainfall-fed terraces, more fertile bottomland soils, high productivity per unit labor input, more available land area and wider crop varieties less subject to temperature extremes. A change in food production rate going from terrace to lowland agriculture involves, for the same labor input, farming area and water supply, the following result from the basic equation:

$$F_{bottomland\ agriculture} - F_{terrace\ agriculture} = exp\left\{\left[\left[\frac{(L_c*-L_c)(Q-Q*)^2(\Phi/\Phi*)}{2(A*-A-\Delta A*)^5}\right]\right]\right\} > 0$$

The difference is largest when L_c is smaller (the case when supply canals originate from a nearby river), $\Phi/\Phi*$ is large (canal water distribution technologies are more advanced, controllable and have longer water supply duration than terrace intermittent-rainfall systems) and $A \rightarrow A*$ (easier agrosystem construction and maintenance on flat terrain with better bottomland soil fertility than on steep mountainside terraces to increase productivity of different crop types in a more temperate climate). From the preceding equation, the RHS is positive implying an increase in food production by transferring *mit'a* labor assets from terrace to bottomland agriculture thus supporting Inka management decisions for best use of labor to increase the productivity of lands under state control. While not all labor was channeled in this manner owing to lack of large bottomland areas close to highland Inka administrative centers, much terrace agriculture continued as it was well adapted to the mountainous terrain and high rainfall levels that prevailed over much of the highland Inka domain.

Cases exist for which Andean societies are characterized by dispersed villages emphasizing their independence from central authority or alternatively, opportunistically adapting to a changing resource base to maintain individual survivability. An example of the former lies in Early Intermediate Period Moche occupation of the Jequetepeque Valley characterized by little central authority (Dillehay 2001); a latter example resides in late post-Tiwanaku V society abandoning the central city under prolonged drought conditions for dispersed local sites near water resources to support farming (Kolata 1991). While structural anthropology theory (D'Andrade 2004) provides general conclusions on social phenomena and societal behavior patterns from observable phenomena, research specific to Tiwanaku societal structure (Janusek 2004; Janusek and Kolata 2004; Kolata 1991, 1996a, 1996b; Ortloff 2009; Ortloff and Janusek 2015) indicates the presence of late governing hierarchical structures in the form of elite palaces and monuments segregated by a perimeter drainage canal from secular urban housing districts (Ortloff and Janusek 2015; Chapter 5). The proposed transition from bottom-up to top-down management of agricultural resources reflects limited group development of early agricultural management to a later formalization by an elite management and ruling class – this based on higher agricultural production efficiency beneficial to all societal classes (Janusek 2004; Janusek and Kolata 2004).

Optimum field system designs

A further question asks how ideas are generated to improve agricultural system design. This posits a science that governs decision making perhaps codified from observations deriving from agricultural experiments (or intuition) that produced either positive or negative results related to field system productivity. Given that an optimum configuration exists for an agricultural system given area ecological conditions, the basic equation is instructive to determine these designs. The closer to optimum field system/irrigation network designs that Andean farmers initially chose (or evolved to) provides insight into their science. Key questions are: What is the optimum way to irrigate a field system by subdivided water-path networks? How close to an optimum methodology was practiced by Peruvian coastal and Bolivian highland societies from early to late times? To address these questions by use of the basic equation, first divide resources as follows: A/n, where n is the number of field plot subdivisions with their required amount of water supply Q/n and supply canals L_c/n. Here sufficient labor resources are assumed distributed as required to service n individual land plots, or equivalently, V/n. Substituting into the basic equation, insight into how land is to be subdivided for maximum access to water supply is given by:

$$\left[(1/F)\ dF/dt\right]\Big|_{divided} \sim n^{3/2}\left[(1/F)\ dF/dt\right]\Big|_{undivided} \tag{9}$$

The interpretation of this result is as follows. As the divisions leading the Equation (9) are A/n farming area division, Q/n water supply and V/n labor, as n increases, the rate of food production dF/dt increases by $\sim n^{3/2}$. This means that the limiting case is that each plant has its own individual water supply, attendant labor and land area. This is equivalent to a plant being in an individual 'flowerpot' which is sensible for greenhouse farms growing specialty plants and plants for genetic research activities. For general agriculture however, following Equation (9), the best way to efficiently section an agricultural area into divided subplots with individual water access involves a high number (n) of smaller land plots each with a proportionate amount of water and labor to optimize production. For increasing n subdivisions, $\{(1/F)\ dF/dt\}|_{divided}$ experiences an increased food production rate. Practically, a limit exists for the number of watered n subdivisions given constraints on the number of exposed water channel surface areas promoting evaporation loss. For typical irrigation systems (Figure 2.12b) of north-coast Peru (Moseley 1983, 2008; Moseley and Deeds1982; Netherly1984; Ortloff 1993, 2009), the optimum field system

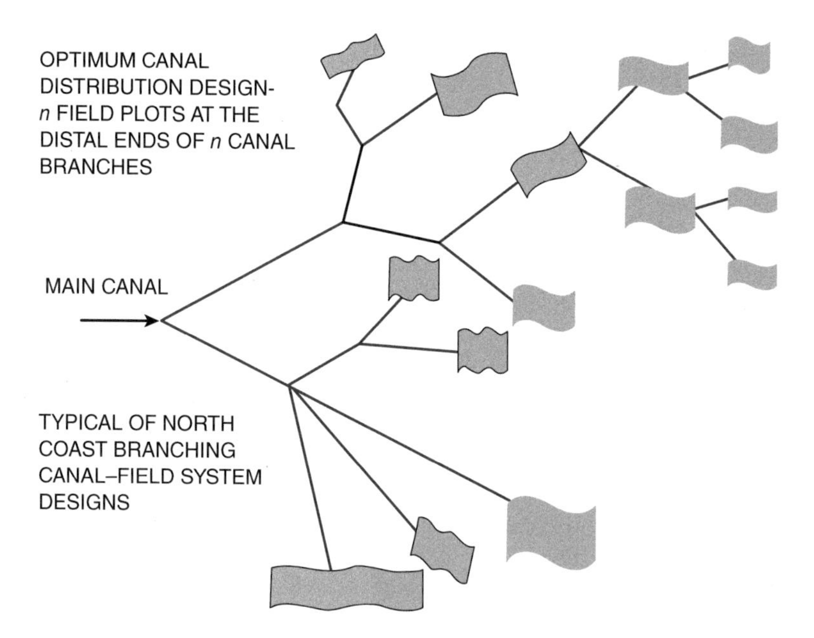

Figure 2.12 Basic equation optimum land partitioning configuration for canal supplied irrigation systems typical of Late Intermediate Period north coast Peru.

configuration consists of many sequential daughter branches from a main canal each serving an accessible field plot. This type of canal branching system was previously noted (Moseley and Deeds 1982; Deeds et al. 1978) in their studies of the Pampa Esperanza canal system adjacent to the Chimu capital city of Chan Chan. Figure 2.12a illustrates their version of the daughter canal branching network leading to individual field plots similar to those shown in Figure 2.12. Here the canal subdivisions are in the form of branches from a main water supply canal as a main canal subdivides into multiple daughter branches that further subdivide and feed water into terminal field plots which further branch into multiple channels each serving elevated berm planting areas. Each field plot is further divided into smaller sections served by individual water channels (Figure 2.12b). Here the optimum water distribution system is provided by an example from nature given by water distribution in a leaf. A leaf contains multiple distribution branches (Figure 2.16) that optimally (least water transfer energy involved) deliver water to leaf subareas.

This conceptual ideal underlies the canal network designs of coastal Peru systems that provided water to land areas in a manner similar to how nature distributes water through elaborate leaf canal systems. It may be suggested that such an observation from nature inspired Chimu thinking on how to design irrigation canal networks efficiently. For field system and irrigation network designs typical of the Chimu society of north coast Peru in the 9th to 14th centuries (Figures 2.9 and 2.12), field system designs existed based on the presence of gradually sloping farming land within alluvial river delta valleys using a river source to provide

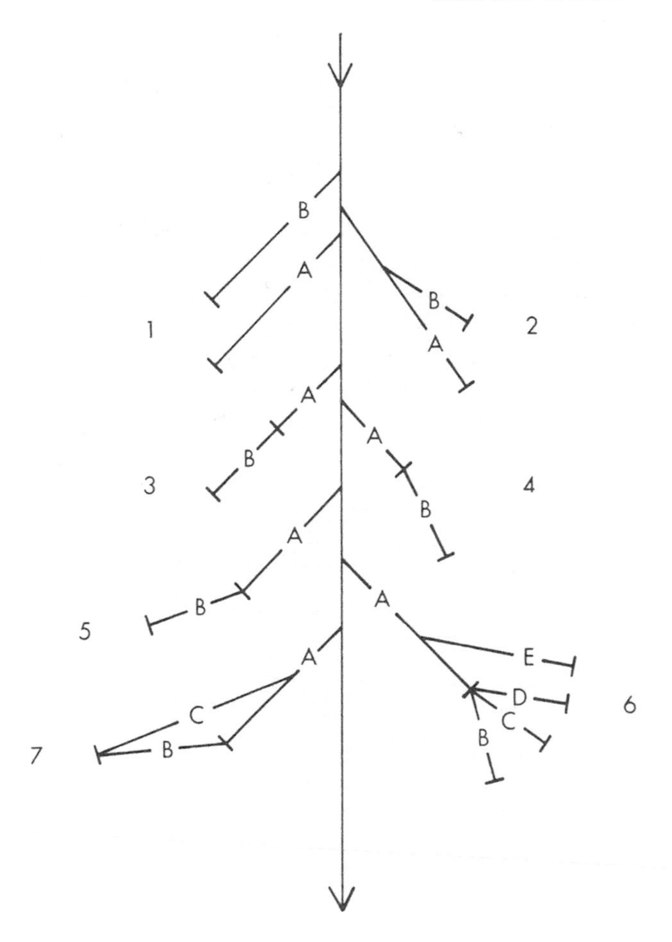

Figure 2.12a Canal branching in the Pampa Esperanza adjacent to Chan Chan (Moseley and Deeds 1982); distal ends of daughter branches associated with field systems shown in Figure 2.12b.

Figure 2.12b Typical Chimu field system sinusoidal water branching to agricultural berms.

irrigation water through canal networks (Ortloff 1993, 2014a, 2009; Ortloff et al. 1982). Figure 2.12 indicates that multiple-branch, daughter canals emanating from a main canal are individually supplying multiple subsections with water channels that provided the optimal way to increase food productivity as the basic equation predicts. The more field plots (high n) with their individual daughter canal water supply, the closer to optimal food production exists. This system reinforces the plant-like fractal efficiency of water delivery to individual field plot areas given the constraint that a land area fraction (ΔA*) is inaccessible due to topographic features that prevent access by canal branches.

This example reinforces a vision of Andean agricultural science that supported intensification involving complex field system design land-water use strategies evolved from observation and field trial results that led to optimum use strategies of land, water and labor resources. An optimum design is represented in a section of the Chimu Intervalley Canal as indicated in Figure 2.17 (Ortloff 2009:51).

Here the canal cross-section shaping changes of this ~1600-m canal section (Ortloff 2009, 2014a; Ortloff et al. 1985) produces a near critical flow at a given flow rate in the canal by manipulating canal cross-section shape, canal width, complex Froude number effects on geometric canal shape expansion/contraction changes and wall roughness. Details of this hydraulic science mastery are described (Ortloff 2009) and this example serves as yet another hydraulic optimization methodology developed in later stages by Andean societies.

For Tiwanaku raised field systems based on multiple agricultural mounds (berms) surrounded by water filled swales (excavated trenches penetrating into the water table about one to two feet), the question arises as to how to optimally distribute sectioned land areas so that each berm mound has an adequate capillary infused water supply to root systems while maximizing the farming surface area of the berms per unit surface area. If, for example, field system designs were to be configured as a series of individual circular mounds then the ratio of the exposed water perimeter surrounding a farming area mound to the internal area is 4/D where D is the circular area diameter. Clearly if D is large then the near center part of the mound has limited access to capillary water from surrounding swales to reach crop roots toward the center of the mound. This presents a mathematical problem to field system designers to figure out how best to maximize farming field plot area configurations with access to a minimum wetted perimeter surrounding a berm so that all of the farming area of the berm has equal access to capillary water supply from surrounding swales to maximize crop productivity and limit evaporative water losses. For cases for which water is supplied by surface irrigation canals, field system designs evolve about the number of sequential subdivisions to effectively use limited water resources to field systems; for agricultural systems that depend upon groundwater, such as raised field systems, somewhat different field system designs are required to maximize crop area per unit land area as the next example case demonstrates.

Contrary to the optimum water supply design to surface canals by a canalized river source, here groundwater lies at an approximate constant depth below the surface of Tiwanaku raised fields estimated to be ~130 km^2 in the Pampa Koani raised-field area (Kolata 1991, 1993, 1996b). While more sub-divided raised-field berms surrounded by swale water increase food production, there is a minimum raised field width sufficient to transfer capillary swale water to root systems and to preserve mound heat-storage capability (Kolata and Ortloff 1989a). Figure 2.13 shows a field system design where n interconnecting water channels distribute water to n raised-field ridges; here the higher n is and the more subdivision berms exist served by swales, the higher productivity results. The key to field system design is then to obtain the maximum farming land area under these constraints to increase yields.

As an example of typical raised field designs, the early raised fields at Taraco on the northwest boundary of Lake Titicaca (Figure 2.14) (Henderson 2011) are shown to conform to a near optimum design reinforcing that Tiwanaku agricultural science involved insight into earlier raised-field designs for productivity optimization. Within the Taraco raised fields are occurrences of imbedded shorter wavelength field systems within larger wavelength field systems indicative of water and heat storage requirements for specialty crops. The partitioning of raised fields had different optimum configurations for different crop types with different water requirements. This type of field area division was also evident at the Tiwanaku satellite farming area at Pajchiri (Ortloff and Kolata1989) where different crop types required different water supply system designs. Similar to the earlier Taraco field system are the later Tiwanaku field system designs (Kolata 1993:184) that mirror the same closely spaced, elongated farming berm land design indicative of some form of technical transfer over time of an optimal

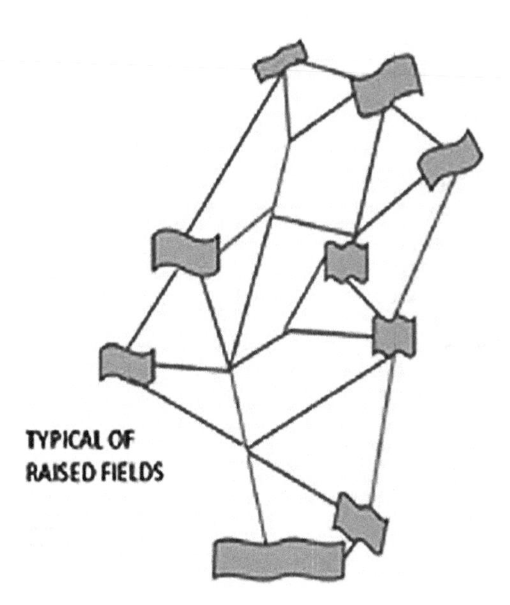

Figure 2.13 Basic equation optimal land configuration for systems typical of Tiwanaku raised field systems, 300 BC–AD 1100.

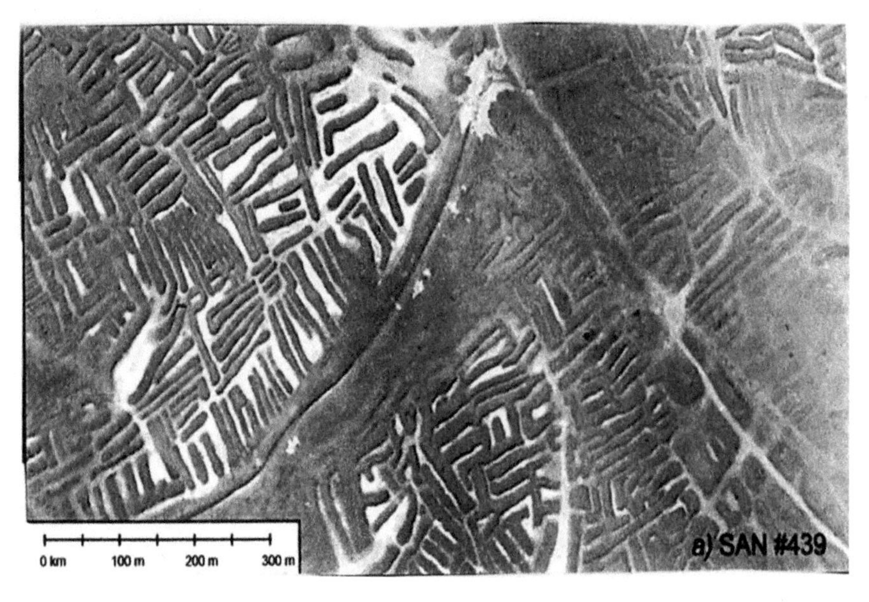

Figure 2.14 Aerial view of Taraco raised-field designs in the northeast Lake Titicaca area of Bolivia.

agricultural strategy (Figure 2.15). While the surface berm configurations were vital to optimize, in some way, agricultural productivity, excavations performed at the Lukurmata area of Tiwanaku indicated that experimentation related to the geometry of raised fields had occurred. Excavated berm profiles at depth below the ground surface showed earlier raised fields had much shorter widths and were more closely spaced together compared to later raised-field designs.

Research conducted on Tiwanaku's raised field agricultural systems in the Pampa Koani region (Tung 2012) and systems under Tiwanaku influence on the northwest regions of Lake Titicaca indicated use of advanced hydrological methodology underlying crop sustainability and yield improvement. Among the advances in agricultural science are usages of heat transfer technology to limit crop destruction by freezing during cold altiplano nights (Kolata and Ortloff 1989a; Ortloff 2014a:86–92) as well as hydrological/hydraulic control mechanisms providing groundwater height control to stabilize raised-field swale water height through seasonal changes in water availability (Ortloff 2014b; Ortloff and Janusek 2015). Additionally, raised-field technology has been shown the most efficient design choice to limit short-term drought effects on crop yield due to continual groundwater supply from intercepted rainfall over vast collection areas continually flowing toward the Lake Titicaca basin (Kolata and Ortloff 1989b) that stabilized the groundwater height in swales between farming berms. Analysis

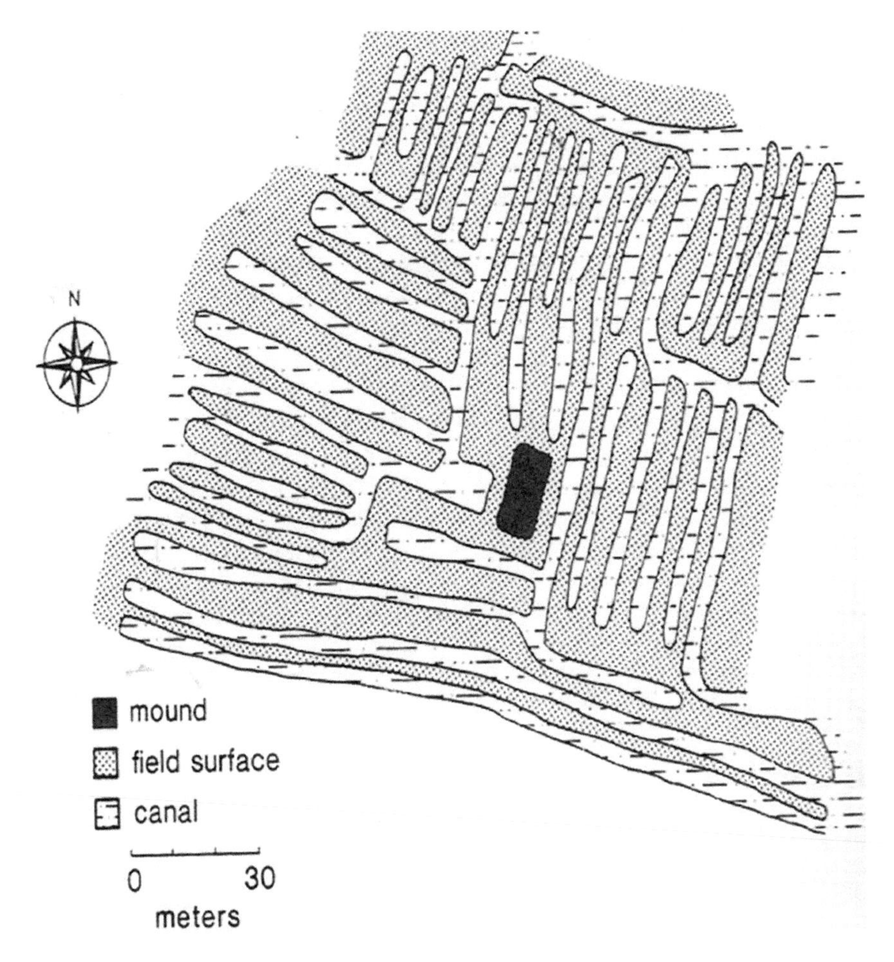

Figure 2.15 Raised-field system geometry from the Lakaya sector of Koani.

of groundwater control mechanisms in the Pajchiri agricultural area (Kolata 1996; Ortloff and Kolata 1989) reveals different berm heights and swale depths appropriate to different crop types. In the Pampa Koani and northwest Taraco regions of Lake Titicaca, different patterns of raised field lengths, widths and orientation are frequently inserted within more regular patterns – each pattern appropriate for the water needs of different crop types (Kolata 1996a; Ortloff and Kolata 1989a). Excavation raised-field berms in the Pampa Koani area indicated stone-base lining and clay layers (Kolata 1996:115–128) to limit cold capillary water transfer from deep groundwater regions into berm interior solar heated swale regions (Kolata and Ortloff 1989a; Ortloff 2009). Due to higher swale water temperatures from solar radiation input, the additional storage heat to berm interior regions limits convection and radiation heat withdrawal during cold altiplano nights to limit freezing damage to crop root systems. In other words, the latent heat removal for water to ice transition within berm interiors during cold altiplano nights is limited by additional heat transfer from elevated temperature swale water capillary heat transfer into berm interiors (Ortloff 2009:129–134). Examination of Taraco raised-field berm patterns (Henderson 2012; Figure 150) as well as Tiwanaku raised fields (Kolata 1993:184; Figure 2.16) reveals a berm shape majority consistency. These figures show that swales are interconnected leading to a continuous water path surrounding berms.

Given that berms must be somewhat narrow to allow swale water to diffuse to root systems, what does this imply about an optimum raised-field design? Consider first a thin rectangle **a** long and **b** wide and **a** >> **b**. The area is **ab** and the perimeter is 2**a** + 2**b**. Now if **b** is made wider to a width permitting swale water diffusion to root systems, then the wetted perimeter is only slightly increases as **a** >> **b**. This design permits a large berm farming area with the smallest wetted perimeter – more farming area is gained only by increasing **a** while maintaining **b** about the same. Following the preceding example of the elongated rectangle, when a typical berm is described as an elongated ellipse with major axis **a** and minor axis **b**, (**a-a** and **b-b**; Figure 2.14), then an **a/b** ratio from 10 to 14 appears to characterize average of berm geometries. Here **a** is the distance from the origin to the major ellipse axis and **b** is the distance from the origin to the minor ellipse axis. When **a** = **b** the ellipse becomes a circle with area $\pi \mathbf{a}^2$. This ratio for an

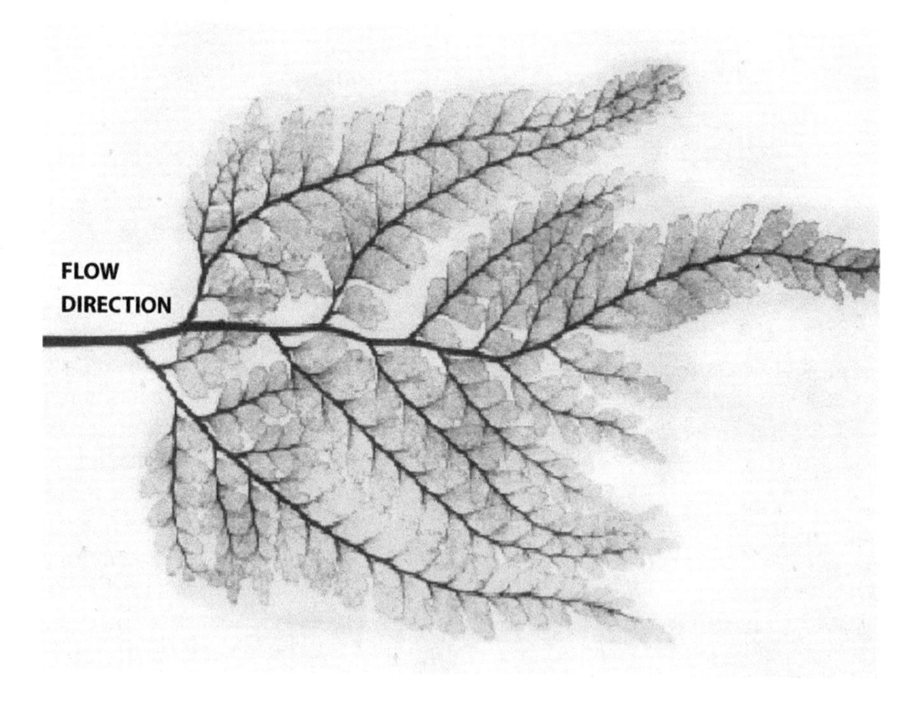

Figure 2.16 Fractal water distributions system in a leaf.

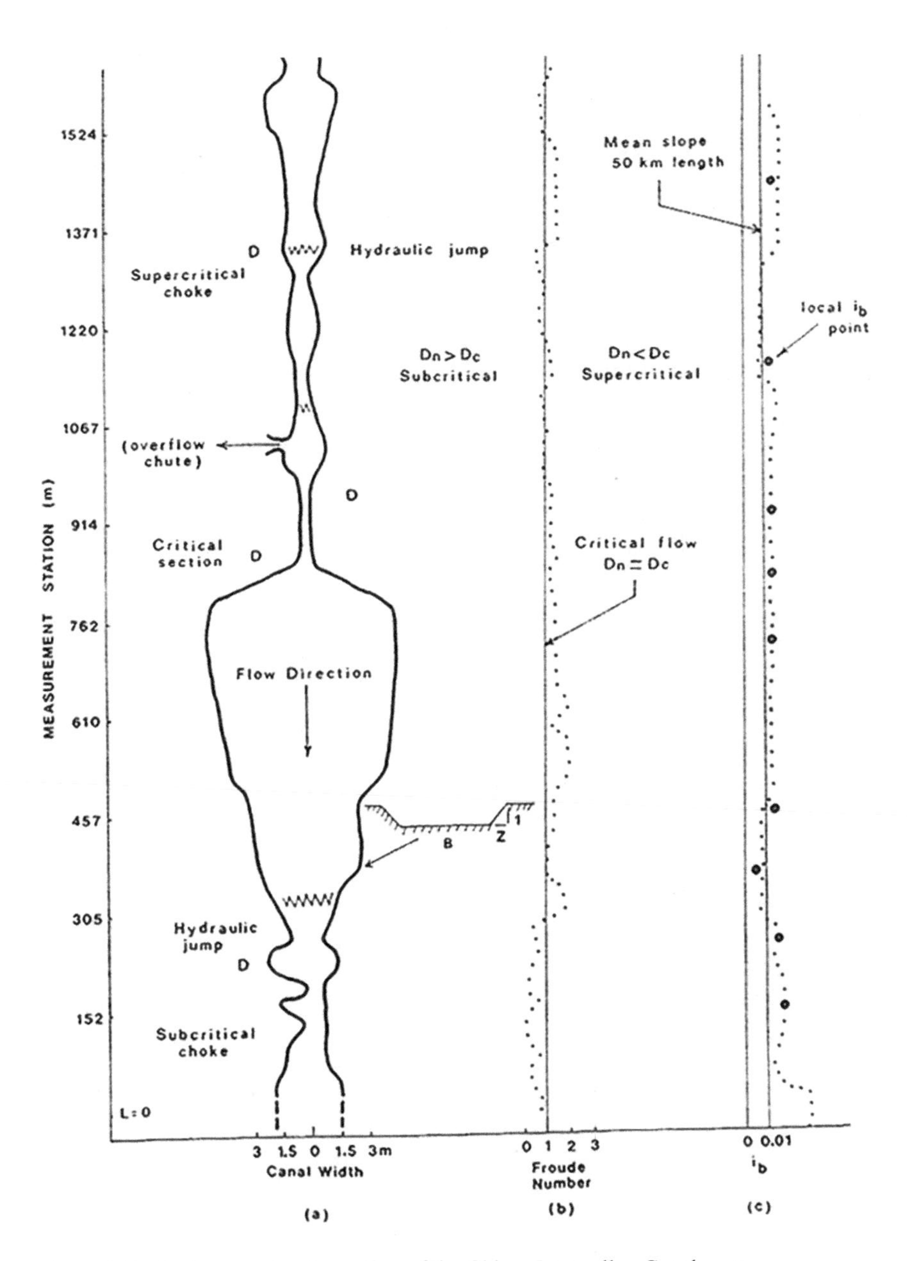

Figure 2.17 Design details of a section of the Chimu Intervalley Canal system.

Source: Ortloff 2009:51

elongated ellipse (**a** >> **b**) is significant in that the ellipse perimeter is a minimum for the given berm surface area (**π a b**) for this class of ellipse. This indicates that the berm pattern configuration yields the minimum wetted berm perimeter and thus as each elongated berm is self-sufficient with a narrow swale width water supply, the berms can be placed closer to each other and will be surround by swale water sufficient to reach plant roots in the centerline of the elongated, elliptical berm. In short, the width of the swales becomes a minor consideration as a narrow swale serves all water supply purposes so that larger swale widths are not necessary. Here the narrow berms (**b** << **a**) provide an easy path for sun-heated elevated temperature capillary water to reach berm interiors to supply water to root systems. This conclusion, in turn, reduces the exposed water surface area of the interconnected swales reducing evaporation loss that helps to locally maintain a constant groundwater profile to maintain constant swale water height. The net effect is that a greater number of closely spaced berms can be watered properly to maintain the crop freezing defense while the greater farming berm area per unit raised field area produces more agricultural yield. In other words, more closely spaced berms similar in design to Figures 2.14 and 2.15 provide more farming area per unit land area while maintaining crop protection against freezing temperatures which is a common nightly occurrence at the ~4500 m altiplano altitude. These advantages are a key indicator of an advanced agricultural science being employed to protect and increase the yields of raised field agricultural systems. The **a/b** ratio of individual berms contains important information related to Tiwanaku engineering practice as their design incorporates a further level of optimization to limit swale water area to reduce evaporation losses and provide additional crop yield benefits.

From the basic equation,

$$\left[\left(A^*-A-\Delta A^*\right)^{3/2}\Big)/\left(Q^*-Q-\Delta Q^*\right)F\right]\left(dF/dt\right)=\left[\left(L_c{}^*-L_c\right)/\left(A^*-A-\Delta A^*\right)^{1/2} + \left(V^*-V\right)/\left|A^*-A-\Delta A^*\right|^{3/2}\right]\bullet\left|Q-Q^*-\Delta Q^*\right|t\left(\Phi/\Phi^*\right)/\left(A^*-A-\Delta A^*\right)^{3/2}\right]$$

for V = V*, Q < Q*, and for L_c considered as the total swale water path length (Figure 2.13) then the question is to obtain a field system configuration where the total L_c is small and A is large where A is considered to be the total berm area per unit field system area. This configuration then permits large berm farming areas to be close together to maximize total farming area per unit field system area. From the basic equation, for small total L_c and large A then dF/dt is a positive constant indicating a large rate of food production. Maximum berm farming area per unit area is achieved when the smallest total length water course serves the largest berm area. This configuration then posits close separation of individual berms each with the minimum wetted berm perimeter that occurs at **a/b** ~ 10–15. Observation of Tiwanaku field system verifies that berms following this rule are configured close to each other (Figures 2.14 and 2.15) confirming optimum designs for both individual berms as well as their separation placement distances in field systems. Summing up results of this later section, analysis indicates: (1) food production is greater when land areas are finely subdivided (large n) and surrounded by narrow swale water channels that can capillary transfer water into berms to supply root systems; (2) for a fixed berm land area, the optimum shape of a raised-field berm element is **a/b** ~ 10–15 because this elongated ellipse shape has a short wetted perimeter length for given berm area and narrow berms shorten the water path to root systems; (3) as each berm is a self-sufficient agricultural unit with its individual swale water supply, then the width of the swales can be minimized thus permitting individual berms to be close together – this permits more land in farming use per unit area.

As elliptical berm area is A = π **a b** and the circumference C is, according to Ramanujan's approximation,

$$C = \pi\left\{3(\mathbf{a}+\mathbf{b}) - \left[(3\mathbf{a}+\mathbf{b})(\mathbf{a}+3\mathbf{b})\right]^{1/2}\right\}$$

Then R = A/C can be shown to be ~4 for typical elongated **a/b** ratios (10 to 15) shown in Figures 2.14 and 2.15. Substitution of different **a/b** ratios is shown to result in lower R values indicating higher perimeter values for a given berm area.

Top-down management implementation of raised-field systems

For 10 < **a/b** < 15, individual berm area is 250 < A < 315 m^2. The question posed to elite management is for a ~20,000 population size, how much berm area is required given a yield/m^2 of berm area? For **a/b** = 15, A = **a b** = 15 $\mathbf{b}^2$ (again assuming a rectangular berm for simplicity) thus the total berm area required is known to feed the population given the agricultural yield per unit berm area. The next requirement is to utilize the most reliable, fertile land area close to the lake where swale water height is constant; this requires a minimum berm width (5 m) with a minimum wetted perimeter for a given berm area such that berms may be placed close together. Given the individual daily food consumption per individual, and the averaged total food productivity per berm times the number of active berms, then the total berm area necessary to support 20,000 city dwellers may be made.

Causality

As noted by Wittfogel (1957a), "thus a number of farmers eager to conquer arid lowlands and plains are forced to invoke the organizational devices that offer the one chance of success: they must work in cooperation with their fellows and subordinate to a directing authority." This observation coincides with the transition from Strategy 1 to 2 described previously using both Tiwanaku and Inka examples (as well as Complexity Theory results. Also, an economic efficiency advantage in terms of increased food production per unit land area was noted between different agricultural strategies and optimized berm geometry and spacing using Equation (8). Presumably some form of calculation involving best usage of land, water, labor and technology available for different water ecologies and landforms preceded enforcement of a later Strategy 2 characterized by top-down central management. Labor efficiency use directed toward maximizing food production directed from a directing authority likely had an economic basis. As to the question of irrigation technology as causal to cultural complexity and state creation, Fagan (2011) states "the main interest of rulers was food surplus which supported the state . . . with the scale of the irrigation system directly proportionate to the scale of management." From Similitude Theory analysis, it can be observed that food surplus was achieved by using optimum field system designs and a distribution of water, land, labor and technology at * values to promote maximum food production. Here technology was a vital factor to accomplish this goal as continual refinement of surveying technology over time was achieved by many of the coastal Andean civilizations (Figure 2.10) utilizing river sourced irrigation canals. In retrospect, from the present analysis, it appears that social complexity coevolves with technical complexity that requires a management system capable of directing cooperative (or forced) labor and has knowledge of technical details to optimize agricultural production to meet increasing population needs with assurance of results. This is in agreement with Lamberg-Korlovsky's (2016) statement "irrigation, and its ever increasing technological requirements and consequent population increase are co-evolutionary processes in the development of cultural complexity" and also the sense of the Orwellian statement that "man is only as good as his technology allows." Previous chapters indicate an extraction of hydraulic lessons from early canal systems that likely indicate the realization of the potential to control agriculture once water technology could be mastered. The interest to see how water control was vital to optimizing food production through sophisticated field system designs specific to local ecological conditions apparently permitted an accompanying management class to exploit this potential new technology to ensure their authority as the intelligence center directed toward the betterment of their subject's welfare and betterment.

Conclusions

Andean agriculturalists improved field system designs with land, water, labor and technology strategies that optimized food production. Underlying their initial choice of a field system design were optimum land/water partitioning configurations close to optimum configurations predicted by basic equation analysis. The science governing these design improvements likely derived from field trials conducted over time to codify advances in technology that improved agricultural productivity – trial improvements from intuitive insights and agricultural experiments with positive and negative results undoubtedly played a role in discovering optimization technologies. Use of the basic equation provides the foundation to replicate the thought process and logical decision making of ancient agricultural engineers albeit in a format vastly different from western science notational conventions. Here economic advantages in the form of increased agricultural production and efficient use of labor and technology guide decision making to produce economic advantages for a society much in the same way as done in modern society. Similarly, refined surveying technology used to increase farmland area provided addition food resources that supported population growth and instilled advances in societal complexity to manage effectively what they created. As a fraction of this population growth is the labor source to maintain the additional agricultural land, an equilibrium condition is achieved in accordance with the *Law of Diminishing Returns* where all members of a population have either voluntary or assigned roles that contribute to the efficient running of the state by an elite managerial class. Use of the basic equation provides a quantitative way to compute gains in food supply through adjustments in land, water, labor and technology by Andean societies; the Similitude Theory methodology demonstrates that basic economic principles apply to decision making. In contrast to the many theories on the route to state formation from ancestral kin group origins, the use of mathematical formalisms (Complexity Theory and Similitude Theory analysis in particular), at least replace speculation (without doubt from its originators as to its veracity) by mathematical formalisms that obey known economic principles. Of course, religious beliefs, past sociopolitical structure conditioning society members, historical past experiences with climate variations affecting sustainability, subsistence shifts, social inequality, gender differences in assigning importance to decisions, intensification of production goals, inherited cultural norms and governance characteristics of societies influence social complexity that provide obstacles (or preferential paths) to the convoluted path toward statehood. But in the context of the discussions presented, a major role is assigned to agricultural development efficiency as one basis for advancing societal complexity. Here economic advantage to a population by means of a sustainable agricultural base is a key factor for progress to occur to higher levels of sustainability and economic benefit for all members of a society. That these economic advantages can be calculated using similitude methodology provides the basis for rational decisions and judgments to be made; however given the vagaries of human nature in making rational judgments even based upon factual material bases, allowance must be made for judgments compromised by biases.

Challenging the path to sustainability, climate change has profound effects on the water supply and productivity or agricultural lands. Similar concerns about excessive rainfall flooding and prolonged drought periods affecting crop cycles affected the ancient societies of the Andean world and led to defensive construction activity to limit destructive flooding and drought effects on their agricultural systems (Chapter 3). For the Chimu canal irrigation systems constructed during the Late Intermediate Period, extended 10th- to 11th-century AD drought together with ongoing tectonic/seismic geophysical landscape change and river downcutting required continual modification of canal (Ortloff 1993, 2009; Ortloff et al. 1985) to continue their productivity. The history of canal development within the Moche Valley Chimu heartland reveals abandonment of early canals having inlets placed far up-river at the Moche River entry location into the Moche Valley later replaced by canals with down-river inlets shown in Figure 2.9 (Ortloff 2009:32). The sequential down-river canal inlet placement from the Moche River was a consequence of tectonically induced river downcutting that successively stranded higher canal inlets and progressively reduced Moche Valley land available for agriculture. Such A* reducing agricultural landscape change originating from tectonic/seismic forces inducing inflation/deflation landscape effects combined with ENSO climate changes seriously affected water supplies to city and nearby farming areas. The result is that the Chimu agricultural environment experienced ENSO and geophysical instabilities that required continual agrosystem modifications to balance declining food supplies to population needs. On this basis, the need to optimize food production and the hydraulic technologies to support it was a necessity as the vagaries of nature occasionally challenged survival of the Chimu and other developed Andean societies. Defensive measures in place in for Tiwanaku raised fields invulnerable the Chimu, the Intervalley Canal construction project to bring in adjacent-valley water resources to the Moche Valley was an application of to short-term drought as water table decline was gradual given gradual changes in mean rainfall levels. For their advanced hydraulic science to match the canal's flow rate to reactivate the main Vinchansao Canal flow rate to provide water to north side Moche Valley agrosystems.

Where geophysical/climate change altered agrosystem continuity, different strategies described in Chapters 3 and 4 to analyze and manage these systems evolved – but as Figure 2.4 indicates, no defense under long-term drought was possible for some coastal societies dependent upon declining highland rainfall runoff into rivers supplying their irrigation systems – or for societies relying on constant groundwater levels (Ortloff and Kolata 1993; Ortloff 2009). As complex issues related to agricultural productivity and intensification existed in ancient Andean societies, intellectual effort was devoted to promote sustainability – this result follows from predictions from the basic equation whose principles were known and used (albeit in some pre-scientific form) by major ancient Andean societies as described in preceding chapters. Augmenting conclusions about hydraulic science thus far presented and applicable to Andean sites, further discussion of hydraulic science for the Tiwanaku, Chimu and Inka societies is presented in later chapters to sum up current knowledge of the advances made by these societies.

Application of similitude methods to the Petra agricultural system

To illustrate a further application of Similitude Theory, the agricultural systems of Nabataean Petra in Jordan are next considered. The agricultural systems have been discussed in the literature (Beckers et al. 2013; Bellwald 2007; Bellwald et al. 2002; Oleson 1995, 2002, 2007; Ortloff 2003, 2009, 2014a, 2014b and basically rely on dams that trap rainfall runoff into *wadis* (streambeds) to recharge groundwater used for agriculture; such systems occupy many of hill areas surrounding Petra. These systems rely on rainy season runoff collection guided on to terraces and may have had early connections to large, spring-fed reservoirs in early phases of agricultural development but show later modifications with the advent of Roman occupation and control starting at AD 106. Starting from the basic equation,

$$\left[\left(A^*-A-\Delta A^*\right)^{3/2}\right)/\left(Q^*-Q-\Delta Q^*\right)\,F\left(dF/dt\right)=[\left(L_c{}^*-L_c\right)/\left(A^*-A-\Delta A^*\right)^{1/2}$$

$$+\left(V^*-V\right)/\left|A^*-A-\Delta A^*\right|^{3/2}\Big]\bullet\left|\,Q-Q^*-\Delta Q^*\right|\,t\left(\Phi/\Phi^*\right)/\left(A^*-A-\Delta A^*\right)^{3/2}\Big]$$

the maximum food production rate (dF/dt) is achieved for Petra's population estimated at 20,000 to 30,000 by large farming areas served by extensive L_c* channels collecting rainfall runoff water flow into *wadis* (streambeds) subject to intermittent rainfall collection (Q < Q*) over short time periods. Farming areas at higher altitudes than the city center of Petra were extensive to support large populations but only a fraction of A* was useful (ΔA* large) due to limited *wadi* channel that could direct limited rainfall runoff water (large ΔQ*) supplies to limited distribution areas within their reach.

While the potential farming area A* and ΔA* were large, available farming land area was limited. Agriculture appears to be concentrated during the rainy season duration implying that food production zones were scattered over large areas with limited productivity for comestibles requiring continuous irrigation. This implies the production of agricultural products that do not need continual watering but are seasonal and whose extensive root systems rely on groundwater (olives, peas, lentils, dates, for example) and agricultural products that have a short maturation times after the rainy season (grapes for wine production, wheat, barley, for example). As the Nabataean empire covered vast areas from the Negev well into Iraq and Jordan, many ecological zones with varying water supplies existed – this may imply that trade in comestible products was an important feature of Nabataean society.

Appendix

A brief exposition of the use of nondimensional groups is presented. While commonly used for engineering applications, the use for archaeological analysis in the present text is unique. Early in the 20th century (in the pre-computer age) as industrialization took hold in major western countries, there was a need to have available reliable engineering formulas serving as the basis for reliable engineering designs. The challenge was to first list all the variables (p1, . . ., pn) governing an engineering problem and then trying to relate these variables in a deterministic, functional form that proved useful for an engineering design. One method to achieve a functional form was to develop correlations based upon isolating one variable (p1) and running tests over ranges of the remaining variables (p2, . . ., pn) to determine "weighting constants C2, . . ., Cn" for each of the p2, . . ., pn variables in the form p1 = f (p2•C2, . . ., pn•Cn) – this method clearly was time consuming, costly and limited in applicability. It was noticed by Buckingham (1914, 1915) that all physical laws must hold in all possible systems of units (English, metric, SI, etc.) and that for physical systems, three primary quantities are mass [M], length [L] and time [T] and that units for all physical variables can be expressed in these terms.

The insight was to develop a methodology to combine individual p1, . . ., pn variables each with combinations of units involving mass, length and time into a reduced number of nondimensional groups of variables – this would reduce the test range based on a fewer number of nondimensional groups. A nondimensional group is a combination of variables each with its appropriate units, such that all units cancel out and the group reduces to a number (which has no units). A simple demonstration of this concept is offered by the familiar Einstein formula $E = mc^2$ where E is energy, m is mass and c is the speed of light. Here dividing each part of this equation by mc^2, the result is $E/mc^2 = 1$ indicating that the units of E, m and c^2 must cancel in this equation as unity (1) has no units. As the units of E are $[M] [L^2] [T^{-2}]$ (or mass times velocity squared) and the unit of mass is [M] and the units of light velocity squared (c^2) are $[L^2] [T^{-2}]$, then the units of E/mc^2 are $[M] [L^2][T^{-2}]/[M] [L^2][T^{-2}]$ which cancel to produce a number – which in this case is (1) unity. The immediate utility of this method for experimenters is that fewer test variables in the form of nondimensional groups can be used rather than individual variables with their appropriate units. Here variables any unit system can be used (metric, English, SI, etc.) to produce nondimensional groups. As many of the nondimensional groups have physical significance, their use in both theoretical and test applications is widespread (Murphy 1950); in fluid dynamics and heat transfer, Reynolds, Prandtl, Rayleigh, Nusselt, Knudsen, Froude and Mach numbers, among others, are widespread nondimensional groups with physical meaning appearing frequently in the literature. For the present application, only land, labor, water, technology and the food production rate are considered in the text; later applications can be extended to include further socioeconomic, political economy variables to widen the scope of similitude methodology in analyzing intensification development.

To illustrate the development of nondimensional group concepts, for example, velocity v has dimensions $[L] [T]^{-1}$, force has units $[M][L][T]^{-2}$ and energy (force times distance) has units $[M] [L^2] [T]^{-2}$. When, for example, two physical quantities with dimensions $[M]^a[L]^b[T]^b$ and $[M]^{a'} [L]^{b'} [T]^{c'}$ are multiplied together, the result is $[M]^{a+a'} [L]^{b+b'} [T]^{c+c'}$. If an equation representing some scientific law is to work in any unit system, its two sides must scale in the same way and their dimensions must be the same. If for example, the time period of a pendulum P is related to the mass m of the bob, the length L of the string and the gravitational constant g then $P \sim m^a L^b g^c$. In terms of consistent units on each side of the equation, then matching units $[T]^1 = [M]^a [L]^{b+c} [T]^{-2c}$ on each equation side leads to a = 0, b + c = 0 and c = -1/2, b = 1/2 so that $T \sim (L/g)^{1/2}$ – which is the well-known relationship obtainable from Newton's laws.

Relevant to the present application, extension of the preceding discussion lies in the concept that all equations representing physical laws must be expressible in a way that *both sides are dimensionless*. This means that if prime quantities are combined in such a way that their dimensions cancel to form nondimensional groups, then functions of these groups automatically satisfy physical laws that hold for all unit systems. This involves an extension of the aforementioned methodology where, for an example case, primary quantities are expressed in an arbitrary functional form as s = f{g, v, t, m, L, ρ, μ} where ρ is density $[ML^{-3}]$, μ is viscosity $[ML^{-1}T^{-1}]$, g is the gravitational constant $[LT^{-2}]$, m is mass [M]. T is time [T], v is velocity $[LT^{-1}]$ and s has length [L] dimensions. Then, in terms of dimensions of these primary variables,

$$L \sim \left[LT^{-2}\right]^{c1}\left[LT^{-1}\right]^{c2}\left[T\right]^{c3}\left[M\right]^{c4}\left[L\right]^{c5}\left[ML^{-3}\right]^{c6}\left[ML^{-1}T^{-1}\right]^{c7}$$

Collecting the exponents of [M], [L] and [T] leads to:

[M]: 0 = c4 + c6 + c7; [L]: 1 = c1 + c2 + c5–3c6 – c7; [T]: 0 = -2c1 – c2 + c3 – c7.

As there are seven unknowns (c1, . . ., c7) and only three equations involving [M], [L] and [T], three unknowns c1, c2, c7 may be expressed in terms of the remaining four unknowns c3, c4, c5, c6. One possible combination is c1= c3 + 2c4 + c5 – c6–1, c2 = – c3–3c4–2c5 + 3c6 + 2 and c7 = – c4 – c6 so that

$$s \sim f\left\{\left(g^{c3+2c4+c5-c6-1}\right)\left(v^{-c3-3c4-2c5+3c6+2}\right)t^{c3}m^{c4}L^{c5}\rho^{c6}\mu^{-c4-c6}\right\}$$

Now grouping together all terms with c3, c4, c5 and c6 exponents, there results

$$s \sim \left(v^2/g\right)\left(g\,t/v\right)^{c3}\left(g^2 m/v^3\mu\right)^{c4}\left(gL/v^2\right)^{c5}\left(\rho v^3/g\mu\right)^{c6}$$

where now all parenthesis (. . .) terms are dimensionless and c3, c4, c5 and c6 can have any value as powers of nondimensional groups remain nondimensional. The appropriate c3, c4, c5 and c6 constants can be determined by experimental testing. This procedure is codified as the Buckingham Pi-Theorem (1914, 1915) and is the basis of the generation of equations derived from the basic equation as developed in the text. Details of this development are presented in Ortloff 2009:134–137.

3 Hydraulic engineering strategies in ancient Peru and Bolivia to manage water supplies for agriculture and urban centers

Introduction

As climate and weather variations played a major role in determining the agricultural strategies of societies in ancient Peru and Bolivia at major sites located in Figure 3.1 (Binford et al., 1997; Ortloff and Kolata 1993), it is of interest to track the adaptive strategies those societies used to combat uncertainties in agricultural production and urban center water supplies. Of most interest are defensive strategies used to counter excessive rainfall and drought periods and their effect on agricultural field systems and urban water supplies over different time periods. Specifically, the effects of ENSO El Niño, La Niña and random rainfall and episodic drought periods were major concerns addressed by defense efforts to counter their adverse effect on agriculture and urban water supplies. The defense strategies of Andean societies are specific to different geographic sectors, ecologies, societies and cultural periods and represent integrated programs that different societies developed to protect large investments in agricultural fields and urban water supply systems. Additional considerations involve agricultural landscape change due to mass wasting brought about by tectonic events, flood events, aeolian sand transfer and shoreline and landscape change effects from flood events (Ortloff and Moseley 2012; Chapter 7) affecting agriculture and marine resource base production. During El Niño flood events, Peruvian coastal rivers carried sand, silt, clay, stones and mud over their banks; heavier sediment particles deposited leaving mounds of stones, sand and silt near the edge of rivers. Interaction of non-Newtonian flood slurry mixtures leaving river mouths interacting with ocean currents has been shown to result in linear beach ridge formations that when dated, provide a time sequence of major El Niño flood episodes (Ortloff and Moseley 2012; Chapter 7). Extreme flood events sent sediment material over riverbanks leaving thin layers of silt and fine sand occasionally interbedded with the clay substrate indicating a wide variety of deposition events depending upon the magnitude of the flood event. For large-scale flood events causing beach ridge formation (Ortloff and Moseley 2012), sequences of beach ridges trapped aeolian sand behind them that gradually infilled coastal bays with sand and flood debris to produce marshes. Generations of marsh vegetation, seeding and dying, and the next generation of plants growing through the matted fibers of parent plants

Figure 3.1 Major site locations mentioned in the text.

led to a sequence of organic deposition layers that transitioned from grass to peat to organic material intermixed with clay and sand to form organic clay below the surface (Pruett 2016). Successive layers of different soil types indispersed with aeolian sand layers deposited between flood events gave evidence of datable geophysical landscape change over time. This geophysical mechanism occurring from millennia of flood events was evident in the coastal reaches of the Supe and adjoining Peruvian north coast valleys where flood-caused beach ridges (Ortloff and Moseley 2012; Chapter 7) blocked debris transfer to the ocean and infilled bays with marsh areas where tortora reeds could flourish – these flood related, bay-infilling landscape events severely altered the marine resource base (fish species, shellfish types and seaweed types used for agricultural fertilizer) from previous norms as well as destroying farm areas adjacent to coastal rivers through land erosion as well as deposition of non-agricultural soils. While few restoration activities were possible to restore lost coastal land back to agricultural production from such events, where possible new inland farm areas were brought into production by canal water transfer from rivers to replace lost near coastal farm areas based on use of innovative technical advances supporting new farming methods and terrains previously thought of as having little agricultural potential. The present chapter describes the many innovations that coastal and highland Andean societies originated to defend lands against reoccurrence of flood and drought damage as well as the new innovations that brought new lands into cultivation.

Figure 3.2 shows the many types of agricultural systems utilized by different societies in different areas of pre-Columbian South America. This figure organizes agricultural systems into categories supplied by surface water sources (rainfall, springs and rivers) and subsurface water systems (groundwater) on the vertical axis, and on the horizontal axis, artificially created and natural farming surface types. For each of the major categories, defense and preservation strategies to protect from excessive rainfall, floods, mass wasting erosion/deposition events and drought exist along with performance and efficiency improvement strategies to increase agricultural yields under more stable and predictable weather patterns.

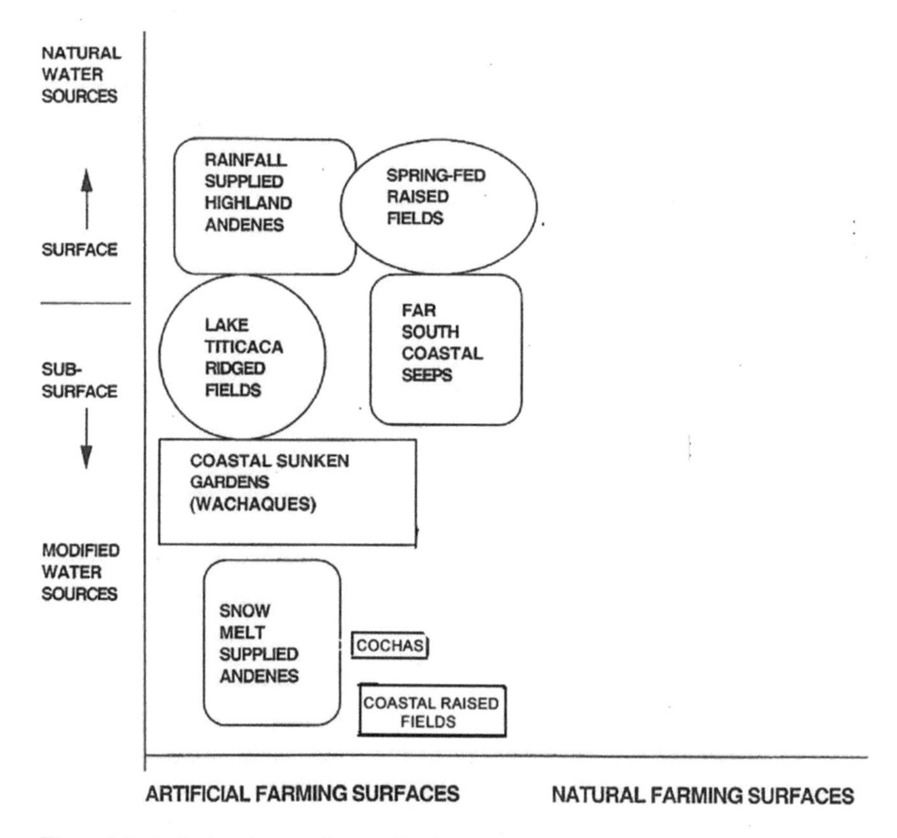

Figure 3.2 Agricultural strategies used in the ancient Peru and Bolivia.

Examples of all strategies put into practice by ancient Andean water engineers are found in the archaeological record and indicate that irrigation farming was a science in itself practiced with considerable foresight and oversight to maintain agricultural productivity despite the many variations in climate and weather that negatively influenced agricultural output.

A listing of defensive strategies follows; each category describes a hydraulic engineering application, examples of site use locations and the time frame occurrence in which events occur.

Canal flow rate controls and use of intravalley and intervalley canals for drought remediation (1)

Examples from the archaeological record include adjustable canal inlet blockages in the form of inserted reed mats and stone barriers at river canal intakes (*tomos*) to regulate flow rates into irrigation canals. Water supply canals frequently contain an elevated overflow weir located on the side of a canal activated when El Niño runoff into canal yield flow rates and flow heights in excess of a canal's flow rate design capacity. Here the side weir conducts excessive water, which originates from a flood event, from a canal to a lower level reservoir, drainage ditch or directly to field systems to alleviate excessive erosion and over-spillage damage to the main canal. An illustrative example of a flood water control construction using a canal width contraction that creates a unit Froude number choke that limits canal flow rate to a prescribed value (Ortloff 2009:111–122). Excess flow beyond this flow rate piles up before the choke to the height of a side weir that then conducts excess water to a lower reservoir or drainage area. This Chimu (or Moche?) structure lies in the Jequetepeque Valley east of Pacatnamu and is an example of a construction limiting excessive rainfall runoff into irrigation canals. A further example case is given using modern hydraulic theory (Figure 3.3) to illustrate how a downstream channel contraction (choke) raises the (subcritical $Fr < 1$ water height prior to the choke contraction zone to allow for drainage by an elevated side weir cut into the canal wall to drain away excess water over the design flow rate value for the canal. This figure summarizes the complexity of flows in expanding and contracting width channels as a function of sub- or supercritical input flow Froude number – while the complexity is describable in modern hydraulic theory parlance, some version of this figure must have existed in an as yet unknown format to Andean water engineers to enable construction of the many sophisticated water structures described thus far in the text.

For excessive rainfall-runoff into a canal section upstream of a contraction section (labeled subscript 1) with a downstream contraction section formed by opposing stones a given width apart (defined as a choke, labeled subscript 2), the passage flow rate in the canal is governed by the choke contraction section width designed to maintain unit Froude number ($Fr_2 = 1$) through the choke width. This defines the maximum flow rate in the choke section (2) and has an effect on the (subcritical) upstream flow (1) to alter its velocity and height when the canal input flow exceeds the design choke flow rate. An illustrative example calculation of

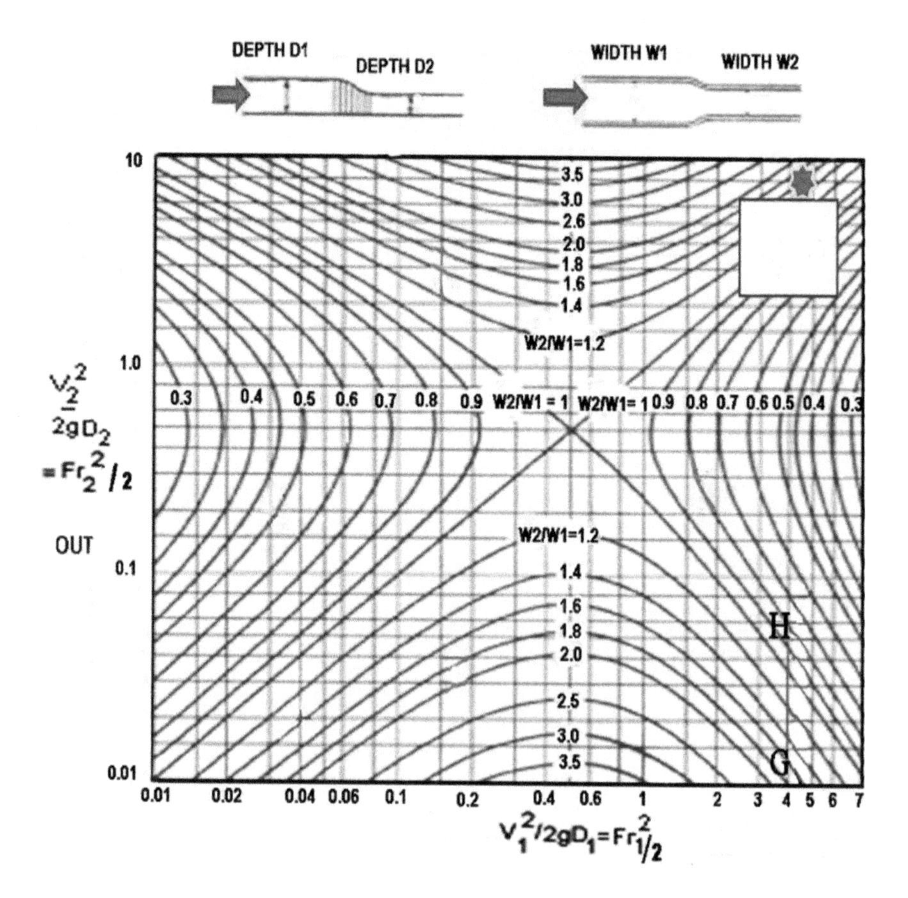

Figure 3.3 Plot (Woodward and Posey 1941) of $Fr_1{}^2/2$ (incoming flow to channel width contraction – vertical axis) vs. $Fr_2^2/2$ (outgoing flow from contracted channel). A vertical line from G to H (G-H) represents the path from an incoming flow with $Fr_1 = 3.0$ to an outgoing flow with $Fr_2 = 0.19$ due to a width contraction ratio $W_2/W_1 = 0.65$. The transition from a supercritical $Fr_1 = 3$ to subcritical $Fr_2 = 0.19$ flow occurs through a hydraulic jump. Left and rightmost chart regions represent W_2/W_1 channel expansion effects; top and bottommost chart regions represent W_2/W_1 channel expansion effects – note the difference in expansion/contraction effects for different incoming flow sub- or supercritical Froude numbers. The channel bottom is assumed flat from inlet to transition sections. For this figure, constant bottom height is assumed.

this effect proceeds following Rouse (1978:140) for the case where a choke channel constriction produces unit Froude number in the choke (2). As the passage choke flow rate is at critical $Fr_2 = 1$ conditions, this limits the flow rate through the choke and is the maximum design flow rate that the canal can support. Excessive inflows into the canal then raise the upstream flow height and lower its velocity. For example, for a trial flow at velocity $V_1 = 8$ ft/sec, depth $y_1 = 6$ ft and width $w_1 = 10$ ft in a rectangular cross-section channel ahead of the choke channel width constriction, the flow proceeds to a $w_2 = 7$ ft choke width with a $z_2 = 1.0$ ft bottom elevation. The gravitational constant is $g = 32.2$ ft/sec^2. Input pre-choke Froude number is $Fr_1 = 0.575$, total head is $H_1 = 0.99$ ft and the flow rate is $Q = 480$ ft^3/sec. In the choke section, the flow rate per unit width is $q_2 = 68.6$ ft^3/sec ft and as the flow in the choke is critical ($Fr_2 = 1$) with the critical depth $y_c = (q_2^2/g)^{1/3} = 5.36$ ft and the total minimum head is $(H_o)_{min} = (3/2)$ $y_c = 7.89$ ft. The total head in the choke section *exceeds* the head of the approaching flow (7.89 ft > 0.99 ft) so a backwater effect occurs to cause an upstream water height rise to change the input head until the flow takes place at critical depth in the choke section or $y_2 = y_c = 5.26$ ft. In other words, the head H_1 must match the H_2 head in section 2 (7.89 ft). This will involve a water height (y_1) rise and a lowering of the velocity in (V_1) to equalize heads. For this to occur, the head in the choke section (H_2) is $(H_o)_{min} + 1$ ft and the head in the amplified water height ($y_{1,\,new}$) upstream section is $Q^2/2g\,(10\,y_{1,\,new})^2 + y_{1,\,new}$. Equating both heads, and solving for the new upstream water height, $y_{1,\,new} = 8.38$ ft. The new upstream water height represents a 2.38 ft rise in height over the original 6 ft water height. The upstream velocity is then $Q/w_1\,y_{1,\,new}$ ~5.7 ft/sec. The placement of an elevated side weir with a height over ~6 ft then conducts excessive height water to a drainage channel. The strategy to limit flow rate and flow height then reduces canal overflow damage and erosion damage to an unlined canal from excessive flow rates. While the example is given for a subcritical ($Fr_1 = 0.575$) Froude number, cases where the (1) flow is high supercritical ($Fr_1 >> 1$) under extreme flood input conditions, the presence of a downstream severe width contraction section causes a hydraulic jump (HJ) to form elevating the water height ahead of the choke. The HJ changes are predictable and for the case of a severe width contraction, the elevation of the jump just ahead of the contraction section is dependent upon the Fr_1 value (Henderson 1966; Morris and Wiggert 1972). Unlike the $Fr_1 < 1$ case, no upstream influence exists in supercritical Fr_1 flow to elevate the flow height as the hydraulic jump is localized before the choke contraction section. For large El Niño events, water velocity in canals can be high indicating a large height hydraulic jump sufficient to activate side weir outflow. The side weir drainage system functions for both sub- and choked supercritical flows under flood input conditions. Examples of sidewall overflow weirs are present in the Chicama-Moche Intervalley and in the Jequetepeque Valley, as well as the Farfan-Pacatnamu Canal systems (Ortloff 2009, sections 1.2 and 1.3). While examples are given in modern hydraulic engineering notation, some semblance of the technology involved must have been known to Moche and Chimu hydraulic engineers albeit in a format as yet unknown to modern investigators.

For Moche and Chimu farm systems in the Jequetepeque Valley, steep offshoot canals from a main canal were found to support supercritical flow. At the entry location to field systems, a vertical plate was inserted into the canal to create a localized hydraulic jump in front of the plate. As the jump height exceeded the plate height, lower velocity water overflow entered channels in the field system limiting erosion and enhancing water absorption into root systems from plants on elevated berms. This use of created hydraulic jumps associated with supercritical flows in steep channels is an indication that Chimu and Moche hydraulic engineers were familiar with differences in Froude number regimes (albeit in a different format as yet to be discovered) and their effects on flows in channels of altered width along their paths. Chapter 7 further explores this conclusion for the Inka Tipon water system.

Mega-canal systems (for example, the Chicama-Moche Intervalley Canal, Figure 3.4) represent a 10th-century AD (Thompson et al. 1995) drought remediation measure to redirect water from a valley with a large flow rate river (the Chicama River) to an adjacent valley of political importance (the Moche Valley, seat of the Chimu capital city of Chan Chan) with limited farming area and the small flow rate, intermittent Moche River. Water from the Chicama River was directed over ~75 km by the Intervalley Canal into the northern Moche Valley main canal (the Vinchansao Canal, Figure 3.4) supplying water to field systems (Ortloff 1993, 2009; Ortloff et al. 1982; Ortloff, Feldman et al. 1985) formerly operational with Moche River water only prior to 10th- to 11th-century AD drought. While the design intent of the Intervalley Canal is clearly to reactivate Pampa Esperanza canals adjacent to Chan Chan with a tailored flow rate sufficient to activate a later phase of the Vichansao Canal (Ortloff 2009), distortions due to seismic and tectonic activity of canal slopes over ~1000 years (Nocquet et al. 2014; Perfettini et al. 2010) prevent comparison of slopes on the order of fractions of a degree constructed ~1000 years ago to present-day slope measurements. A canal in front of Chan Chan containing silt deposition from irrigation water use was recently

Figure 3.4 The Chicama-Moche Intervalley Canal and its link to the Moche Valley Vichansao Canal.

measured to have an uphill slope thus confirming tectonic landscape distortion over a ~1000-year time period. Claims of nonfunctionality of the Chimu Intervalley Canal flow to the north Moche Valley area based on present-day slope measurements are not consistent with tectonic distortions of the original intended Intervalley Canal slope measurements put into place by Chimu engineers. Global positioning networks confirm that a long segment of the northern Peruvian coast (5°–10° S) is currently moving coherently 5–6 mm/year east-southeast relative to the main continent (Nocquet et al. 2014). Such displacements can result in subtle and spatially variable surface deformations without requiring major earthquakes or fault scarps (Perfettini et al. 2010). It is plausible that current uphill canal segments of the Intervalley Canal were caused by localized crustal bulging or tilting sometime during the past ~1000 years for the Intervalley Canal and other intravalley and intervalley canals on the Peruvian north coast. The question as to whether the Intervalley Canal supplied water successfully to the Moche valley is still unresolved.

Elsewhere the Lambeyeque-Supe-Leche Intervalley Canal system and the Lima Intervalley Canal complex (Kosok 1965) provided canalized irrigation water diversion from valley rivers together with extension of existing segments of intravalley canals to provide irrigation water to arable land areas not previously in production. A further subsidiary function of intravalley canals was to redirect water to fields close to valley necks to recharge the lower valley groundwater aquifer – the resulting higher water table limited irrigation canal seepage and evaporation from longer canals serving downvalley field systems and provided groundwater for city wells (Figure 4.8) within Chan Chan (Ortloff 2009) and *wacheque* farming pits that were served by groundwater elevation. For normal operation of irrigation systems in predictable weather/climate patterns, additional field areas adjacent to, and located in between the source and destination valleys of intervalley canals, were supplied by water from intervalley canal offtakes to supplement intravalley agricultural land area and food supplies. Areas along the Chicama-Moche Intervalley Canal exhibit many field areas of this type.

Recent work in the upper Nepeña Valley (Poma 2018: Imgs. 01, 04, 11) for dual and multiple daughter branches of the ~45 km long open channel flow Huiru Catac Canal ascribed to Middle Horizon times, with later Recuay and Inka modifications, shows a 700 m drop over a measured 25 km length. The canal is sourced by multiple high-altitude lagoons (Laguna Capado and Tocanca) starting at 4500 m altitude. Due to the steep average slope involved (0.028), portions of the channel were infilled with stones to slow water velocity while other canal segments contained dams to effectively lower water velocity to limit canal bed and side wall erosion. Canal cross-section shaping varies from trapezoidal to rectangular possibly to limit water height within canal walls. Unlike the *amunas* associated with Supe Valley Caral (Chapter 7) where high altitude lagoons provided water through ground faults to support high groundwater levels, the Huiru Catac canal system provided irrigation water through open channels to adjacent field systems.

Water diversion channels (2)

Channels of this type were used at the Middle Horizon Tiwanaku Pampa Koani and Lukurmata raised field farming areas southeast of Lake Titicaca (Figure 3.1) to intercept and shunt excessive rainfall runoff from hill slopes directly into Lake Titicaca through large water collection canal systems (Ortloff and Kolata 1989a, 1993; Ortloff 1997; Kolata and Ortloff 1989b). While intercepted rainfall on to fields helped to provide surface and groundwater to the swales adjoining the Pampa Koani raised field berms, excessive water input increased the swale water height beyond its functional use height to sustain root crops. Here swales were normally excavated about ~0.5 to 1 m into the groundwater level to provide the correct moisture level for major root crops planted on berms with heights of ~1.0 m above swale water height.

Tiwanaku drainage canals with elevated outlet weirs set above the channel bottom directed excessive channel floodwater above a given height through the weirs into adjoining channels draining directly into Lake Titicaca to rapidly drain flood and excess rainwater to prevent amplification of groundwater and swale water height. The water height in swales is equal to the groundwater height and to modulate and control the groundwater level with respect to berm raised field planning surfaces, channels collecting and conveying water from upstream ponds and water collection zones had overflow weirs set a certain height above the channel sidewall activated when floodwater height exceeded the bottom height of the weir. This design led excess water to flow directly into Lake Titicaca through a connection channel and served to regulate groundwater height by draining away excess water. Swale water height was important to provide percolated water to plant root systems on berm surfaces. Further regulation of excessive groundwater height was controlled by aquifer seepage water flowing into drainage channels that led water to Lake Titicaca – this feature was possible as all swales were interconnected thus providing drainage access to the lake. The combination of these effects (runoff weir drainage channels to limit water addition to raise groundwater height and drainage channels within the raised fields to maintain swale water height), provided groundwater height control sufficient for major crop production during both dry and rainy seasons. During the dry season, river water could be diverted to distribution channels to maintain groundwater height; this design feature permitted multiple harvests to occur with a stable and constant seasonal groundwater swale height. There are applications of rainfall runoff collection channels for Moquegua Valley mountain region terraces (Figure 3.7) occupied by Wari and Tiwanaku presence to divert excessive rainfall into terrace water supply canals leading to downhill spillage channels draining into valley bottom fields (Moseley 2008). Possible use of Moche Early Intermediate Period Great Trenches (Ortloff 2009:29) observed in the Moche Valley for flood water diversion to the ocean is a further example of flood defensive measures – here further research on this feature needs to be done to clarify its role.

North Coast Chimu coastal *mahamaes* and *wachaques* (sunken gardens) and Chan Chan city wells (Figure 3.8) (Moseley and Deeds 1982) were activated by irrigation canal water seepage from field plots added to the existing groundwater profile originating from sierra watershed rainfall infiltration and river water seepage as examples of Chimu groundwater recharge technology. It is noted that during incipient drought stages, that new *mahamaes* and *wachaques* were dug closer to the coast to intercept the declining groundwater profile; survey of the areas west of Chan Chan reveals a sequence of sequentially abandoned pits that served temporarily to provide food resources to city inhabitants.

Tiwanaku use of spring-fed canals in the Lukurmata area north of Tiwanaku to deliver water to localized raised field water troughs provided water with different nutrient biochemical composition (Carney et al. 1996; Ortloff 1997, 2009) for specialty crops and provide a further example of groundwater recharge for a specific purpose.

Sunken gardens, wells, dams, spring systems and reservoirs (3)

Wetland agriculture in Preceramic times (10,000–4000 cal BC) in the form of raised fields at the site of Huaca Prieta-Paredones (Dillehay 2011, 2017; Dillehay et al. 2005) located in the Chicama Valley coastal zone is, to date, the earliest indication of organized agriculture to support the local population. The Paredones complex was apparently the main farming area of the site whose agriculture was supported by high groundwater levels from rainfall runoff interception and Chicama River infiltration input into the aquifer supporting the farming area phreatic zone. Examples of sunken gardens include north coast Chimu and earlier north

Figure 3.5 Two Preceramic Period Caral pyramids.

and south coast societies' use of excavated pits dug to groundwater phreatic level (*mahamaes*) and pits excavated close to springs (*wacheques*) for agricultural water supply during low river runoff and drought periods (Moseley and Deeds 1982; Moseley 2008). Sunken pits intersecting the water table and collecting rainfall (*cochas*) operational in late Tiwanaku V times (Ortloff 1997, 2009) in the Bolivian Pampa Koani area were used to support agriculture and serve as examples of last-resort water systems during drought periods. Use of wells interior to Chan Chan (Figure 3.8) and located close to the coastline where the water table was close to the ground surface and whose depth could be lowered to intercept a declining water table in drought periods served as a Chimu drought defensive measure during the 10th- to 11th-century AD drought. Abandonment of higher level sunken gardens as the 10th- to 11th-century drought progressed and replacement by new sunken gardens closer to the coast where shallower depth intersected the higher

Figure 3.6 The Huaca Major pyramid at the site of Caral.

Figure 3.7 Moquegua Valley terrace agriculture supplied by channeled rainfall runoff/ snowmelt water source.

Figure 3.8 Chimu Chan Chan well within one of the eight walled compounds.

water table proved a defensive measure to continue agriculture during extended drought that severely affected the Peruvian north coast in that time period. A subsidiary use of natural wetlands for agriculture on the Peruvian north coast relied on land depression areas close to the water table (West 1971) and use of elemental raised field systems noted in the Santa Valley. The use of reservoirs penetrating groundwater level with lead-off canals to field systems sustained Supe Valley agriculture at Preceramic Period Caral (2500–1430 BC) and adjacent valley sites (Ortloff and Moseley 2012) to provide water year-round to field systems – this is an example of groundwater use for agriculture dating back to the Preceramic Period.

Runoff interception trenches and runoff canal shunts to rivers/lakes/pools (4)

An example of use of these techniques is seen in the collection and channeling of rainfall runoff flows into channels connecting directly to Lake Titicaca to limit runoff interception and seepage into groundwater; this technique regulates local groundwater height in the swales surrounding raised field mounds. This regulation activity controlled the height of the phreatic moisture layer above the groundwater layer to serve different crop types with different root system depths having different water needs.

Further examples originate from the canals used to drain excess groundwater into channels to regulate moisture available for root systems in raised field plots (Ortloff 1997, 2009). The perimeter canal surrounding the ceremonial center of Tiwanaku (Ortloff and Janusek 2014; Ortloff 2016a) previously designated as a "moat" served to accelerate ground drying by runoff collection and aquifer seepage collection transferred by canal and groundwater seepage to the nearby Tiwanaku River to promote improved environmental and hygienic conditions for city inhabitants. Chapter 5 provides details of other urban water management discoveries at the city of Tiwanaku. Chimu use of interception trenches uphill from canal systems near Farfán in the Jequetepeque Valley to intercept rainfall runoff originating from mountainous terrain (Ortloff 1997) to direct flow into *quebradas* (flood erosion gullies) limited erosion and overflow damage to major canals (Ortloff 2009).

Terrace agriculture (5)

Examples of terrace agriculture (*andenes*) from Middle Horizon Tiwanaku, highland Middle Horizon Wari and Late Horizon Inka (Figures 3.1, 3.3, 3.7 and 3.9) societies' use of stepped planting surfaces on mountainsides provided the option to move agricultural zones to higher elevations to intercept higher rainfall amounts during normal rainfall and especially during low rainfall and drought periods. Use of high elevation canals sourced by snowmelt water during the warmer dry season and rainfall runoff water during the rainy season provided additional sources of irrigation water beyond rainy season rainfall (Chepstow-Lusty et al. 1996:831) to lower level farming areas. Examples of mountain ridge channels being alternately

Figure 3.9 Inka terrace agricultural system (*andenes*).

directed from one side of a ridge to the other side provided water for terrace (*andenes*) agriculture on both mountain sides; examples exist in the Moquegua valley uplands controlled by Tiwanaku colonies. For situations where raised fields were flooded from high Lake Titicaca water levels (a ~1.0 to 2 m lake level rise can flood upwards of 20,000 hectares of low-slope land bordering Lake Titicaca), use of terrace systems provided an alternative to continue agricultural production for the Tiwanaku during lake flooding events. Use of multi-terraced hillsides above river bottomland in the upper Supe Valley area together with spring-sourced canals irrigating field systems (Ortloff and Moseley 2009; Ortloff 2009) provided a defensive strategy by varying the types of agricultural systems to ensure agricultural continuity through weather/climate changes; the use of a variety of agricultural systems is noted at the preceramic site of Caral in the Norte-Chico region of coastal Peru. Figure 3.5 shows the one of the monumental temples at preceramic Caral indicative of the advanced nature of site architecture and its associated water system.

Specialty sunken gardens (6)

Sunken gardens (*wacheques* and *mahamaes*) were used by late and post-Tiwanaku V altiplano societies up to and including Inka occupation of altiplano territories to supplement food supplies as drought extended past AD 1100. Excavated basins (*cochas*) located near streams or in areas distant from Lake Titicaca margins

Figure 3.10 The Moray agricultural system.

where the water table remained high throughout dry periods from incoming groundwater from vast watershed areas provided a limited agricultural base to support localized populations emigrating from drought stricken Tiwanaku city collapse. The Tiwanaku colony site of Pajchiri demonstrated field systems at different heights above and below the water table for different crop types with different root depths that required different water requirements. The Inka created large, terraced concave circular pits (Figure 3.10) for specialty agricultural products at Moray (Wright et al. 2006) with temperature and humidity variations to determine best growing conditions for specialty crops – this was likely a unique example of an experimental horticultural facility used by the Inka. Later thoughts on Moray by Wright et al. 2006 indicate that it served as only a ceremonial center. Chimu *mahameas* and *wachaques* provided exploitation of local water resources largely impervious to short-term drought conditions to extend limited agricultural production of specialty crops. Recent research at Tiwanaku (Ortloff and Janusek 2014) indicates that *cochas* and variants of raised fields were integral to the urban housing areas of the city during its later Tiwanaku V phases and had water supplies from spring-sourced canals.

Agricultural reserve land and alternative field system fallowing strategies (7)

Examples include Tiwanaku raised field usage area shifts to high groundwater areas as diminishing rainfall interception and drought depressed the subsurface groundwater profile. At the Titicaca lake edge, the groundwater profile intersects the receding Lake Titicaca shoreline water boundary. Although much effort went into groundwater height control by the Tiwanaku through use of drainage and water diversion canals, long-term drought inevitably lowered Lake Titicaca height with a corresponding depression of the groundwater profile in areas close to the lake edge. Distant from the lake, the groundwater profile remained high from infiltrated rainfall into vast upland watershed areas in past times gradually percolating as groundwater toward the Lake Titicaca basin. These distant areas were farmed in combination with *cochas* to provide limited agricultural produce to supplement pastoral livestock sources of food as drought progressed past AD ~1100. Agricultural field area shifting allowed fallowing of non-used raised field areas to increase later productivity during regular rainfall, non-drought conditions that prevailed for many centuries of Tiwanaku occupation of the Pampa Koani raised-field area.

Groundwater collection into subsurface transport channels supplying agricultural field systems and amuna systems (8)

Underground galleries penetrating the water table collected seepage water and channeled it to surface fields in the Nazca area of far south coastal Peru. Examples of this method of water extraction for agriculture are described in the literature (Schreiber and Rojas 2007; Lane 2017). Here rainfall was scarce and rivers deeply downcut due to high tectonic uplift rates preventing the possibility of river sourced irrigation canals. The water table supplying the galleries derived from rainfall interception and absorption into groundwater at distant mountains.

Use of canalized groundwater pools/reservoirs fed by springs was effective in mitigating short-term drought for the Supe Valley site of Caral (and other adjacent valley sites) as the groundwater table showed little variation in height throughout the year due to the large infusion of groundwater originating from mountain watershed areas percolating water into the lower valley groundwater table through subsurface fault zones. The Supe Valley is unique in that it maintained a high water table throughout the year and thus was a natural destination for preceramic societies taking advantage of the valley's agricultural potential. Some speculation exists that highland lakes and lagoons transferred water into Supe Valley groundwater through surface and subsurface channels (*amunas*) built by local site inhabitants but more likely water from highland lakes was the source of groundwater transfer that supported valley bottomland farming via a geologic fault in the Supe valley area.

Canal and river water/groundwater seepage utilization (9)

An example originates from the north coast Chimu utilization of irrigation canal water seepage to recharge interior Chan Chan wells (Moseley and Deeds 1982). This additional conservation use of canal and field system seepage water added to the existing groundwater base derived from sierra rainfall interception sources and helped maintain numerous groundwater-penetrating wells within the royal compounds of Chimu capital city of Chan Chan. Moquegua Valley Chirabaya coastal groundwater seepage agriculture (Clement and Moseley 1991) in the Ilo area of southern Peru was based on collection of deep groundwater seepage on cliff face troughs directed by channels toward adjacent field plots. This mode of irrigation produced flows on the order of a few liters per second and thus could only support limited agriculture. This system provided agricultural continuity through drought periods as the water table had a low subsidence rate and was able to maintain limited agriculture while surface river flow rates diminished to the point where surface irrigation was no longer possible. Some of these earlier systems continued into Spanish colonial period times where olives and grapes used for wine production were the major crop types; surveys of the lower Moquegua Valley reveals several wine production facilities left over from colonial times with large storage containers dated from the mid 15th and 16th centuries. Some limited opportunistic north coast raised field agriculture at Casma, Viru and Moche Valley coastal areas was observed as a result of amplified water table height after El Niño rains provided moist soils suitable for limited agriculture (Moore 1988b, 2005).

High sierra reservoir water storage for canal delivery to coastal valleys (10)

Examples exist from high altitude (> 2000 m) Cordillera Negra mountain lagoons interconnected with canals to deliver water to highland valley and terrace agricultural systems. Water from these sources augmented groundwater levels in coastal valleys by means of water seepage through geologic fault zones and/or percolation through bedrock-limited, shallow soil layers (Avila 1986). The 50-km-long Hairu Catac Canal (possibly of Inka construction) irrigating the Iambe, Moro and San Jacinto sierra areas and Nepeña Valley foothill areas are examples of this type of water storage and transfer system. The use of a major dam structure in the Nepeña Valley uplands (Baudin 1961:62–70) is a further example of water storage for later transport to valley farm areas by canals. The probable water source supplying the Supe Valley high water table comes through geological faults transferring sierra infiltrated and runoff water to the lower coastal valley aquifer to create the high water table prevalent throughout the year that underwrote sustained canal agriculture for Caral and other Supe Valley Preceramic sites (Ortloff and Moseley 2012). It is noted that to supplement Supe Valley bottomland agriculture in preceramic times, a water transfer canal from the Supe River was built on an infilled terrace to conduct water to the inland site of Chupacigarro; whether further canals of this type were in existence to bring water to urban Caral has some evidence (Ortloff 2009) but severe erosion of regions separating the upper and lower portions of the city may prevent resolution of this issue. Use of lakes and ponds that penetrated Supe Valley groundwater levels (Figures 1.18 and 1.18a) provided drinking water at remote Supe Valley locations as well as creating moist farming area for water intensive crops such as cotton and specific fruit and vegetable varieties noted from plant seed remains taken at the Caral site.

Canal water/diversion of runoff to modify groundwater profiles (11)

Tiwanaku Pampa Koani raised-field agricultural systems on the Taraco Peninsula as well as field systems at nearby Pachiri exemplify this procedure which was used to regulate water table height for different crop types with different root system moisture requirements. Use of the Tiwanaku Pajchiri farming area's multiple elevation canal systems to supply field berms set at different heights depended upon regulation of water table height for different crop types (Kolata and Ortloff 1996). Seepage from irrigation canals surrounding the Chimu capital city of Chan Chan provided groundwater to amplify the water table height within the city for purposes of maintaining numerous wells (Figure 3.8) with city walled compounds.

Snowmelt water collection channels directed toward mountainside terraces (12)

Examples are found in upper highland Moquegua Valley drainages in late and post-Tiwanaku V times as well as for high sierra Middle Horizon Estuquiña and Wari terraces that utilized seasonal snowmelt water supplies channeled to agricultural terraces (Figures 3.7 and 3.9). This additional seasonal water supply provided for multi-cropping vital to sustain urban and rural valley settlements. Further examples from the central Andean Peruvian Patacancha Valley illustrate this strategy (Chepstow-Lusty and Bennett 1996; Chepstow-Lusty et al. 1996). This technique was used in elevated temperature periods and seasonal temperature elevation periods as a water source (water storage as ice) for terrace agriculture. Inlets to canals supplying water to lower-level terraces were relocated accordingly as the snowmelt boundary varied with precipitation, altitude and temperature.

Vertical and lateral agricultural zone shifts (13)

Relocation of Tiwanaku raised field agriculture on the margins of Lake Titicaca to terrace agriculture on nearby hill slopes when Lake Titicaca height excursions from excessive rainfall completely covered the raised fields provides an example of agricultural zone shifts. Possible reuse of very early phase Tiwanaku terraces during LH Inka occupation times by populations occupying mountainous and hill areas overlapping or outside of the Tiwanaku homeland area represents a strategy shift toward the use of terrace agricultural systems given the ample amount of rainfall in the altiplano environment in post-drought times past AD ~1200. The presence of terraces and adjacent raised fields in close proximity to the city of Tiwanaku offered options to shift agricultural areas between Pampa Koani raised fields and nearby highland terrace areas depending on the variation of local water table height in raised field swales as influenced rainfall intensity and duration, Lake Titicaca height and groundwater arriving downslope to the lake from distant watershed areas. Lateral shifting and extension of Tiwanaku agricultural raised field zones toward receding lake edge areas during the dry season and extended drought periods and expansion of raised field areas distant from the lake edge during extended rainfall periods added flexibility to continue agriculture to allow for changes in rainfall intensity as well as dry period duration.

Canal intervalley and intravalley transport/distribution canals (14)

The Chicama-Moche Intervalley Canal (Figure 3.4 shows the ~20 m fill-terrace banks upon which the canal was supported in the Quebrada del Oso region between the Moche and Chicama Valleys. The Lambeyeque-Supe-Leche Canal system and the Lima Complex (Chillon-Rimac-Lurin Canals) (Kosok 1965) had use as a drought remediation measure by distributing river water from large flow rate rivers to reactivate desiccated field systems in adjacent valleys with smaller, intermittent flow rate rivers during extended drought periods. During extended wet periods, water transfer from large rivers to adjacent valley canal systems provided extension of the growing season (Ortloff 2009) as well as for additional land areas to be brought into production. Intervalley canal water irrigated field systems located alongside the canal path and provided an efficient use of water close to the canal water source by limiting seepage and evaporation water losses associated with long canal paths; as for water delivery to the Pampa Esperanza area of the Moche Valley, further research is needed to confirm this connection. Long intravalley canals within the Peruvian Lambeyeque Valley (Racurumi, Taymi Canals, Kosok 1965) served to redistribute water from rivers to field areas only reached by a separate canal dedicated to access difficult-to-reach areas.

Canal hydraulic efficiency improvement features and flood damage controls (15)

Chimu hydraulic efficiency improvements in the Chicama-Moche Intervalley Canal design came through examination by modern hydraulic theory analysis methods that involved manipulation of canal wall Manning roughness coefficient (n), bed slope (i_b), canal cross-sectional shape (hydraulic radius R_h) and canal width expansion/contraction effects on flows for sub- and supercritical flows. Analysis of the open channel flow hydraulic design principles used on a 1.5-km section of the canal near Quebrada de Oso (Ortloff et al. 1985; Ortloff, Feldman et al. 1982, 2009) indicated knowledge of modern open channel water control principles to achieve a canal design for a specific flow rate (~4.3 m^3/sec) while having flow delivered to a destination location in the Froude number (Fr) range $0.8 < Fr < 1.2$ – a design close to modern open channel flow practice (Morris and Wiggert 1972). The canal exhibits, over a measured 1.5 km length, variations in n, i_b, R_h and sub- and supercritical contraction/expansions of canal geometry to control flow and maintain near critical slope conditions to yield a flow close to unit Froude number. Here critical $Fr \approx 1$ flow, where $Fr = V/(g\ D)^{1/2}$, represents the maximum water transport rate per unit width of a canal, where g is the gravitational constant, and D is equal to the wetted canal cross-sectional area divided by the wetted perimeter. Here exact $Fr = 1$ flow is to be avoided due to water surface instabilities causing canal overbank flow and increased erosion of unlined canal sections. The canal design sustains a Froude number close to, but either somewhat higher or lower than $Fr = 1$ as is typical in modern transport canal design practice. The canal design incorporated a hydraulic width contraction elements (Woodward and Posey 1941) before the (now missing) aqueduct crossing the Quebrada de Oso that converted flow from supercritical to subcritical over the aqueduct (line G-H, Figure 3.3) – this velocity reducing effect had the effect to limit erosion of the unlined sidewalls of the aqueduct. The Figure 3.3 graph holds for a flat bottom and vertical channel side walls as the channel transitions from inlet to contraction and expansion sections. As flow leaves the aqueduct, further n, i_b and R_h changes restore the flow back to a low supercritical value. Further hydraulic features of the canal (Ortloff 2009) indicate an ingenious way to use a vortex resident in a concave part of a canal wall whose size increases with increasing channel velocity. As flow rate and velocity increase from flood water input, a hydraulic jump is formed in the contraction zone between the vortex outer boundary and the channel wall that elevates water height that then flows out of an elevated drainage side weir. This design feature regulates excessive flow rates over the design value from El Niño flood runoff into the canal. Analysis reveals that changes in i_b, n, R_h and sub- and supercritical Froude number effects were effectively utilized to produce a near modern design (Ortloff 2009) but it remains for future investigators to reveal the format of hydraulic knowledge discovered and utilized by Andean water engineers.

Use of half-hexagon cross-section canal shapes for minimum flow resistance using straight-sided canal walls (Morris and Wiggert 1972; Henderson 1966) was observed in the Chimu Pampa Huanchaco canal system (Ortloff et al. 1982; Ortloff, Feldman et al. 1985) adjacent to Chan Chan. Use of streamwise canal shape changes to regulate water height to activate overflow weirs served to limit El Niño flood damage; examples of this design feature are found in Jequetepeque Valley canals and the Chicama-Moche Intervalley Canal. Use of a choke contraction section in a canal to allow a given design-maximum flow rate to pass though the choke (at Fr = 1) causes a water height change upstream of the choke for any excess flow rate over the maximum allowable design flow rate. The excess high height flow then activates an elevated side weir eliminating and directing the flow to a lower drainage area – this effect reduces canal damage from erosion and overspillage such as would occur during El Niño flood events. An example of this technology is found in the Jequetepeque Valley used to supply water to field areas south of the site of Pacatnamu (Ortloff 2009:117–121).

A further example of flow control derives from changes to existing canal designs related to lowering seepage and evaporation losses for long length transport canals. This is accomplished by a canal lining or letting an impervious silt layer accumulate on canal bottoms. For the Chicama-Moche Intervalley Canal, stone plate linings exist in given curved sections of the canal where water turbulence and rotational velocity excursions are high to limit sidewall erosion. Further Intervalley Canal modifications are in place by replacing a long U-shaped canal path along the walls of a deep interior valley by a constant slope, straight canal segment joining the entry and exit points directly. The shortened aqueduct canal path lessened the length of the canal by about ~1.0 km thus reducing flow resistance, water seepage from the canal bed and evaporation water loss.

Lomas farming (16)

This type of farming derives from coastal fog condensation water supply to lichen pastures on Peruvian north central coast areas. Winter fog (*garua*) condensation starts lichen growth – this begins the food chain starting with snails using the lichen as a food source then followed by predator species feeding on lower life forms in the food chain. This coastal zone served early local residents of the area by providing a food source composed of larger predator species drawn to the area for its diversity of plant and animal life as a food source. While the lomas systems were utilized in the early Formative Period to provided agricultural plant products as well as edible protein sources (Lanning 1967), the degree to which human intervention prevailed in later time periods to control and amplify its potential as a food production area remains a research topic.

Marine resource base adaptations during drought (17)

For drought-affected coastal regions with access to marine resources (fish, shellfish and edible aquatic plants), emphasis on exploiting marine resources provided a protein supplement to plant carbohydrate food supplies. Several Peruvian north central coast preceramic societies organized valley sites with coastal villages providing marine resources and inland sites largely invested in industrial and comestible crop production. Trade between coastal and inland sites formed a cooperative economic model seen in the Supe Valley and to a lesser degree, in adjoining north central coast societies (Shady 2009, 2007; Ortloff and Moseley 2009; Sandweiss et al. 2009; Moore 2005). While this cooperative model served well to provide a balanced carbohydrate and protein diet for valley populations, later Formative Period coastal societies resorted more heavily on exploiting marine resources to supplement food supplies from irrigation agriculture.

Adaption toward sierra pasturalism (18)

As a 10th- to 11th-century AD highland drought response, a shift from agriculture (due to recession of groundwater in raised field swales) toward herding and pasturalism occurred – this new emphasis on alpaca, llama and vicuña herding was possible due to well-watered grassland pasture areas (*punas*) prevalent at higher altitudes – here snowmelt water provided supplemental water to augment rainfall. This shift to pasturalism provided a protein food source to augment produce from drought-affected raised field and terrace areas. In many cases (Murra 1982a), major highland societies maintained colonies in different ecological zones that were able to export food varieties back to homeland areas to maintain their viability during both normal rainfall periods and drought periods. While partially effective in supplying cities with imported food items, this means of support was marginal to support large populations and as drought intensified, spatially distributed colonies in different ecological zones also declined from drought and were unable to provide an adequate agricultural support base for their homeland cities.

Raised fields (19)

Early evidence of Precramic Period use of raised fields dating from ~5000 to ~1500 BC at the Huaca Prieta-Pardones site located along the coastal Chicama littoral (Dillehay 2017) exploited high groundwater levels in natural sunken valley depressions for farming. This technical achievement was likely the earliest example of this type of agriculture noted in Preceramic Period Peru. Examples from the Bolivian altiplano from the Tiwanaku society are low vulnerability agricultural systems largely resistant to short-term drought due to the continuous arrival of groundwater from watershed zones distant from Lake Titicaca combined with slow groundwater subsidence from surface evaporation. These fields originated about AD ~600 when industrial-scale agriculture prevailed in later phase of Tiwanaku development – excavations to ~2 m depth in the Pampa Koani area show that experimentation took place earlier to decide the best geometric design of ridges and swales for maximum agricultural productivity and heat storage capability. Earlier raised field versions had shorter wave lengths between berm ridges than those observed in later phase constructions. Raised fields are constructed by excavating swale trenches deeper than groundwater level so that the swales penetrate the top of the saturated groundwater level. Typically raised field mound widths vary in the 4–10 m range and lengths from 10 to 100 m; mound heights are usually 1.0 to 2.0 m above the swale water surface and swale water depths are on the order 0.5 to 1 m but both berm orientation and berm geometry vary for specialty crop regions. Raised field swale water is able to store radiative heat from the sun during the day to prevent freezing of root crops during cold altiplano nights (Kolata and Ortloff 1989a, 1996; Ortloff 2009) thus preserving crop spoilage from freezing. Analysis of raised field geometry at Pampa Koani and earlier northern field systems indicate that raised field berm geometry was optimized to produce to maximize food production per unit land area – this feature is described fully in the Similtude Theory, Chapter 2.

City environmental control through use of a drainage moats to accelerate ground drying and regulate the deep groundwater profiles for monument foundation stability (20)

Use of a depressed, encompassing perimeter canal (described previously as a "moat" in the literature) surrounding the ceremonial center of Tiwanaku (as discussed in Chapter 5) collected rainfall runoff and seepage from phreatic and deep groundwater layers to promote rapid soil drying throughout rainy and dry seasons thus improving environmental and health living conditions for city inhabitants. By aquifer seepage into the moat during both wet and dry seasons and water supply through a newly discovered moat-intersecting canal fed by distant springs (Ortloff 2014b), the groundwater level was stabilized throughout seasonal weather cycles. The year-round stabilization of the deep groundwater level prevented deep aquifer dry-out and collapse which had the beneficial effect of maintaining the bearing strength of foundation soils underlying large monumental structures within the moat boundary thus reducing their settling and deformation (Ortloff and Janusek 2014; Ortloff 2014b). This the "moat" characterized as a separation barrier between elite and secular parts of the city had additional hydraulic and hydrological features that included hygienic benefits, rapid drainage benefits to dry site soils during and after the rainy season as well as structural benefits to maintain large monument stability.

Hydraulic jump control to minimize canal sidewall and bed erosion (21)

An example of Tiwanaku water engineering utilizing hydraulic jump control comes from the Lukurmata area north of the city of Tiwanaku. Rainfall runoff collected into a steep, unlined channel (Ortloff 2014b, Ortloff and Kolata 1989a) flows into Lake Titicaca bypassing a lakeside temple to protect it from flood damage. As lake height is high during the rainy season, the intersection of a high velocity, supercritical channel flow with calm lake edge water produces a large, turbulent hydraulic jump progressively eroding the unlined outlet soil bank in the upstream direction. To avoid upstream migration of the hydraulic jump and erosion of the fill channel, Tiwanaku hydraulic engineers skillfully varied channel outlet geometry to essentially produce a hydraulic jump of zero height thus canceling the jump (Ortloff 2009; Kolata and Ortloff 1989**b**). Modern literature (Bakhmeteff 1932) describes a similar technology of hydraulic jump cancellation resulting from channel expansion to create critical flow conditions at the jump location; essentially a hydraulic jump is created of zero height. Tiwanaku engineers anticipated this technology discovered by western science by some ~1000 years.

Open channel flow engineering in the Chicama-Moche Intervalley Canal (Ortloff 2009: Figure 1.21) and the Jequetepeque Canal previously described demonstrates that water engineers knew how to create hydraulic jumps to transition flow from super- to subcritical flow to avoid canal erosion over unlined canal sections and to locally elevate water to activate side weirs to dispense water flows over the canal design flow rate to limit erosion of unlined canal sections. Although modern hydraulic engineering methodology can reveal hydraulic features intentionally incorporated into Andean canal designs, it remains to be discovered as to how Andean water engineers became aware of hydraulic phenomena and then successfully utilized this knowledge in their water system designs. The Inka site of Tipon (Chapter 6) reveals Inka knowledge of sub- and supercritical flow characteristics as well as advantages of creating critical flows in channels – this level of hydraulic knowledge previously unnoted in the literature is a new addition to the accomplishments of Andean societies.

A further illustration of a channel constriction choke is found in agricultural field systems in the Jequetepeque Valley adjacent to the major water supply canal. The major canal elevated over the field systems supports steep lead-off canals from the main canal to channels supplying water to lower field systems. As water velocity is rapid (supercritical) in steep canals due to gravitational forces, a choke is provided to create a hydraulic jump prior to the choke – the hydraulic jump elevates the water height and slows it down as it passes through the choke constriction. The resulting low water velocity permits slow water entry to all channels between the field system's multiple elevated berms' elevated planting surfaces supporting plant root systems. Thus evaporation losses are reduced as water is directly available to root systems much in the manner of modern methodology. This field system design, typical of Peruvian north coast field systems, reduces evaporation losses as stagnant water in channel is reduced to minimum amounts. The choke shown in Figure 3.11 indicates indigenous knowledge of flow-type transition from supercritical flow on the steep canal incline to subcritical flow through the canal constriction.

Examination of the Inka Main Aqueduct at Tipon (Ortloff 2018 and described in Chapter 6) indicates an aqueduct design flow rate to limit over-bank spillage in a steep-to-mild slope channel section of the aqueduct. For flows in excess of the selected design flow rate, aqueduct overbank spillage would occur at the base of the steep channel section from the hydraulic jump water height change. Thus knowledge of hydraulic jump velocity-induced height change was apparently known to Inka water engineers to design the wall height of the aqueduct channel. Chapter 6 details the modern technology used that involved hydraulic jump technology to compute the flow rate of the main water supply aqueduct at Tipon.

Figure 3.11 Channel constriction (choke) designed to slow entrance flow into a field system inlet; site located in the north coast Peruvian Jequetepeque Valley.

Canal geometry analysis as a source of irrigated crop water supply (22)

Recent work (Ertsen and van der Spek 2009) investigates the role of canal longitudinal extent and cross-section geometry on the seasonal carrying capacity of water that can be supplied to crop areas using examples from the northern Peru Lambeyeque Valley. The maximum flow rate of the Chicama-Moche Intervalley Canal used to activate a later phase branch of the Vinchansao Canal supplying water to Chan Chan's field systems to reactivate the drought stricken Pampa Esperanza field systems outside of the Chimu capital of Chan Chan (Figure 1.23) has been determined in detail by previous analysis (Ortloff 2009). Details of technology to determine the matching flow rate in the Moche-Chicama Intervalley Canal to the acceptance flow rate of the Moche Valley Vinchansao Canal indicate a yet further element of Chimu water technological achievements. Here prediction of canal flow rates proved useful to determine water supplies to different crop types as well as the maximum extent of field areas supplied by the carrying capacity of different canal designs. New work summarized in the *Encyclopedia of Geoarchaeology* (Gilbert 2017) analyzes the role of geophysical landscape change on the developments of ancient and modern societies' agricultural systems and social development.

Hydraulic engineering technical advances (23)

Analysis of the Inka Tipon water system near Cuzco (Ortloff 2018, Chapter 6) indicates knowledge of different effects of subcritical, critical and supercritical input flow regimes in a supply channel to produce a desired water flow velocity and height change in downstream expanded or contracted channel. This is demonstrated by applications in the Chimu Intervalley Canal, the Jequetepeque Canal and the Tipon water system described in Chapter 6. Although described in terms of modern hydraulic nomenclature, these effects were understood by Andean water engineers through controlled experiments and/or observations from nature. This degree of hydraulic engineering knowledge centuries before formal discovery in western science indicates the presence of research facilities or test facilities to codify results for use in canal design; although such facilities have not been identified as of yet, the prevalence of sophisticated water systems at many Andean sites advances the premise that hydraulic engineering and the facilities to support new advanced water system designs existed.

Drought/flood defensive measures in different geographic regions and time periods (24)

While modern agricultural systems import produce from different worldwide climate zones throughout the year and have redundancy of crop types in different domestic and worldwide locations to maintain steady supplies, ancient agricultural systems possessed few of these features to sustain large populations and thus had extreme vulnerability to climate and weather induced collapse of their agricultural systems. On this basis, the measures listed in the preceding sections give indication as to the concern of ancient societies to protect their agricultural systems to the degree that their technology and innovative resources permitted. The catalog of defensive measures listed in Table 3.1 gives direct evidence of memory and response of past catastrophic climate and weather events and constitutes further evidence of the importance of such events in shaping the technological responses to protect agricultural fields systems and vital canal and aqueduct systems against excessive rainfall and drought episodes.

Sacrifices (25)

Although outside of technical defensive means to control destructive climate and weather change events affecting agricultural continuity, use of child sacrifices on a large scale were made by the Chimu late in the AD 1000–1471 time period toward the close of an exceptional rainy period lasting from AD ~1250 to 1450 (Romey 2019:54–77) on the north coast of Peru. Extensive excavations (in progress) in the Huanchaco town area north of Chan Chan and the city of Trujillo have exposed 269 child and 466 llama sacrificial burials thought to represent appeasement to deities to cease the long 10th- to 11th-century AD drought period followed by increased by El Niño flood events in the AD 1250 to 1450 time period. While the Chicama-Moche Intervalley Canal attempted to remediate 10th- to 11th-century extensive drought by bringing water from the Chicama River to the dessicated Moche Valley field systems, apparently major, continuing flood events in later times damaged the agricultural field system base adjacent

Table 3.1 Defensive, expansive water engineering measures

EH (and Preceramic, Formative)	*EIP*	*MH*	*LIP*	*LH*	
North Sierra					
+ Rainfall	—	2, 6	—	—	—
– Drought	—	—	6, 13, 19	—	—
Southern Highlands					
+ Rainfall	—	—	2, 5, 6, 12, 14, 23	6, 14	6, 7
– Drought	—	3, 4, 5, 7, 8, 12, 13, 14	4, 6, 7, 8, 14	6, 7, 14	19, 20, 21, 22
North/Central Coast					
+ Rainfall	2(?), 21	2(?), 15	—	1, 5, 15, 17, 23, 26	—
– Drought	3, 4, 6, 9, 7, 11, 12, 17, 18, 20, 21, 22	4, 9, 17, 18, 21(?)	—	1, 3, 4, 7, 10, 18	—
South Coast Valleys					
+ Rainfall	—	3	—	—	—
– Drought	—	9, 10, 18	7, 9, 10, 18	7, 10, 18	10

to the Chimu capital city of Chan Chan (Moseley and Deeds 1982) – but not beyond repair to later continue limited agriculture. Continued threats to Chimu agricultural sustainability over long time periods apparently led to the ultimate sacrifice thought sufficient to finally reverse the long term of negative events that challenged their survival. Further evidence of child sacrifice is noted in the Inka society where elaborate burials on mountain heights were made to ensure continued favor of deities (Reinhard 1992). Clearly such events presupposed the presence of divinities that control the fate of societies beyond what human intervention, inventiveness and defensive technologies can control or influence – an observation that persists to the present day.

In Table 3.1 the + Rainfall notation denotes defense strategies exercised in excessive rainfall periods to protect vital agricultural systems; this category also includes efficiency improvements in water supply technology which have been observed in the archaeological record. The – Drought notation represents drought defense strategies and innovations observed in the archaeological record. Table 3.1 is divided into different spatial divisions of the Peruvian and Bolivian landscape. These spatial divisions represent the Peruvian highland sierra (as far north as Wari occupation areas around Arequipa in Peru), altiplano regions characteristic of Lake Titicaca highland Bolivia and the Peruvian north and south coast valleys in different time periods from early Preceramic/Formative times (2500–1600 BC) to Late Horizon (LH) times (AD 1400–1532).

Preceramic Period (2500–1600 BC); Formative Period (1600–300 BC), Early Intermediate Period, EIP (300 BC–AD 700); Middle Horizon, MH (AD 700–1000); Late Intermediate Period, LIP (AD 1000–1400); Late Horizon, LH (AD 1400–1532). The — notation indicates insufficient data available.

Summary and conclusions

The adaptive strategies and time sorted in Table 3.1 indicate conscious effort by different Andean societies to control water supplies for agriculture and urban use using a variety of engineering technologies specific to different geographical locations and agricultural systems in different time periods. While water engineering and delivery efficiency improvements accompanied long-term favorable climate trends, other engineering responses provided defense measures against ENSO El Niño flooding and long-term drought. For changes in weather and climate affecting agricultural productivity, additional responses in the form of modifications to existing systems, or in dire drought circumstances or as a response to population pressure, design and building of intervalley canal systems to bring additional water supplies to new farm land areas proved the only solution. The defensive responses given earlier required long-term observation and recording of past weather and climate patterns and changes in climate that portended future disasters unless large-scale measures were put into place to counteract decreases in water supply and concomitant changes in societal sustainability. Once recognition of threats to water supply were confirmed by observing and recording weather and climate trends over time, a leadership plan evolved either by elite management or communal action to organize decisions to implement planning to ameliorate future crisis events affecting the well being of their society; origination of such management structures arising from practical needs is addressed in Chapter 1. Further logistics to recruit labor and provide material and rations for extended field operations were planned to perform required tasks and originated a management structure according to derivations in Chapter 2. From personal exploration of the Intervalley Canal inlet structure in the Chicama Valley, a management site was discovered which contained stacks of crudely made unfired clay plates – these obviously part of a logistics plan to distribute and feed workers at remote locations to improve labor efficiency. Management was in effect at a high level of control in the Chimu society as further demonstrated by the multiple compound structures observed at Chan Chan each of which was devoted to an individual ruler and his retinue.

Table 3.1 provides a library of agro-engineering responses to destructive long- and short-term climate and weather variations and demonstrates that climate and defensive technology development were important factors in design, operation, supply logistics, administrative planning and construction to ensure sustainability of agricultural fields vital for societal survival. While many of the technology advancements listed came from response to agricultural threats, many derived from technical improvements used to increase irrigation water flow rates to fields and/or reduce water seepage and evaporation losses by canal redesign and relocation – all to improve agricultural yields. In total, the listings indicate an active and creative engineering mentality existing in Andean societies to conceptualize optimum irrigation system designs commensurate with environmental and water availability conditions and then to put plans into action by organizing construction labor, supply logistics for workers and after construction, maintaining project operation by experienced farming and technical personnel. As might be anticipated over the extended time period for Intervalley Canal construction (~50+ years), new innovations were likely made improving its operational efficiency over time by appropriate engineering redesigns. In total, ancient Peruvian and Bolivian water systems appear to reflect many hydraulic and hydrological technologies brought about from years of innovation and nature observation that rival modern civil engineering practice.

4 Climate change effects on sustainability of ancient Andean societies

Noting the influence of and response to climate change on the dominance and dissolution of major Andean societies, a means to quantify cause and effect relationships can be expressed in the form of a sustainability factor equation (Q) (Ortloff and Moseley 2009) derived in this chapter. While minor climate change and weather effects affecting the urban and agricultural water supply of a society that influence societal sustainability can be largely countered by technological advances (Chapter 3), major climate change results in radical realignments and transitions of existing polities' sociopolitical economies. From derivation and use of the derived sustainability factor Q equation, examination the Uhle-Rowe divisions of Andean history into Horizon and Intermediate Period categories can be shown to reflect changes in climate patterns that influenced the agronomic base of major Andean polities. The discussion to follow summarizes new information related to major Andean polities to demonstrate the applicability of the Q equation to highlight and explain the underlying effects of climate and technological innovation on Andean historical patterns. Several key parameters are next reviewed that compose the Q equation and are discussed in detail in the following discussion.

Runoff ratio (R)

During drought, coastal farmland that normally supports an abundance of crop types under normal climate and weather patterns can only sustain a reduced number of crop types with low water demand. As a result, crop yields are reduced in response to lower water supplies. Highland sierra rangeland in contrast can support pastoralism and crop types adapted to cold extremes based upon elevated rainfall amounts at higher altitudes. One reason for differences in sustainability of coastal and highland agriculture under drought is related to soil types that influence rainfall infiltration, runoff and water transport rates from highland rainfall sources compared to coastal soil types. Highland soils subject to higher rainfall amounts typically retain water within their porous structure until saturation is reached; beyond this level, rainfall runoff collected into rivers occurs to supply Peruvian coastal valley irrigation systems. Channeled runoff into rivers is further subject to evaporation, seepage and subsurface porous aquifer retention effects resulting in imbalance in the rainfall/runoff delivery rate from sierra to coastal areas. Here the important point is that sandy Peruvian coastal farm areas depend entirely on intermittent rainfall runoff from highland sources collecting into rivers that supply coastal irrigation systems. Given that Peruvian coastal areas experience limited precipitation, their agricultural systems are totally dependent upon highland sierra rainfall runoff. As sierra rainfall sources collecting into rivers supply adjacent Peruvian coastal valley irrigation systems, evaporation and ground infiltration seepage exists from irrigation transport canals as a further source of coastal water supply decline that subtracts from water amounts infiltrated into groundwater – thus coastal Peruvian agriculture dependent upon excavated pits to the groundwater phreatic layer (*wacheques* and *mahamaes*) require deep excavation to reach deep groundwater levels. Expansion of Peruvian coastal agricultural field systems as the means to increase agricultural yields to match population growth also results in increasing evapotranportation losses from increased exposed canal surface areas – yet another constraint on agricultural productivity is that water lost in this manner is not infiltrated to groundwater to maintain a high water table and phreatic water supply to crop root systems. As drought is a key player in deciding the fate of Andean societies' agricultural systems, it is necessary to examine the sources and origins of drought to establish their intensity and duration in different ecological zones – with this information, the susceptibility of different agricultural system types experiencing different vulnerabilities to drought can be exposed.

To illustrate the effect of drought stress and the hydrological relationships between rainfall, soil types, runoff and flow losses, an example from the Moquegua Basin in southern Peru is given. The Moquegua Valley lies on the Pacific watershed to the west of Lake Titicaca. The valley river system is 139 km in length, with headwaters reaching slightly above 5100 masl. Along the Moquegua coast precipitation is negligible but increases gradually in the interior with altitude. However, the quantity of rainwater only exceeds saturated retention values in 19% of the basin above 3900 masl. Between 3900 and 4500 m, average rainfall is about 360 mm/year, about 260 mm of this amount is retained in saturated soil so that only 100 mm is available as runoff (ONERN 1976). In this zone, a 10%, or 36 mm, decline in rainfall to 324 mm decreases runoff by 36% from 100 mm to 64 mm. Given a specific soil type with a given retention capability, a 15% decrease in rainfall results in a 54% decrease in runoff. In the elevation zone between 4500 and 4900 masl, rainfall averages 480 mm/yr and a 10% or 15% rainfall reduction results in runoff reductions of 21.8% and 32.7% respectively. The asymmetric disparity between rainfall and runoff reduction diminishes as precipitation increases. Comprising less than 3% of the Moquegua Basin, precipitation in the zone of alpine tundra above 4900 masl is principally in the form of snow and ice (ONERN 1976). The runoff contribution is unknown because an unknown amount of moisture is retained in glaciers and snowfields. Nonetheless, for the upper river basin as a whole, rainfall declines of 10% to 15% result in runoff reductions of 25% to 40% or more. Significantly, the asymmetric relationship between rainfall and runoff also works in reverse. Increased precipitation rapidly saturates the soil which then discards water and amplifies runoff. This effect was prevalent in the first two centuries AD when precipitation rose by 20% to 25% and the runoff by 72% to 90%. Drought stress is exacerbated by the fact that once rainfall saturates the soil and excess water is released, some surface runoff is lost to evaporation and seepage. Due to these factors, the Moquegua River loses about 4% of its flow per kilometer in the arid sierra at elevations around 2250 masl. Other than during spring floods, the river channel does not normally carry surface flow at elevations below 1200 m thus limiting agriculture to terrace farming (*andenes*) in mountainous areas. From this discussion, it is apparent that farming areas at different altitudes can have differences in sustainability given different water retention characteristics. Farming in the coastal section of the drainage depends on springs fed by subsurface groundwater flows originating high in the river basin. The relationship between highland rainfall and coastal spring flow is highly asymmetrical because subsurface water flows through porous geological strata and porous soil deposits have different hydraulic conductivity and saturation values. Although these values are poorly known, there are indirect indications that coastal spring flow may have dropped by 80% at the start of the AD ~1100 drought (Ortloff 1996, 2014b).

These calculations are approximations for the Moquegua Basin. Other soil adsorption and precipitation values characterize other drainages. Nonetheless, relationship between rainfall and runoff is nonlinear and drought always exerts a greater asymmetric effect on runoff based coastal farming than on rainfall farming at higher mountainous elevations. The runoff not directly channeled into rivers is diminished en route to coastal zones by further infiltration into increasingly more porous soils (adding to the local water table profile) as well as by evaporation losses. The resulting coastal river hydrographs track the availability of coastal irrigation water over time and variations exist between coastal valleys as functions of local soil geomorphology, topography, evapotranspiration, agricultural productivity potential, temperature and specific humidity history. The net effect is one of a nonlinear, but generally similar, relationship between unit amounts of input sierra rainwater at different altitudes and times, with time lags of deliverable water to coastal irrigation systems. Even after arrival of irrigation water to coastal field systems, approximately 85% is lost to evaporation from soils leaving only 15% to be absorbed by plant root systems. The effectiveness ratio of water supplied to water absorbed by root systems further skews the nonlinear relationship between sierra water supplied by rainfall and the useful amount available for agriculture at lower elevations. It may be concluded that in dry periods, rain at high altitudes may sustain some form of agriculture in these zones, but coastal agriculture derived from runoff into rivers from the same watershed will experience a severe deficit of irrigation water. In terms of quantifying the runoff effect, the Runoff Ratio (R) is defined as the average net runoff rate from an area divided by the average rainfall delivery rate to the same area. Here R = 0 denotes zero runoff and R = 1 denotes that all delivered rainfall to an area converts to runoff, implying a total saturation condition. Intermediate R values are thus a key component related to the sustainability Q of a society related to changes in the rainfall amounts and its effect on agricultural systems due to extended drought.

Vulnerability index (1-V)

Figure 3.2 shows a plot of the main agricultural strategies practiced by Andean civilizations. A vulnerability index is defined relative to available rainfall levels. The first entry, raised field agriculture (Index 1), was widely practiced by the Middle Horizon (AD 300–1100) Tiwanaku III-V around the southern and western periphery of Lake Titicaca. Tiwanaku groundwater based agricultural systems are largely invulnerable to short-term drought due to the continuous arrival of subsurface water from earlier rainfall events in the immense watershed collection zone around the lake. Because groundwater transport velocities are on the order of a few centimeters per month, groundwater from distant collection basins may have originated as infiltrated rain that occurred many years earlier. Groundwater based raised field systems comprised of long elevated berms surrounded by water filled swales at local groundwater level are likewise relatively invulnerable to seasonal excessive rainfall because elaborate field drainage systems built by the Tiwanaku shunt water directly into Lake Titicaca, thus limiting infiltration into the water table. Raised field areas close to the edge of Lake Titicaca, however, are inundated and lost to agriculture as lake level rises from episodes of heavy rainfall into the lake as well as surface runoff.

Because collection basin rainfall rates and Lake Titicaca height vary with seasonal and climate-related rainfall/runoff fluctuations (Binford, Kolata and Brenner 1997), the raised field water table height sufficient to maintain the necessary raised field swale depth and phreatic zone for agriculture shifts in time and laterally in position within the extensive lacustrine field systems (Ortloff 1996; Kolata et al. 1989b). The Lukurmata and Pampa Koani raised field system areas north of Tiwanaku were farmed in localized subzone areas supplied by spring water added to elevated groundwater profiles in swales to supply agricultural root systems – remaining areas were allowed to fallow to reconstitute nutrients for farming. Careful regulation of the groundwater profile by use of drainage and supply canals thus permitted different crop types with different water needs for their root systems to provide maximum crop yields. Prolonged drought over many years can destroy the heat storage features that provided frost damage protection under diurnal and seasonal temperature variations (Kolata and Ortloff 1996a) and decrease the height of the water table in raised field troughs necessary to sustain agriculture. The raised field systems can, nevertheless, be optimized to highland climate conditions and cycles to produce high crop yields as demonstrated by modern resurrection and use of these systems (Kolata 1996b). Chapter 2 provides yet further insight into the sustainability of raised fired systems – here the geometry of agricultural berms and their placement is shown to be optimum to generate maximum yields per unit land area.

The next least vulnerable agricultural system to rainfall fluctuations is a variant of raised field systems – sunken gardens (Index 2) denoted as *wacheques*. These systems are pits excavated to the water table phreatic zone and are mostly found as a last resort drought response used when the water table continues to decline out of reach of plant root systems and canal-based irrigation system are no longer possible to sustain. Sunken gardens are common in the AD 1100 post-collapse settlements around Tiwanaku and *wacheques* are also found in Peruvian north coastal valleys in response to the continuance of the pan-Andean late Middle Horizon drought into the Late Intermediate Period. Although groundwater based agriculture has a slow response to drought due to the slow decline of the water table height from reduced rainfall penetrating the evaporation zone as well as resupply from infiltrated rainfall from areas distant from lake Titicaca), these systems have no defense against excessive rainfall as groundwater level rises and drainage paths are nonexistent as higher levels of lake water flood lake edge raised fields.

The next level of vulnerability (Index 3) is represented by agricultural terraces (*andenes*) widely used by Inka, Wari and post-Tiwanaku highland altiplano societies. Terraces are mostly supplied by rainfall and provide a well-drained agricultural system effective during rainy seasons. Other terrace agriculture variants are supplied by snow melt and spring flow channeled water during periods of low rainfall and elevated temperature. As rainfall diminishes, these systems gradually become marginal for agricultural production unless supplied by channeled water from higher rainfall elevation areas or snow/ice deposition areas subject to seasonal elevated temperature melting.

Next in increasing order of vulnerability (Index 4) is canal-fed irrigation as practiced primarily by north and south coast Peruvian polities. Systems of this type are viable for flooding in the presence of highland rainfall exceeding saturation conditions, i.e., high R values. Similarly, extended drought reduces runoff to rivers thus compromising irrigation agriculture. As such, if coastal agriculture flourishes, then highland agricultural areas have an excess of water supplies due to the nonlinear input/delivery R relationship. The highlands appear always to have demonstrated less vulnerability to agricultural stress regardless of the level of rainfall, provided the technology to use the available water was adequately developed in each ecological zone and drainage technology to control the water table height was in place. On this basis, theories arise about drought periods influencing lowland agricultural vulnerabilities and population sustainability compared to highland societies and the dominance of these societies in different time periods.

Generally, long canals are more vulnerable (Index 5) than short canals due to greater seepage and evaporation losses, tectonic/seismic distortions and higher technological demands for low-angle surveying and construction (Ortloff 1995, 2009). Yet more vulnerable (Index 6) are the coastal seeps that supply agriculture, mainly in northern Chile and in the Ilo area of the Moquegua Valley (Ortloff 2009). Such systems rely on groundwater seepage to coastal bluffs over long underground distances and are only marginally productive compared to all other water delivery systems.

Survival of the more vulnerable agricultural systems is questionable past critical drought rainfall reduction levels and extent of the drought and extinction of highly vulnerable systems is inevitable whenever reconfiguration to lower vulnerability systems is impossible. In the presence of yearly rainfall and runoff variations, high vulnerability systems must have superior technology, innovation and modification features to maintain agricultural production. A Vulnerability Index (1-V) is next defined such that the largest values of the index denote the least vulnerable agricultural systems. The Vulnerability Indices are defined as follows: for Index 1, (1-V) = 1; for Index 2, (1-V) = 0.6; for Index 3, (1-V) = 0.3; for Index 4, (1-V) = 0.2; and for Index 5, (1-V) = 0.05.

Technology change under climatic duress

Andean agricultural systems have a long history of evolution and improvement over time from a variety of agricultural strategies (Ortloff 2009). An initial choice of an agricultural system compatible with local ecological conditions is made by early inhabitants. System evolution proceeds through observation of agro-production changes in response to field system design changes and water supply variability. Key requirements are the preservation of the system and its efficient functioning under seasonal weather fluctuations and those arising from large-scale climate fluctuations. Therefore, the ability to modify an agricultural system to maintain sustainability in anticipation of climate-related changes in water supply is part of the original design concept of the system (Chapter 3). Agrosystem modifications must be performed more rapidly than the time it takes for the deleterious effect of a climate variation to manifest, i.e., *agricultural* system modifications are only effective when long-term climate changes are initially observed and system modifications carried out in anticipation of effects of a long-term negative trend. Therefore, the Vulnerability Index (1-V) of an agricultural system depends upon the sustainability of an initial agricultural field system design choice in the face of weather and climate variations, as well as the ability to modify technology (T) in time (t), denoted as (dT/dt), faster than the climate induced creation/evolution rate of a climate related disaster (dD/dt).

Agricultural sustainability model: the Q equation

Key parameters influencing agricultural sustainability are next combined to produce a trend equation from which increases or decreases in each term imply a net increase or decrease in agricultural sustainability (Q) of a society. The quantities in the equation (Q, R, S, P, V, Y, dT/dt and dD/dt) are nondimensional values normalized to the maximum reference state for each variable. A large value of Q connotes agricultural sustainability, while a small value denotes the opposite. From preceding discussion, a model equation, *based upon agricultural parameters only* (i.e., excluding implied or induced social, political, economic and/or governmental system effects) is postulated as:

$$Q = S + Y\,A\,R + (1 - V) + P'(2 - P'\,(dT / dt) / (dD / dt)$$

Where R is the Runoff Ratio ($0 < R < 1$), S the agricultural storage capacity ($0 < S < 1$) normalized to S_{mx} where S = 0 represents zero crop storage and S = 1 represents total storage of all unconsumed crops. The quantity Y A R represents the main comestible crop yield per unit land area times the total land area times the available water supply to the area normalized so that $0 < Y A R < 1$. Here the lower limit trends toward poor crop yields over a small land area with poor water supplies while the unity limit indicates the best crop selection over the largest possible agricultural area sustained by irrigation. The term, $P' = P/P_{mx}$ is defined as the population density where P_{mx} is the maximum population sustainable by the in-place agrosystem. If $P' = 1$, then $P'(2-P') = 1$ at the maximum population level balanced with the food supply, i.e., $\partial Q/\partial P' = 0$. If $P = 0$, then $P'(2-P') = 0$, indicating that a very small population exists (such as may occur after a natural or man-made disaster). Thus: $0 < P'(2-P') < 1$.

As before, (1-V) is the agricultural Vulnerability Index ($0 < V < 1$) for the agricultural systems shown in Figure 3.2. For the remaining terms, $0 < dT/dt < 1$ represents the time rate of technology (T) change to surmount a long-term negative climate effect (excessive rainfall, drought, temperature variations, climate induced landscape changes) on agricultural production. Here the maximum dT/dt value is assumed to be unity to represent a technology growth rate typical of most advanced agriculture-based societies and can be large due to technical innovations. The dD/dt derivative may be large for rapidly evolving disasters such as El Niño events reducing Q dramatically in a short time period.

The $1 < dD/dt < \infty$ term is representative of the rate of change of disaster-producing climate factors. If $dT/dt \div dD/dt = dT/dD$ is large, the rate of development of technology to defend against climate-induced changes in water supply exceeds the rate of disaster evolution on the same time scale and a positive effect on agricultural sustainability Q results. If an El Niño flood event occurs beyond any defense mechanisms' ability to provide protection (Chapter 3), then $dT/dt \div dD/dt$ is a small number indicating no contribution to agricultural sustainability. If, however, a climate-related disaster evolved at the same rate as a defensive technology, then $dT/dt \div dD/dt$ can be a positive constant and Q shows increased sustainability. If $P' < 1$, then the labor force to make rapid dT/dt corrections is not available and Q decreases. If $P' > 1$ in the presence of a declining agricultural supply, the agricultural resources are inadequate to feed a large workforce over time to ensure rapid dT/dt changes to increase Q. Thus, only a population balanced with agricultural supply (including storage) promotes large sustainability Q values. The relative value of Q ($0.2 < Q < 4$) (increasing or decreasing) applied to highland and coastal societies at different time intervals then gives indication of underlying factors behind the sustainability Q of a society based on agricultural parameters in different time intervals. Overall, from the Q equation, sustainability is enhanced when the runoff ratio/water supply (R), land area (A) in cultivation and crop storage (S) are all high, a stable population is balanced with agricultural output ($P' = 1$), the system vulnerability $(1 - V)$ is low, the technology innovation rate dT/dt exceeds or equals that of the disaster evolution rate dD/dt, and soil productivity (Y) per unit of water input is high. High Q indicates a successful, well-managed society with foresight to maintain a sustainable agricultural base despite weather and climate variations. Low Q indicates gaps in the perception of threats that will cause an agricultural system to fail or operate in a marginal manner. Of course, for extreme, long-lasting negative climate variations such as long-term multi-year drought, Q must ultimately drift to smaller and smaller values indicating that sustainability is no longer possible.

Andean historical patterns

In the Uhle-Rowe chronological sequence, each time division (Horizons and Intermediate Periods) is characterized by a dominant polity (or polities) with distinct political and economic structures, governmental systems, architectural and settlement patterns, ceramic and religious iconography and agroengineering practices. Frequently, one dominant trait characterizes the period. During Horizons, one society exerts overarching influence over vast territories. During Intermediate Periods, dominant regional states may exert control primarily through branching government structures capable of integrating adjacent territories into the same ideological and political template or in certain cases, by conquest of adjacent societies to garner their resources agricultural and otherwise.

The Early Horizon (EH) time period is characterized by highland Chavín influence diffused into Peruvian north and central coast radiation centers showing similar, but locally interpreted, artistic traditions in iconographic, ceramic and textile traits. The expansion of Chavín influence from highland sources appear to have been religion based. Minor south coastal societies (Paracas Cavernas) arose at this time with regional influence.

During the Early Intermediate Period (EIP) major coastal architectural and agricultural complexes were begun by the Moche dominating the Peruvian north coast. Lima polities were preeminent on the central coast, and Paracas and later Nasca polities were established on the south coast and characterized by limited centralized administrative control. The minor north coast Recuay and Huarpa societies of the central highlands arose in this period but had only local extent and influence with minor irrigation works compared to the major costal and highland polities. During the Middle Horizon (MH) a shift back to highland dominance occurred with late Tiwanaku (Phases IV and V) and Wari polities dominating much of the southern and central Andean coastal and highland regions through political, economic, military and religious influence. Large agricultural complexes in the form of raised fields constructed by the Tiwanaku and terraces constructed by the Wari, were in place in conjunction with secondary administrative centers such as the Tiwanaku centers of Omo, Pajchiri, Lukumata and Wankarani and the Wari centers of Pikillaqta, Cajamarquilla, Viracocha Pampa, Cerro Baúl and Wari Wilka. The Late Intermediate Period (LIP) is characterized by a shift back to prominence of coastal societies with the Chimu occupying a north coast zone from the Chancay to the Lambeyeque Valley. The Chimu incorporated a complex of administrative centers (Farfán, Manchan, Purgatorio) with older ceremonial centers (Pacatnamú, Chotuna) in north coast valleys adjacent to the Moche Valley where the capital of Chan Chan was located. The idea of centrally administered, multi-valley agroengineering complexes directed by satellite administrative centers sharing common political, social and religious practices appears to be a feature of this period. Ica culture was dominant in the Peruvian south-central coastal area at this time with the minor intermediate highland Recuay and Cajamarca societies having only regional influence. Military conquest and dominance of highland and coastal polities by the Inca state occurred in the Late Horizon (LH). It appears coincidentally that the EH-EIP-MH-LIP-LH chronological sequence somewhat corresponds to a geographic alteration of prominence between highland and coastal polities. Because the effects of climate on agricultural systems had some role in the sustainability of Andean civilizations, these effects are discussed in terms of the Q equation.

Late horizon and intermediate period sustainability

The climate change history reflected in the Quelccaya, Huscaran and Titicaca Lake cores indicates a major, long-term drought starting in the late EIP at AD ~600. This drought had a role in the decline of the coastal Moche and Nasca polities at their traditional sites, while the highland Tiwanaku and Wari polities began their rise to prominence observed later in the MH period. A drought-induced lower Runoff Ratio (R) affected coastal zones disproportionately. Coastal canal-based irrigation systems have high vulnerability (low 1-V) because they are runoff-dependent. Known coastal agricultural storage facilities were minimal and population apparently was in balance with pre-drought agricultural resources (i.e., balance is taken to mean that agricultural resources were adequate to continually sustain population size). Yields for irrigation-based agriculture were high consistent with high R. While technology to modify and defend agricultural systems was limited in early EIP times, a slowly evolving drought crisis developed, and $dT/dt \div dD/dt < 1$ resulted as technical innovations alone could not overcome extreme long-term drought even at sites with high soil productivity. The net EIP result was a Q decline of coastal polities – mainly the Moche polity on north coast Peru. While large populations provide labor resources, unless technology is present (or can be rapidly developed) to utilize these resources, then large populations adapted to food supply levels developed during adequate water supply periods suddenly become a liability when drought onset is rapid. With reference to the Q equation, drought reduces agricultural sustainability and Q decreases during drought for vulnerable Index 4 and 5 systems characteristic of the north and south coast EIP where low R and S prevailed. Some cultural transfer of the Moche tradition to northern coastal valleys (and integration with local societies in that area) and the creation of new Sicán centers such as Pampa Grande occurred in late EIP and early MH times, indicative of the need to restore Q to higher levels, primarily by utilizing high agro-technology levels (dT/dt) in combination with the canal-interconnected, higher flow rate (R large) rivers (such as the Leche, Chicama and Lambeyeque rivers) with vast, fertile land areas (large Y A R). Highland Tiwanaku and Wari societies achieved high levels of sustainability during late EIP and MH times due to elevated highland rainfall rates. Because of design features that imply large dT/dt, the low vulnerability Tiwanaku raised field systems and Wari terrace systems flourished under both high rainfall and intermediate-term drought conditions. The highland Wari and Tiwanaku polities have high R, S, Y, 1-V, with balanced P′. Highland rainfall was adequate during the late part of the EIP, so that Wari terrace agriculture flourished; however, toward the end of the MH diminishing rainfall levels undermined the productivity of these systems as sustained, multi-century drought took hold after AD ~1000 leading to collapse and dispersal of these societies noted at AD ~1100. With respect to the Q equation, highland altiplano Tiwanaku in the late EIP and early MH were characterized by low vulnerability (V) raised field systems, large storage facilities (S), high yields from raised field agriculture (Y A R large), large water supply, high dT/dt but slowly increasing dD/dt as drought began to reduce annual rainfall levels by 5% to 10% from previous norms. Although a large population existed that theoretically could be utilized to modify the location of the agriculturally productive raised field zones in the Lake Titicaca area, the raised field design was not easily modified as it would require decreasing the elevation of raised field berms to accommodate the lowering groundwater level associated with the water height in the swales. Given that the raised fields covered ~75 to 100 km^2 from different estimates, this repair task was clearly not humanly possible even with the large population of urban and rural Tiwanaku. The net result was a high Q value that indicated high sustainability through the MH until the deepening drought began AD ~900–1000 in the early LIP. At the end of the MH, highland polities underwent slow collapse and population dispersal due to effects of long-term drought, while coastal polities (Chimu, Sicán and Lambeyeque) flourished. This can be explained by observing that while R, S and 1-V were low in similarly afflicted coastal areas, dT/dD, Y and P′ were high, reflecting the development of advanced canal technology with sufficient labor to implement major agroengineering projects altering canal placement and design (Ortloff 2009) to accommodate drought effects (dT/dt large), and the availability of marine-based food supplies to supplement land food resources. This later phase of Chimu society saw the dispersal of Chan Chan Moche valley population to satellite sites in adjacent coastal valleys as well as conquest of northern coastal valleys with large land areas and adequate water resources (particularly larger rivers with longer time period high flow rates) as detailed in Ortloff (2009). Coastal valley canal systems can be easily modified (Ortloff et al. 1985, 2012) for reduced water supplies; this is evident with the reconfiguration of intravalley and intervalley canal systems (e.g., the Chicama-Moche, Motupe-Leche-Lambeyeque and, Chillón-Rimac-Lurín systems (Kosok 1965; Ortloff 1993, 2009; Ortloff et al. 1982) that redistributed available water from larger rivers over long distances between valleys to large field system complexes in valleys with low water resources. While water supplies were adequate, low-slope surveying accuracies (large dT/dt) extended canals to larger cultivatable areas (Ortloff 1995, 2009). When water supplies declined due to long-term drought, canal replacement and cross-sectional area reshaping for hydraulic efficiency improvements provided an optimum strategy (increasing dT/dt further) to distribute available water supplies brought in by inter- and intravalley networks (Ortloff 2009). Military conquest by the Chimu of northern valleys, including absorption of the Sicán of the Lambeyeque Valley AD ~1375 added to land areas and water supplies available for agricultural exploitation. The potential to transfer main agricultural zones to higher flow rate north coast valleys is a variant of "storage capability S" or simply an increase in R. Therefore, sustainability of coastal societies under declining water resources is aided by their ability to alter irrigation systems (technology and placement), while highland raised field and terrace systems could not be easily modified and are thus susceptible to long-term drought extinction.

To illustrate this point, while the Tiwanaku raised field systems have high (1-V), and water table height and agricultural area shifts were controllable in the presence of extended drought, an option to lower all ~130 km² of the Tiwanaku-Lukurmata-Koani raised field planting surfaces to accommodate the late MH drought-induced sinking water table would require a vast labor input over many years to achieve marginal benefit. As an example, Figure 4.1 illustrates the time dependence of Tiwanaku raised field agriculture. For population size estimate limit $P'(t_1) \approx 20{,}000$ and $P'(t_2) \approx 40{,}000$ and R(t) between 1 and 2 limits for sustainable population and agriculture at AD ~900 (top plane) to support the population, the sustainability factor Q lies at ~3.5 (point B, Figure 4.3).

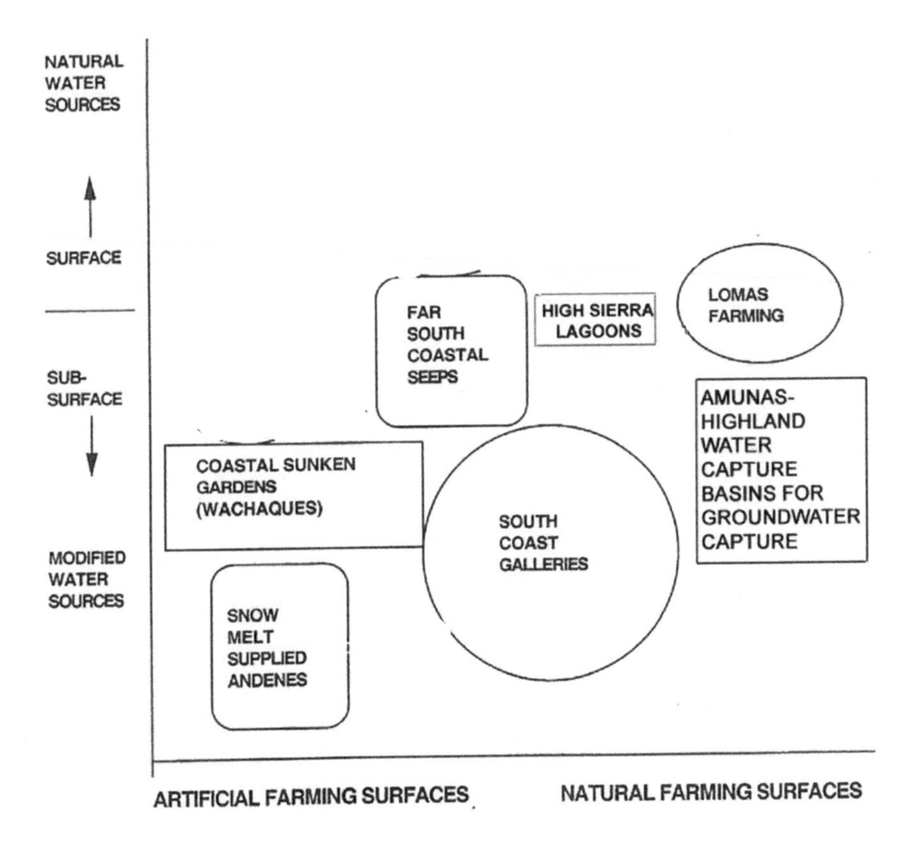

Figure 4.1 Main agricultural strategies of Andean societies.

With increasing time into the 10th century AD, drought initiates leading to a contraction of raised field productivity. Here the near Lake Titicaca groundwater level declines precipitously as lake level decreases due to lower rainfall interception amounts and lowered groundwater supply. Swale water levels decrease to the point where near lake agriculture is no longer tenable over vast raised field areas and agriculture has been moved to distant field systems from the lake edge where the water table remains high due to the incoming source of groundwater from intercepted rainfall deposited over the long pre-drought time period at distant areas from the lake edge. Although this strategy entailed some agricultural resources available to city population, it was insufficient to maintain previous population levels. The archaeological record shows that population dispersal from urban Tiwanaku occurred to scattered sites that could support local family groups around *cochas* (excavated pits to intersect the local water table for limited agriculture) as well as Tiwanaku site abandonment to remote area that had water and land resources to sustain elements of the previous population. One such site was found on the eastern slope of the Andes indicative of the dispersal range. Figure 4.1 indicates as the drought commences starting from about AD ~900 with a decrease of about 5–10% less rainfall per year over the 10th to 12th centuries AD, that finally at about AD ~1100, and agricultural yield is no longer at levels to sustain a large urban population as the sustainability Q value slowly declines to a level where city abandonment occurs with dispersal of the previously high population levels. Figure 4.2 indicates that several of the drought remediation measures described in Chapter 12 and represented as R(t) that supported larger P′(t) increases played a role in extending and improving agricultural yields in centuries preceding inception of 10th-century drought; and as such, despite these defensive measures, the inability to deepen swale troughs over vast raised field areas several kilometers from the lake edge as drought intensified proved fatal to the Tiwanaku society. Largely ignored in the climate change and water supply conditions affecting large populations are the effects of temperature change T(t) accompanying climate change. Figure 4.4 represents an analog scenario where certain bacterial species inimical to human health through degradation of water quality have pathogens that destroy crops; here species concentration increases with temperature. Temperature increase effects on human health multiply as resident bacterial populations increase with time. Here temperature changes affect both human health as well as invasive bacterial species that may affect food supply. Research (Thompson et al. 1989, 1995) suggests that such a temperature rise accompanied the 10th- to 11th-century drought.

If then certain bacterial species inimical to human health under humid conditions flourish, then health considerations can further challenge population sustainability. In Chapter 5, discussion is focused on rapid drainage of rainfall by means of a moat drainage canal within the urban center of Tiwanaku – this feature helped to dry household interiors rapidly and thus reduce respiratory and mould related health problems associated with warm and damp living environments. If indeed the moat feature was designed to promote health benefits under climate change effects that promoted bacterial health problems, then this feature

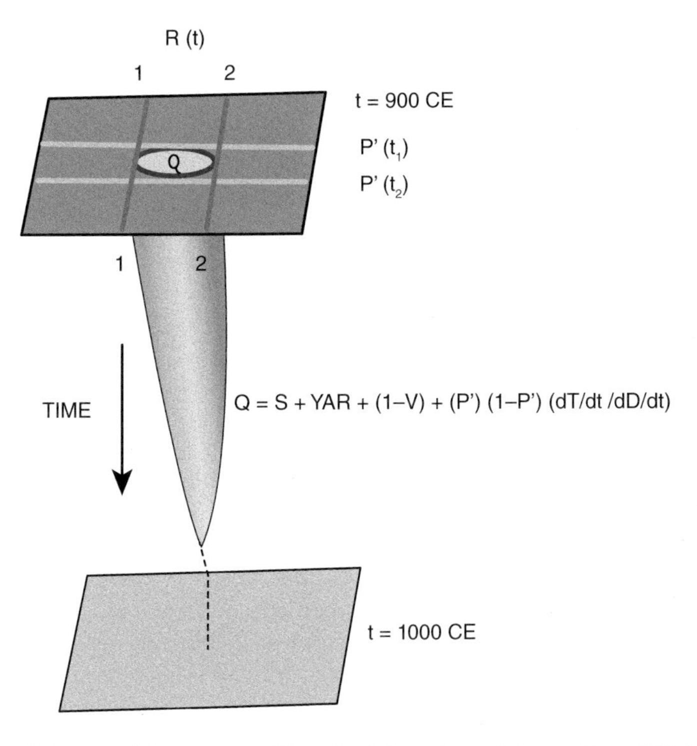

Figure 4.2 Diagram of the progress of the urban Tiwanaku society over time. P′(t) describes the population estimates $P'(t_1)$ ~20,000 and $P'(t_2)$ ~40,000 at ~600 CE; between R(t) represents water availability between 1 and 2 limits to sustain raised field agriculture. The descending time (t) axis indicates contraction of raised field agriculture to near extinction at ~1100 CE.

is a first realization of a hydrological structure intended for that purpose noted in the Andean world.

While a large Chimu labor force on the coast could have been productively employed in canal modification and intervalley connection projects and expansion into large land areas with adequate river resources, the large highland labor force at Tiwanaku was not sufficient to modify ~130 km^2 of raised fields to respond to the declining water table level induced by sustained drought. Therefore, higher dT/dt was possible for coastal irrigation systems due to more easily modified canals, use of intervalley canals that redistribute water, and relocation of agricultural production centers to adjacent water-rich valleys. Highland altiplano agricultural systems were limited in design modifications (low dT/dt after initial construction of raised fields) to react to long-term drought. Although highland agriculture based upon mountainside terraces elevated farming zones to take advantage of higher rainfall to ameliorate drought conditions, Y decreases due to poorer upslope soils and decreasing farming area with height reduced the agricultural potential. None the less, construction of several terrace structures adjacent to the Tiwanaku raised field areas somewhat utilized the large Tiwanaku workforce to help sustain agricultural output but only in a limited, marginal form compared to what was available from the raised field systems under normal pre-drought conditions prior to AD ~900 drought inception. For coastal zones, the lower runoff R available was better utilized due to higher Y from fluvial-deposited soils, more effectively utilized labor resources, and high dT/dt from various defensive strategies (Chapter 3), despite the higher the vulnerability index of canal systems.

Eventually, coastal agricultural systems had the potential for quicker recovery as normal levels of rainfall resumed in the late LIP. The water supply advantage to highland systems (large Q, R, S, 1-V, Y, A, P′, high dT/dt and low dD/dt) was again manifest in the LH to the advantage of post-Tiwanaku societies and ultimately, the Inca society. The Inka (*mit'a*) policy of population relocation to revitalize high and lowland agricultural centers saw the end of many conquered polities operating in their previous political-economic and sociopolitical modes.

Figure 4.3 indicates that Q values for the coastal southern Moche branch (i.e., Moche polities further south than the Jequetepeque Valley) and Nazca polities have descending Q values that reach minimum values in the 6th- to 7th-century AD drought episode (Thompson et al. 1994, 1995, 1989; Shimada et al. 1991). The southern Moche polity was reliant on runoff channeled into coastal river valleys while the Nazca relied on subterranean gallery quanat water supplies and thus both coastal polities were more vulnerable to drought stress than highland societies. Similarly, the later LIP Chimu polity centered in the Moche Valley appears to expand its agricultural base during drought periods through migration and conquest expansion to northern provinces with large water and land resources. The Sicán polity of the far north Lambeyeque Valley, with its strong Moche identity in architectural and iconographic patterns and traditions, is absorbed by Chimu conquest. While details of the highland Wari polity and its MH expansion are fragmentary, little is seen of classic highland Wari central and satellite sites past the 11th century AD indicating that both coastal and highland societies were subject to a long-term drought that largely influenced levels of agriculture evidenced in prior centuries (Kurin 2016). While post-Tiwanaku sites occupied by dispersed Tiwanaku city inhabitants resorted to limited raised field usage far away from the lake edge, the use of dispersed *cochas* was in evidence for smaller population groups as the city of Tiwanaku was abandoned. Expansion of the coastal Chimu by conquest and colonization occurred in order to exploit the larger agricultural land areas and larger water resources of northern river valleys. This pattern of expansion into northern coastal and sierra territories is observed by the Wari polity during the Middle Horizon perhaps for similar reasons based upon obtaining new territories and colonies under their control with different and less vulnerable agricultural systems that expanded their ability to maintain an agricultural base less dependent upon heartland terrace systems. While the Chimu capital of Chan Chan in the Moche Valley experienced loss of agricultural lands and a heroic attempt to revitalize these lands by construction of the ~75 km long Intervalley Canal to bring water from the Chicama Valley into Moche Valley to revitalize

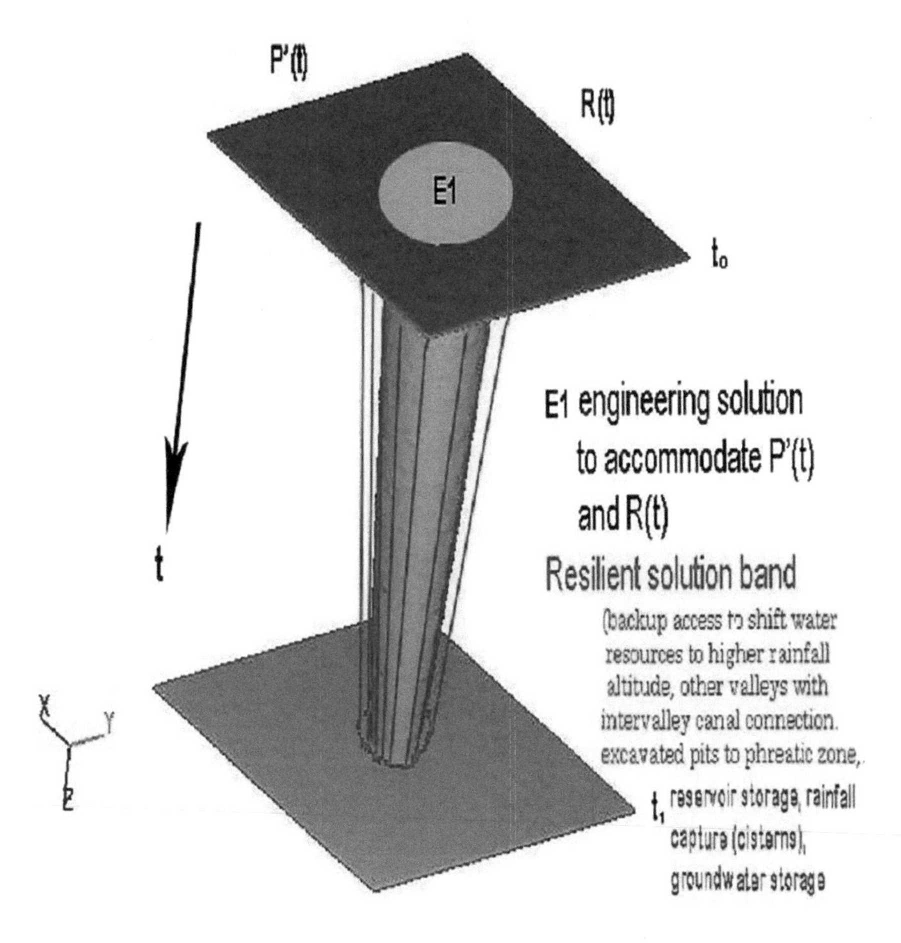

Figure 4.3 The sustainability Q diagram denoting Q values for different Andean societies from Early Intermediate Period (EIP) to Late Horizon (LH) times.

existing canal systems (although the question remains unsettled if the Intervalley Canal actually delivered water as far as the Pampa Esperanza field area adjacent to Chan Chan), expansion into northern valleys provided a degree continuity and revitalization of the centralized Chimu polity through northern coastal valley administrative centers to oversee agricultural production for Chimu colonies. While limited agriculture existed in the Moche Valley through late phase small canal systems, almost all of the extensive field systems adjacent to Chan Chan were lost as drought progressed. Expansion of agricultural systems in northern valleys ultimately recovered as the rainfall increased past the 12th century AD – this related to worldwide climate changes in the Medieval Warm Period.

The highland Inka, through conquest and aided by bountiful rainfall in their home and conquered territories in the 12th century, were beneficiaries of positive

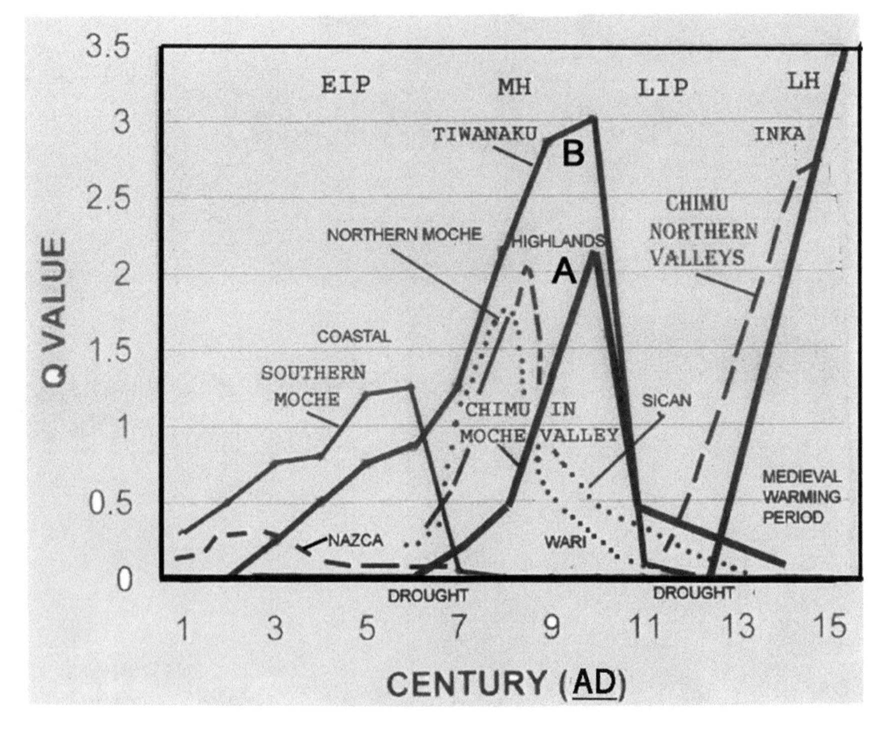

Figure 4.4 Engineering defense solutions represented by an increase in productive water resources R(t) supporting increases in population P′(t) provided high sustainability Q at least up to initiation of 10th-century drought.

climate change effects. Each successive Inka ruler expanded conquered territories and resettlement of conquered peoples to industrialize agriculture and storage systems to the highest levels achieved by prior Andean polities. With conquest and incorporation of Chimu territories in the late 14th century AD, the northern Chimu disappear from the archaeological record as their cultural symbols, practices, cities and settlements are subsumed and erased by Inka overlords to fit the Inka political economic model. Figure 4.3 indicates that climate variations play an important role in deciding the fate (Q) of coastal and highland polities.

To this point, the Q equation has been used to show the collapse and/or transformation of ancient societies as related to major drought/flood episodes (Figure 4.3). The following discussion investigates structural changes in societal political development and reconfiguration as observed from Andean archaeological history. From Figure 4.3, the coastal polities that transform through 7th-century drought/flood events are the southern Moche with a major administrative center in the Moche Valley; the site of Galindo represents the termination of Moche V valley presence and the introduction of new architectural norms (rectangular compounds, absence of pyramid mounds and elaborated temples, changes in burial practices, large concentrations of food storage facilities, etc.) suggesting influence seen in later Chimu architecture in the Moche valley. Although occupation of the Moche site continues for several decades past the major El Niño event, induced sand inundation ultimately causes site and field system abandonment. As coastal polities were dependent upon runoff supplied rivers for irrigation agriculture, drought had a more severe effect than for highland polities closer to rainfall sources. For Tiwanaku dependent upon spring and groundwater raised field agriculture, drought effects had a time delay on agricultural productivity as groundwater levels depend upon intercepted rainfall and subsurface transport from rainfall interception events of past centuries. While coastal polities decline under drought conditions, highland polities thrive – as Chapter 4 indicates for early Tiwanaku phases but later collapses due to extensive period drought ultimately stranding the raised field systems.

The presence of forts in northern sector coastal and sierra areas controlled by the Moche, the loss of valleys from Huarmey to Virú and the contraction of the southern frontier back to the Moche Valley characterize societal changes catalyzed by flood/drought incursion in the 7th century. Moche society is spatially divided into Upper Moche (characterized by metalwork and fragmented townships lacking a centralized administrative and religious center and occupying valleys from Jequetepeque northwards to Lambeyeque) and Lower Moche with emphasis on elaborate ceramics, elaborated ritual centers and a cohesive central administration located at Moche with common religious norms and iconography in valleys south of Jequetepeque and extending as far south as the Santa Valley. While the southern division centered in the Moche Valley was abandoned after the Galindo site was abandoned, the northern branch was intermixed with local independent societies and a presence established in the Lambeyeque Valley at Pampa Grande that ultimately adopted a modified iconography and urban space usage pattern from prior norms while retaining major pyramid structures and elite housing compounds together with major agricultural expansion through elaborate canal systems. The later phases are sufficiently different from Cerro Blanco Moche roots to be relabeled Sicán with a capital at Batan Grande in the Lambeyeque Valley. Concentrations of urban workplaces devoted to metallurgy, craft production, textile manufacture, comestible processing and irrigation agriculture management signal a move toward secular pursuits deemphasizing religion and ritual compared to previous times. As opposed to administrative centers in valleys under Moche control enforcing Moche religious and governance canons, it appears that post-7th century, administrative control is centered in Lambeyeque while outside of this valley, local *señorios* and *parcialidades* control agricultural zones with some degree of autonomy from state-sponsored religious, iconographic and ritual practices. This may imply that practicality and secularism replaced reliance upon deities and religious rites for societal continuity and that the new importance of irrigation technology and multi-valley irrigation canal systems was the real key to survival.

From Figure 4.3, the Tiwanaku polity had a documented collapse in the 11th century primarily as an outcome of an extensive drought period and the inability to modify raised field systems to accommodate a declining water table. Similarly, a decline in mean rainfall levels in this time period had negative effects on the terrace and valley bottom agriculture of the Wari centered in the Ayacucho area of highland Peru. Population dispersal from the core city of Tiwanaku to areas with water supplies sufficient for small group survival laid the foundation for the consolidation of emergent societies (Lupaca, Colla, Pacajes and other societies to dominate territories on the western and eastern borders of Lake Titicaca) in subsequent centuries. Again fragmentation of monolithic state structures with a centralized administrative center controlling vast territories degenerated into localized population centers defending their local resource base with defensive forts and structures guarding their local agricultural fields and water resources; this focus on security undoubtedly incorporated memory of drought devastation in prior centuries and underwrote the need to protect individual, localized agricultural zones that provided survival for local populations. This new focus decentralized former overarching administrative structures concentrated in Tiwanaku and its altiplano and Moquegua Valley satellite centers and led to resurgence of ethnic groups concentrated in homeland areas to form new administrative centers with local governance. The original ethnic Moquegua Valley populations (largely Huacane) saw incursion and then withdrawal of Tiwanaku presence as the valley's agricultural value decreased under drought conditions. As Tiwanaku had an eclectic population mix composed of different ethnic groups brought together in a city environment to participate in economic and urban community life, once the agronomic base was threatened by multi-century drought, a reversal of attractions that brought people together in an urban setting reversed itself leading to either reestablishment of original groups back to original homelands or out-migration of groups with common ethnic ties to areas more favorable to survival with land and water sources. From Figure 4.3, Wari Q values are seen to increase while those of the Moche decline even under 7th-century drought conditions. This is attributable to the R effect where coastal polities experience a disproportionate runoff decline and highland polities rely on higher rainfall levels by moving terrace agriculture to higher elevations utilizing snowmelt-supplied canals to irrigate terraces.

Figure 4.3 shows the progress of the Chimu polity from inception at AD ~800–900 in the Moche Valley to final closure by Inca invasion (and conquest in the 14th century AD. At the Moche Valley capital of Chan Chan, a succession of royal elites ruled from nine separate compounds over an empire extending from the Chillon Valley on coastal Peru to northern valleys close to the present-day border of Ecuador in later phases of empire. Based on maximum exploitation of irrigation agriculture in valleys under their administrative control through satellite centers, the pan-Andean drought of the 10th to 11th centuries AD brought closure to reliable food supplies from seasonal, intermittent rivers characteristic of valleys south of and including the Moche Valley. To maintain agriculture, military expansion into northern valleys with larger land areas and river water sources resulted; the Chimu invasion of the northern Sican territory in the Lambeyeque Valley at AD ~1375 effectively incorporated lands and resources into the Chimu Empire (C1, Figure 4.3). Thus the 10th- to 11th-century AD drought had the effect of expanding Chimu control into conquered northern valley lands while maintaining overarching administrative control from Chan Chan through its local valley administrative centers. Presumably, redirection of exported agricultural products back to Chan Chan along coastal road networks helped sustain the center's prestige as a control and craft center involved in the manufacture of luxury goods that by exchange to local administrative center's managerial elites, guaranteed continuity of the economic system. As the empire became more expansive over new areas and incorporated conquered populations under new rules and rulers, coherence only possible with fewer branches of the administrative tree presented problems as different valleys had different responses to the ongoing drought in terms of available production and decisions regarding export quotas and diversion to the needs of local population. Thus gaps between administrative plans and reality occurred challenging the future coherence of empire in the 11th to 12th centuries; past this period, gradually improving rainfall and runoff amounts characteristic of global climate change gave resurgence to the empire's extended resource base now including almost all of the north coast. Thus expansion by conquest as an earlier drought response to ameliorate incipient agricultural shortfall production of the 10th to 11th centuries ultimately proved fortuitous as the return of higher rainfall levels in later centuries only increased the resources available for further population growth at higher living standards – all to the benefit of the elite ruling class's power and authority for having made right decisions appreciated by the beneficiaries of these decisions. While the Chimu Empire grew in wealth and productive territory, it became a target for the Inka to assume ownership of all the Chimu had accomplished – history records conquest of the Chimu in the 13th century and their assimilation into the empire goals of the Inka (C2, Figure 4.3). Elsewhere due to higher rainfall amounts, north sierra Cajamarca transitions through drought episodes successively while coastal Lima society appears to phase out due to 7th-century drought.

In terms of predictions of dire changes in societal structure proceeding from 21st-century global warming effects, lessons from Andean history are instructive as within this historical time period numerous agricultural crises were brought about by climate change effects. The success of responses largely depended upon options to modify field and water supply systems to accommodate new water availability conditions; here the Chimu, after valiant efforts to sustain Moche Valley agriculture by canal recutting and use of Intervalley Canal waters, opted for conquest and resettlement into northern valleys with more land and water resources. With these resources and increasing runoff for irrigation past the 12th century AD, final conquest and integration of Sicán polities was completed in the 14th century. The ability to easily modify canal system placement and design in desert sands to adjust to changing water availability levels and sources provided ongoing use of declining land areas within the Moche Valley while expansion into northern valleys provided increased sustainability based on exploiting other ecological environments rich in agricultural resources. Based upon 21st-century prognostications and examples of climate-induced resource wars, the Chimu first solved their agricultural resource problems by use of canal design and modification technology to extract more product from declining land and water resources (as innovations such as drip irrigation, genetically modified crops, pesticides/herbicide use, water conservation, crop hybridization and crop rotation, etc,. are in use in the 21st century); when technical limits were reached, the Chimu version of resource wars followed with conquest of ecologically rich lands. This scenario is currently being played out in the 21st century with rich Asian and Middle Eastern nations leasing/buying vast areas of third world Africa to guarantee agricultural resource zones for their populations. Were it not for some degree of world censure for military and economic conquest, the picture may quite resemble that of the Chimu and military conquest.

For the Tiwanaku, their fate was tied to raised field agricultural systems not easily modified under long-term drought conditions so that city collapse and population dispersal was inevitable. For the highland Wari, terrace agriculture could be sustained by moving farming areas to higher elevations where rainfall was more abundant; however, loss of land area, less fertile soils and cold limited productivity. Similar in strategy to the Tiwanaku, dispersed colonies centered on water resources resulted. The lesson for the 21st century is not to depend solely upon a fixed design agricultural system that relies on one water source only – if climate change alters the water supply to that area then the total system cannot be easily modified to accommodate an alternate imported water source. The thought to consider is alternate water supply systems to known fertile land areas. If, for example, declining rainfall diminishes agricultural output, then piping or channeling water from another water source would be required to restore output. As an example, if US Midwest grain fields undergo drought, and groundwater pumping is inadequate to maintain productivity, then a massive canal network from the Great Lakes may be necessary to revitalize at least a fraction of formerly productive lands.

For the Nazca, *quanat* systems based upon groundwater extraction proved reliable for many centuries but lack of recharge due to long-term drought proved fatal to their survival. As much 21st-century agriculture across the world is now dependent upon groundwater extraction, and rainfall infiltration recharge does not match extraction due the declining depth of the groundwater surface past the evaporation zone, such systems are unsustainable under long-term drought. Here the fate of the Nazca shows the outcome of dependence on groundwater systems and their vulnerability to long-term drought.

Climate change lessons from the andean historical record

As current media commentary is focused on consequences of global warming, books with titles *Global Warring* and *Climate Conflict* predict near-apocalyptic visions of enormous and specific geopolitical, economic, and security consequences and, as such, the world of tomorrow looks chaotic and violent. Other writers call climate change an existential threat that could usher in state failure and internal conflict in the economically vulnerable societies of Africa and Third World areas. Dire prognostications are not limited to the more vulnerable societies but include the developed world with its dependence upon reliable supplies of energy, food, water, markets, communication, transport and infrastructure to provide and underwrite economic stability with a degree of predictability. Even with years of economic history and subject lessons, chaotic patterns would still emerge that unsettle stability and with the additional variable of climate change; these instabilities can be magnified and serve as catalytic agents for unraveling societal order. Yet for all these prognostications, limited factual material exists to support the destiny of the contemporary world to perceived climate change effects as much lies in the realm of computer predictions and idea extrapolation into conditions not yet experienced by 21st-century civilization. Collective societal perceptions and responses to crises bring forward questions as to whether human psychological response remains affected by the cultural conditioning of different civilizations in different ages to yield predictable responses to survival challenges.

Chapter 2 introduces Complexity Theory derived from biological system analysis methodologies. Figure 4.5 illustrates the multiplicity of events that can originate from an initial species state (or several coexisting species states) given external influences that can alter these states, eliminate populations or even originate new variations on original species populations.

The social analog would entail different vying theories competing for collective communal acceptance by all society members based on economic and labor-saving advantages apparent to all society members. En route to this final state are external influences (climate change affecting the food resource base, foreign invasion, etc.) as well as internal influences (diverse religious and custom beliefs by faction groups, labor shortages to implement collective action plans, political opposition to rule by elites, reluctance to change sociopolitical, socioeconomic and political science established norms, etc.) that disorganize the path toward a stratified state organization where all members accept their assigned roles under an agreed-upon leadership class. As examples from Old and New World societies illustrate, once this collective agreement societal structure state is achieved, then societal advancement follows.

A backward look to examine the historical record coupled with climate change records is useful to infill information as to systematic social, political and economic responses to past climate crises. For example, drought in the Central Asian steppes led to barbarian westward migrations that ultimately challenged Ottoman dominance of Middle East societies and collapsed state governance and rule; sustained drought influenced Maya classic period collapse; Little Ice Age climate change closed out Greenland European settlements; the Ming Dynasty in China is thought to have collapsed from drought AD ~1644 and collapse of the Khmer Kingdom at Angkor in present-day Cambodia by successive drought cycles AD ~1340–1360 and AD 1400–1420 are well-known examples of climate change affecting historical patterns; further examples of drought influencing cultural transformations in both Old and New World settings are found in Manzanilla (1997). A more contemporary United States example is the 1930s Dust Bowl drought event triggering a two million person migration event influencing political restructuring and survival attitudes of the survivors to later appreciate and then

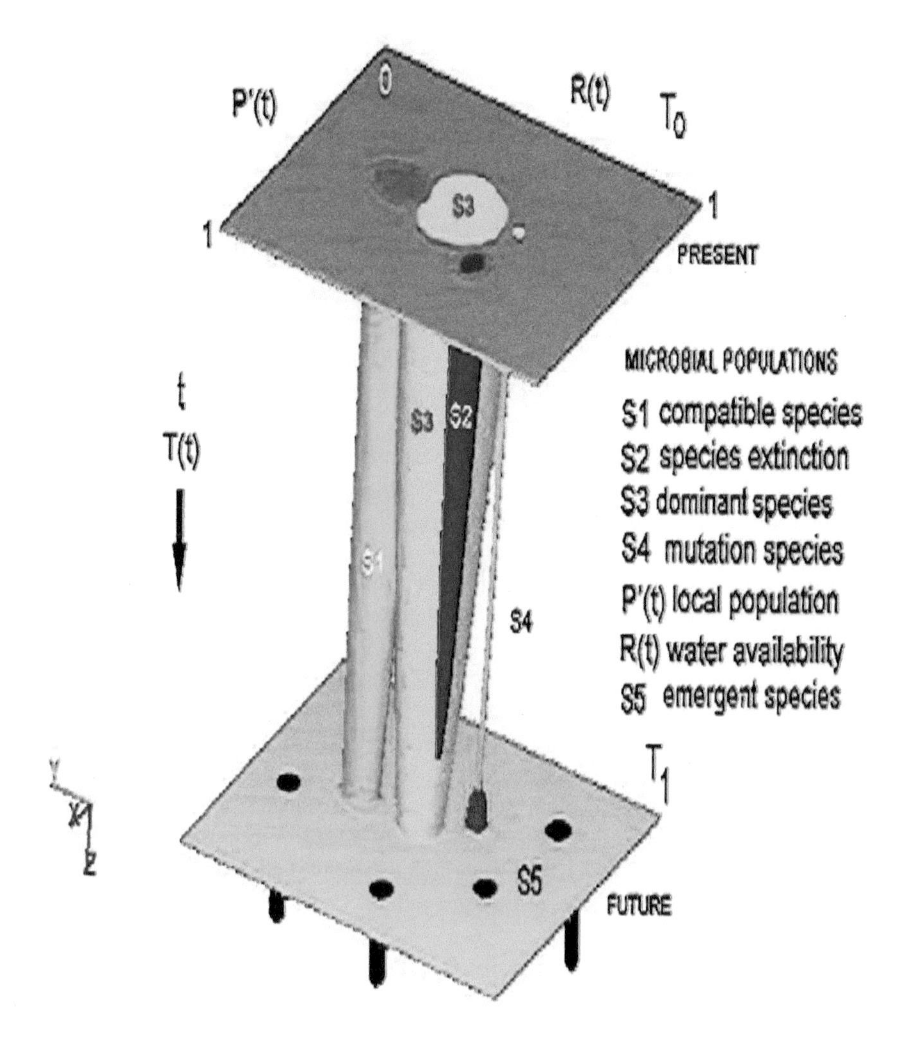

Figure 4.5 Hypothetical temperature T, water availability R(t), time t influence on bacterial species population P′(t) growth/decay histories that may influence population continuity and survival.

demand government programs to restore cohesion, stability and enlightened governance back to a fragmented 1930s society experiencing the Great Depression. While climate change can be a catalytic factor in societal change, governmental structure and its ability to respond positively to disasters, ethnic conflict, outside intervention, vulnerability of agricultural systems to climate change by innovation and creativeness for problem solving are further determinatives of the future.

A survey of climate change effects on Andean historical patterns reveals that major drought effects in the 7th century AD initiate termination of the Nazca society and, in the case of the Southern Moche, relocation to northern Peruvian coastal locations with more favorable conditions for agriculture is evident from current research findings. For the drought reaching its peak in the 11th century AD, the northern Moche-Sicán as well as the Chimu heartland in the Moche Valley underwent collapse of their agricultural systems; the highland Tiwanaku and Wari societies follow suit in their heartlands with dispersal into regions with more adequate water resources and collapse of colonies tied to the central capital city. While dispersal may involve occupying lands of other groups by assimilation, coercion and migration peaceful or otherwise, in the case of the Chimu, military conquest of northern valleys occurred with military forays into the Jequetepeque and Lambeyeque Valleys in the 11th to 13th centuries. The vulnerability of different choices of agricultural system for different societies in different regions appears to influence the outcome: here difficult-to-modify or defend against drought systems predetermine the fate of societies – particularly the Tiwanaku's raised field and the terrace systems of the Wari. For easily modifiable systems, such as those in coastal areas, the innovative capabilities of the Chimu combined with conquest and absorption of new lands and water resources provided sustainability until military conquest by Inka armies in the 14th century. Thus a myriad of responses is observed from the archaeological record to climate variations – as to whether this pattern is a portend of responses ahead for the globally warmed 21st-century societies remains unanswered as this history is yet to be written. But as human nature appears independent of place and time as witnessed by recurring cycles of world history, the Andean past may have elements of our future.

5 New discoveries and perspectives on water management and state structure at AD 300–1100 Tiwanaku's Urban Center (Bolivia)

Introduction

The AD 300–1100 pre-Columbian site of Tiwanaku located on the high altiplano of Bolivia demonstrated an advanced use of hydrologic and hydraulic science for urban and agricultural applications that is unique in the Andean world. From recently discovered aerial photos taken of the site in the 1930's, new perspectives of the water system of the ancient city, beyond previous interpretations of a major channel as a dividing 'moat' between ceremonial and secular parts of the city, are now possible from new discoveries of a network of water channels not previously known. Surrounding the ceremonial core structures of urban Tiwanaku was a large encompassing drainage channel that served as the linchpin of an intricate network of spring-fed supply and drainage channels to control both surface and groundwater aquifer flows. The drainage channel served to (1) collect and drain off rainfall runoff into the nearby Tiwanaku River to limit flood damage; (2) accelerate post-rainy season ground drying by collecting aquifer seepage from infiltrated rainwater into the drainage canal to promote health benefits for the city's population; (3) provide water from a newly discovered spring-fed channel to two subterranean channels to flush human waste from elite structures within the encompassing drainage channel to the nearby Tiwanaku River; (4) maintain the groundwater level constant through both rainy and dry seasons to stabilize the foundation soil underneath massive pyramid structures to limit structural deformation; (5) facilitate rainy season water accumulation drainage from the floor of a semi-subterranean temple into the groundwater layer to rapidly dry the temple floor and (6) provide drainage water to inner city agricultural zones. The water control network in Tiwanaku city is analyzed by Computational Fluid Dynamics (CFD) modeling of transient surface and groundwater aquifer flows to illustrate the function of the drainage canal in both rainy and dry seasons.

The hydrological regime of Tiwanaku

Of interest are the engineering accomplishments of ancient civilizations. In the discussion to follow, the pre-Columbian AD 300–1100 city of Tiwanaku located in the high altiplano (~4500 masl) region of Bolivia demonstrated use of advanced hydrologic principles to maintain city drainage during the long rainy season through a complex network of surface and subsurface channels. This water control system involved both groundwater stability control coupled to spring-supplied surface channels to perform health and monument maintenance benefits to city inhabitants and, as such, was a unique demonstration of hydrologic engineering not seen at any other pre-Columbian South American site. Research at the site has revealed that the city's ceremonial center, composed of monumental architecture and elite residential compounds, was circumscribed by a large drainage channel (Figures 5.1, 5.2 and 5.5, W-D-V-X). While previous researchers interpreted one purpose of the drainage channel as a boundary separating sacred and secular urban areas of the city, the encircling channel additionally served as a hydraulic regulator enhancing aquifer drainage to promote rapid post-rainy season soil dry-out as the dry season progressed. Water collecting in the drainage channel from aquifer drainage, rainy season runoff and flow from canals intersecting the

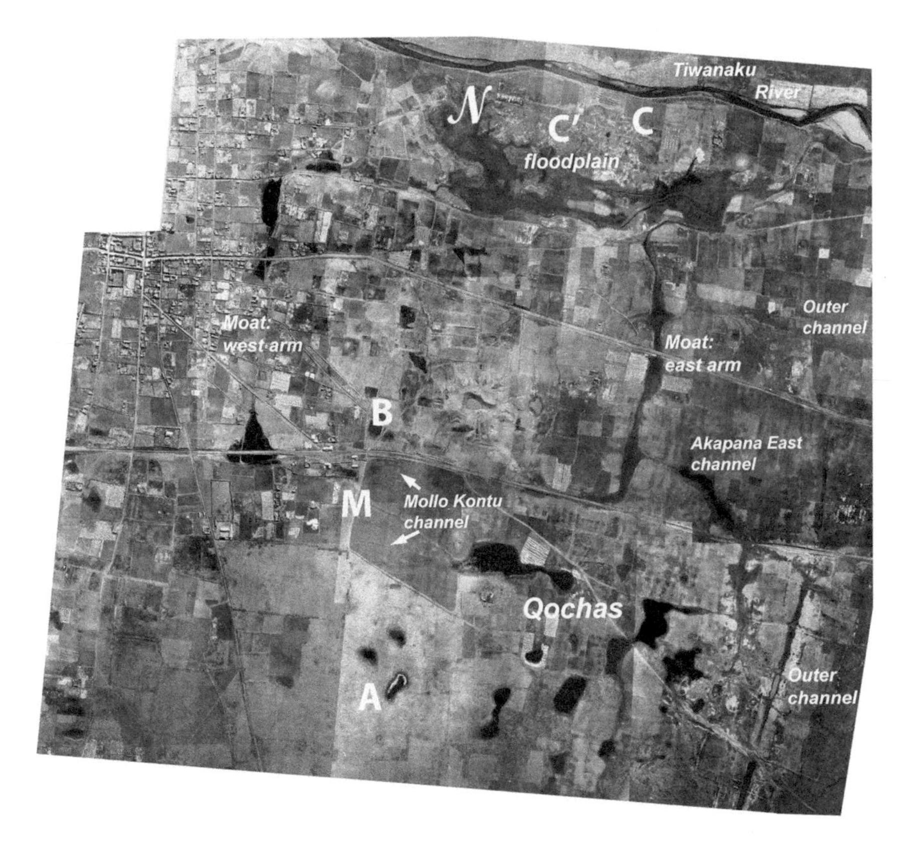

Figure 5.1 Aerial photo view of the inland drainage channel surrounding the ceremonial core of Tiwanaku indicating the intersecting Mollo Kontu (M) canal, qocha regions and the Tiwanaku River to the north.

Source: Alan Sawyer aerial photographs provided to author in 1986, use permission granted.

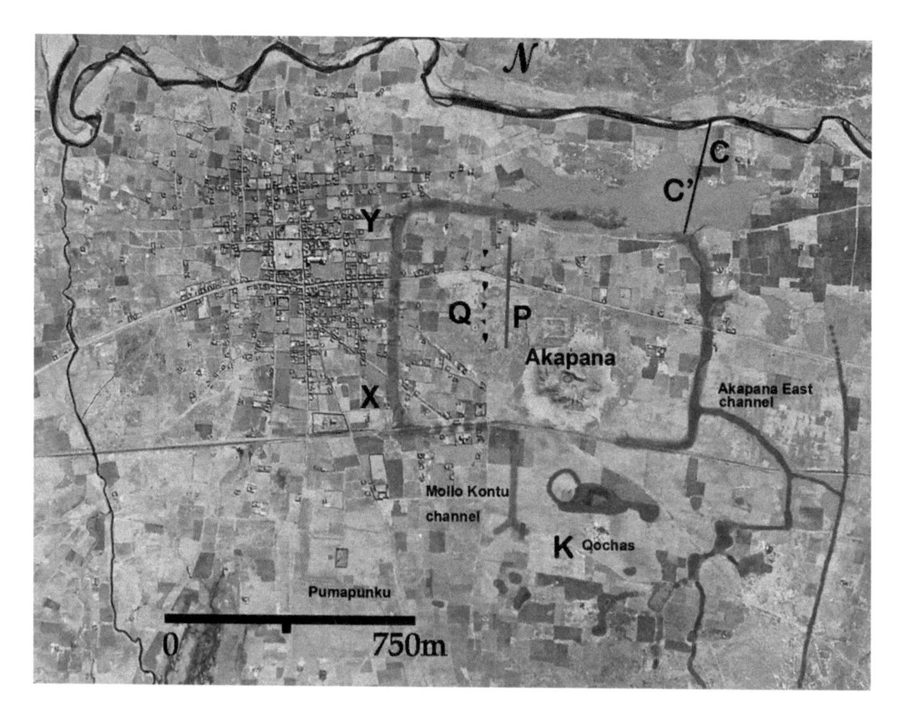

Figure 5.2 Details of the intersection of the Mollo Kontu canal with the southern arm of the drainage channel.

Source: Alan Sawyer aerial photographs provided to author in 1986.

drainage channel rapidly exited through connecting canals to the Tiwanaku River to limit groundwater recharge during the rainy season.

During the dry season, continued aquifer seepage and flow from adjacent canals recharged the groundwater to maintain and stabilize its height through seasonal changes. With the depth of the drainage channel set at approximately the height of the stabilized groundwater profile, the saturated drainage channel bed limited further seepage and promoted rapid water transfer of accumulated water directly to the Tiwanaku River. One net effect of the stabilized groundwater level through rainy and dry seasons was to maintain the bearing strength of soil under large monuments within the ceremonial center to limit structural distortions (Peck et al. 1974). Water accumulating in the drainage channel from aquifer drainage and channeled spring water flow provided flow through dual subterranean channels to flush human waste delivered into the subterranean channels from elite compound structures to maintain hygienic conditions in the residential compounds. The multifaceted hydrological aspects of the drainage channel thus served city environmental and hygienic conditions through rapid soil drying in city housing areas while promoting structural stability for its many monuments as well as aiding in rainy season drainage from the Sunken Temple (Figure 5.5, F) floor. While groundwater control mastery is apparent in the urban setting, additional research on Tiwanaku raised field agriculture indicates similar advances in use of groundwater control technology not previously reported in the literature.

To demonstrate the seasonal interaction of surface and aquifer water flows, a porous media aquifer CFD model is utilized for computational fluid dynamics analysis (Figure 5.5) to demonstrate the inland drainage channel's role as a hydrological control element vital to the city's sustainability during wet and dry seasons.

Settlement history

The ancient (AD 300–1100) city of Tiwanaku, capital of a vast South American empire, has been the subject of research starting from early 20th-century scholars that continues to the present day (Créqui-Montfort 1906; Bandelier 1911; Denevan 2001; Means 1931; Bennett 1934; Posnansky 1945; Ponce 1961, 2009; Kolata 1986, 1993, 2003; Erickson 2000; Ortloff 1996, 2009, 2014a, 2016; Ortloff and Kolata 1993; Browman 1997; Janusek 2004, 2008; Bentley 2013). The city, located at the southern edge of the Lake Titicaca Basin in the south-central portion of the South American Andes at an (altiplano) altitude of ~4500 masl incorporated an elite area bounded by an encompassing drainage channel that supported temple complexes, palace architecture and a seven-stepped monumental pyramid (Akapana) designed to serve ceremonial functions and provide residential structures for Tiwanaku's rulers. Outside of this center lay a vast domain of secular urban housing structures. An intricate network of canals acting in conjunction with the drainage channel performed several hydrological functions: rapid ground drainage during both wet and dry seasons to promote health advantages for the city's 20,000–40,000 inhabitants; flood defense to preserve the ritual center and surrounding urban structures and, most importantly, height excursion control and stabilization of the deep groundwater base underlying the site. This latter hydrologic function prevented dry-out collapse of the deep aquifer that preserved soil bearing strength that limited subsidence of foundation soils underlying monumental structures adjacent to the drainage channel's boundary. The management of water systems within the city demonstrates hydrologic engineering expertise consistent with that found in Tiwanaku's raised-field agriculture (Kolata and Ortloff 1996) and demonstrates Tiwanaku hydrologic engineering mastery.

Newly discovered aerial photographs taken in the 1930s provide data to interpret the extent of, and insight into, the hydrologic function of the drainage channel. The early photographs reveal traces of the drainage channel's north and south arms in addition to a newly discovered) north-south canal (Mollo Kontu canal M (A-B), Figures 5.1, 5.2, 5.3 and Figure 5.5 in the CFD model) intersecting the southern drainage channel arm.

The drainage channel collected flow from adjacent canals, rainfall runoff and infiltrated rainfall seepage from the saturated, near-surface phreatic layer of the aquifer as well as from the deep groundwater aquifer (as the drainage channel depth intersected to top portion of the deep groundwater layer) to transfer water

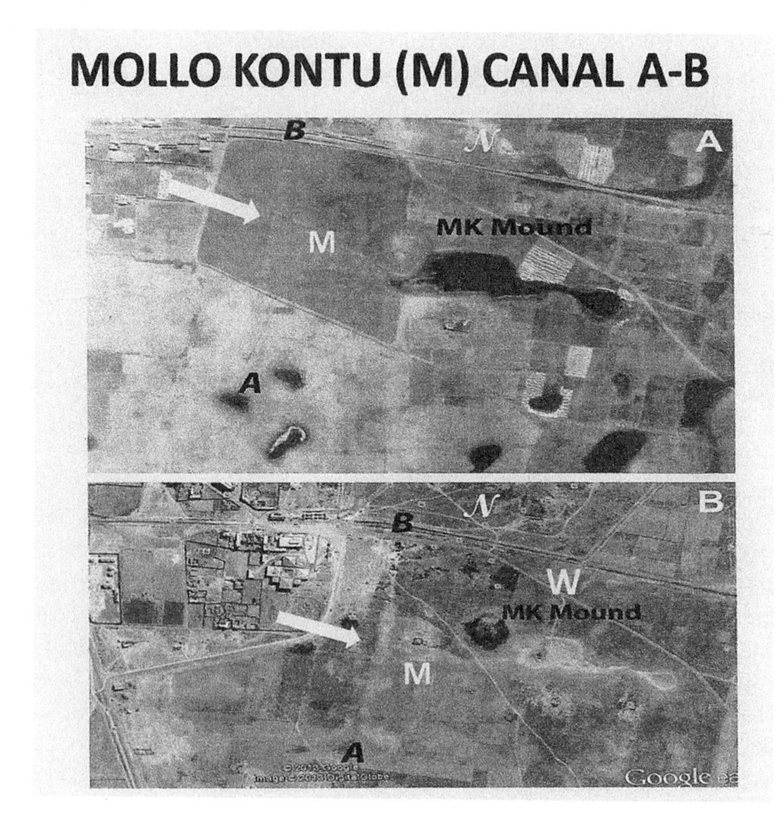

Figure 5.3 The Mollo Kontu canal (M) and surface features from Google Earth satellite imagery.

into the nearby Tiwanaku River during the rainy season to prevent deep groundwater recharge. During the dry season, phreatic aquifer seepage into the drainage channel plus water from the intersecting Mollo Kontu canal maintained the deep groundwater aquifer level relatively constant while surface evaporation and recession of the near-surface phreatic aquifer served to rapidly dry the ground surface promoting health and livability benefits for city inhabitants. The design intent of the builders of the drainage channel envisioned control of the deep groundwater level through wet and dry seasons to maintain the physical integrity of monumental structures by preventing the dry-out collapse of the deep aquifer underlying the main ceremonial core as water constantly remained in aquifer pore areas. Given a stable upper boundary of the deep groundwater layer throughout the year, physical strength properties of foundation soils were maintained thus limiting structural distortion and settling of the massive platforms of the Kalasasaya temple and Akapana pyramid (Figure 5.5, R) within the ceremonial center. Additionally, with the stable groundwater layer well below the floor of the Sunken Temple (Figure 5.5, F), rainy season aquifer drainage was facilitated into the drainage channel. Thus, beyond the drainage channel's role in creating a ritual and social boundary between the ceremonial center and secular residential city districts, its engineering design contributed many practical benefits to living conditions for city residents throughout wet and dry seasons.

Tiwanaku hydraulic analysis

To demonstrate the drainage channel's hydrologic functions, multiple data assemblages used to construct a Computational Fluid Dynamics (CFD) hydrological model (Figure 5.5) include results of archaeological mapping and excavation (Kolata 1993, 2003; Janusek 2003, 2004, 2008; Janusek and Earnest 2009; Couture 2002; Couture and Sampeck 2003), Google Earth imagery and the 1930s aerial photos taken over the site of Tiwanaku. These aerial photos reveal the site decades before modern urbanization and monument reconstruction began and were taken at a time of year when many features held water thus providing a clear view of Tiwanaku's hydrological features. From these photographs, the outline of the drainage channel is shown in Figure 5.1 as the dark encircling boundary to the ceremonial center. The curvature of the drainage canal V-D-W-X shown in Figure 5.5 is derived from earlier observations (Figure 5.18) of the canal (Bandelier 1911; Bennett 1934; Posnansky 1945; Doig 1973:352) made before years of erosion and soil deposition infilling that continues to the present day.

The east drainage canal arm (denoted Moat: east arm, Figure 5.1) averages 5–6 m deep and ranges 18–28 m in top width. Subterranean canals originating from the drainage channel's south arm were drainage conduits for Tiwanaku's monumental and elite residential structures (Couture and Sampeck 2003). Since the south arm of the drainage channel is shallower in depth than the north arm as determined by ground contour measurements (Kolata 2003), a fraction of the water that accumulated in this arm flowed down-slope through the drainage channel's east and west arms toward the Tiwanaku River while a portion of accumulated water in the south arm flowed into the dual subterranean channels (Figures 5.4 and 5.5) underlying the ceremonial center region. Given the two degree declination slope of the subterranean channels toward the north, water accumulating in the drainage channel's bottom during the wet and dry seasons provided flush water cleaning for the Putuni palace's waste removal/drainage facilities. The north canal section (Figures 5.1 and 5.5) is vital for understanding an additional feature of the drainage channel's hydrological function. North of Tiwanaku's northwest monumental complex, the shallow alluvial plain drops sharply downward toward the Tiwanaku River's marshy floodplain. One portion of the east arm of the canal turns west and disappears into the floodplain (Figure 5.1, C') while canal (C) continues north toward the Tiwanaku River. One portion of the west drainage channel arm led into the marsh north of the Kalasasaya (Figures 5.2 and 5.5) while an ancillary arm continued northeast toward the river. The north portion of the drainage channel

Figure 5.4 Photo and plan view of the excavated portion of the subterranean Putuni channel P.

thus divided into several branch canals that intersected the floodplain and drained accumulated water from the north arm of the drainage channel. Water not directly shunted to the Tiwanaku River through canal C (Figures 5.1 and 5.5) drained water in the floodplain's aquifer that directed seepage into the Tiwanaku River. The floodplain area thus served as a productive agricultural system for the urban center by utilizing its available water surplus.

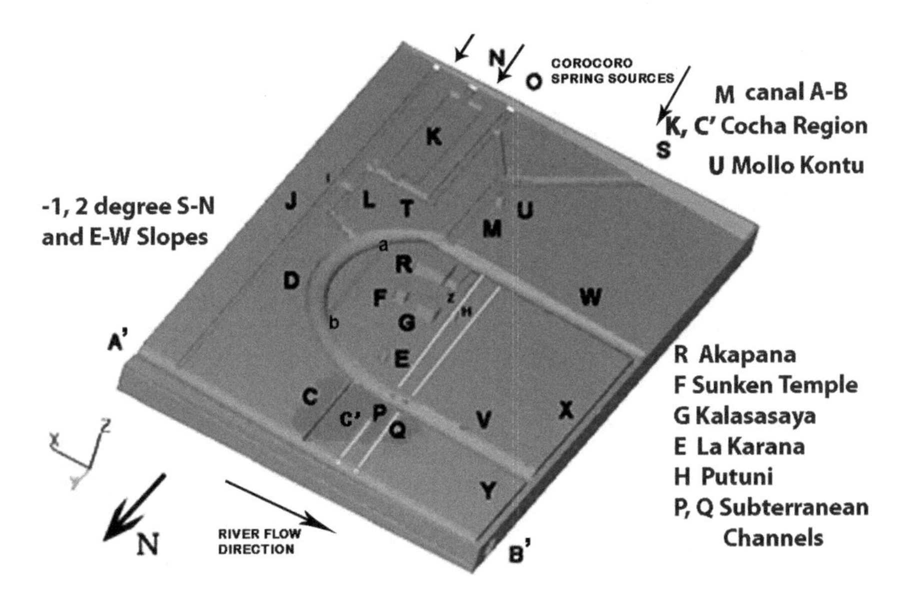

Figure 5.5 Representative CFD model of hydrological and architectural features of the Tiwanaku urban center. Model is a best-estimate representation of the 9th- to 12th-century AD site geometry scaled from aerial photos, historical sources and ground survey; line a-b represents a later drainage path interpretation compared to earlier curved versions by Posnansky 1945; Bandelier 1911 used for the model.

The drainage channel in the urban hydrological network

Where the groundwater surface emerged from depressed areas, springs formed. Several canals in the southern portion of Tiwanaku were engineered to utilize this canal water input. The westernmost Choquepacha area's canal (Bruno 1999; Janusek 1999, 2003, 2004, 2008; Kolata 2003) derived from a natural spring on a bluff southwest of Pumapunku. The spring was fitted with a basin that included several incised canal stones carved to convey water (Posnansky 1945, II: Figure 42). Combined with the output of an adjacent stream that drained a marshy area of the terreplein, the Choquepacha area supported extensive terrain amenable to pastoral grazing and farming immediately to the west of the Tiwanaku urban area (Bruno 1999). Other features relate directly to the hydrological function of the drainage channel. The first feature is a north-south Mollo Kontu canal that supplied water from springs originating from the southwest portion of the site near the Pumapunku complex into the southwest portion of the drainage channel (Figure 5.1-M, 5.2-M, 5.3-M and 5.5 S-M). The second feature is an interlinked cluster of sunken basins, or *qochas* that occupied the southeast or Mollo Kontu portion of the site (Figure 5.1). *Qochas* are pits excavated into the phreatic aquifer layer that capture and store rainwater and serve to expand planting surfaces and pasturage while creating micro-lacustrine environments that attract waterfowl (Craig et al. 2011; Denevan 2001; Erickson 2000; Flores Ochoa 1987). Figure 5.1 depicts a series of canals dendritically linking the *qochas* to one another with a branch connecting to the Akapana East canal (L, Figure 5.5) that drained into the east arm of the drainage channel. The third major feature is a long, narrow, outer canal (J) on the east side of Tiwanaku (Figures 5.1, 5.2 and 5.5). While the role of this canal is unclear, its southern portion is straight and follows an alignment that mirrors that of the Pumapunku complex to the west; its northern portion shifts course and bounds the east edge of the site. The east canal (L, Figure 5.5) links with the drainage channel (Figure 5.1, Figure 5.5, W-D-V-X) by connector canal I indicating that the outer canal was part of an encompassing urban hydraulic network. The areas immediately east of the canal contain Tiwanaku's residential sectors and include Ch'iji Jawira, a barrio of ceramic producers that depended on a constant water supply (Janusek 2003; Rivera Casanovas 1994, 2003). Immediately east of the Ch'iji Jawira sector is a low brackish marsh; from this marsh, the outer canal (J) likely provided fresh water from springs for Tiwanaku's easternmost residential sectors and drainage of excessive canal flow during the rainy season.

The east and west arms of the drainage channel directed water around the monumental complex toward the Tiwanaku River to the north (Figures 5.1, 5.5, A'-B'). The C' floodplain was an integral part of Tiwanaku's larger hydraulic network that served to facilitate drainage of both groundwater seeping from drainage channel arm D-V (Figure 5.5) and rainwater runoff during rainy season peaks. Intricate surface canals and dual subterranean stone-slab constructed canals (Figures 5.4 and 5.5) provided additional drainage within the area bounded by the drainage channel. The elaboration of surface canals on the interior floor of the Sunken Temple (Figure 5.5, F) and areas outside the Kalasasaya (G) (Ponce 1961, 2009) indicate a drainage connection to either (or both) the drainage channel and the subterranean channel P (Figure 5.5); additional drainage by seepage into the stabilized low groundwater layer also helped to keep the temple floor dry throughout seasonal changes. The Akapana pyramid (R, Figure 5.5) incorporated an intricate, stone-lined canal network that routed water from the uppermost level down through successively lower platforms and finally out through several portals in its basal terrace (Kolata 1993, 2003) then to several open surface basins draining into vertical pipes (and/or the drainage channel) that conveyed water into the subterranean canals P and Q (Figures 5.4 and 5.5) into drainage channel arm V (Bennett 1934:378–385). Canal P (Figures 5.2 and 5.5) provided flushing water to remove human waste from the Putini residential compound for conveyance to the Tiwanaku River. Water was temporarily pooled in the sunken courtyards near platform monuments rendering them lakes for ritual events and reservoirs for controlled water distribution.

Excavations between Putuni and Kerikala complexes indicated structures within the ceremonial core region that articulated with Tiwanaku's subterranean drainage network. This area housed high status groups until, at approximately AD 800, the construction of the Putuni platform repurposed the space to support recurring state-sponsored ceremonies (Couture and Sampeck 2003; Janusek 2003:146; Kolata 1993, 2003). Located 2.5 m below the current ground surface, canal P (Figures 5.4 and 5.5) consisted of sandstone slab masonry with vertical side slabs approximately 1.0 m high and horizontal slabs about 0.8–0.9 m across (Couture and Sampeck 2003; Crèqui-Montfort 1906; Janusek and Earnest 2009; Kolata 1993; Ponce Sangines 1961). Several vertical pipes consisting of multiple stacked, perforated stone disks conducted surface water from features within the Putini into the lower subterranean channel P (Janusek and Earnest 2009). Water from the drainage channel's southern arm (V) supplemented by water from canals L and M (Figure 5.5) together with seepage water from both phreatic and top portions of the deep aquifer was used to flush waste water through subterranean channels P, Q located in the west portion of the monumental core.

Models of water management at Tiwanaku

To demonstrate insights related to the hydrological function of the drainage channel, use of Computational Fluid Dynamics (CFD) (Flow Science 2018) is made for cases that address seasonal variability in water input. Here the equations governing aquifer percolation (Bear 1972) were numerically solved by finite difference methods to show transient water transfer within the aquifer for two seasonal water availability cases. Case 1 considers effects existing at the termination of a rainy season on Tiwanaku's canal systems and city open surface areas. The rainy season in the south-central Andean altiplano generally runs from November through March. Case 2 considers effects of limited water input from springs and aquifer seepage into the deep groundwater layer during the April through October dry season. Data from aerial photographs, Google Earth imagery, contour maps and ground survey provided the basis for the computational model (Figure 5.5) to demonstrate hydrological features of the drainage channel and its encompassing hydrological network. A porous soil model of the subsurface aquifer is used to demonstrate hydrological responses of the canal network and drainage channel for the two cases. The Figure 5.5 model surface and subterranean features are on the same scale as Figures 5.5, 5.2 and 5.3 and represent best estimate 9th- to 12th-century AD water supply and distribution network canal paths inferred from photographic and ground survey data. The canal inlets shown (Figure 5.5 – J, N and O) are sourced by canalized Corocoro springs located south of the modeled area as are canals (S-M) leading from the Pumapunku area. Key monument architectural and hydrologic features in Figure 5.5 are:

A-B the Tiwanaku River, flow direction A to B (east to west)
C: drainage channel to A-B
C' the drainage and agricultural complex supplied from drainage channel C
E: La Karaña residential complex
F: Sunken (or Semi-subterranean) Temple
G: Kalasasaya
H: Putuni

I: Connecting channel between canals L-K-L and J
K: multiple interconnected *qocha* region supplied by canal N arm D
L: Akapana East canal, which drained *qocha* region K toward drainage channel
M: Mollo Kontu canal linking supply canals O and S to drainage channel arm W
N: Supply canal to *qocha* region K
O: Connecting canal to canal M
P and Q: subterranean channel pair with declination slope of two degrees to the river; channel P runs underneath the Putuni H
R: Akapana seven-stepped truncated pyramid
S: branch canal to Mollo Kontu canal M
T: lateral transverse canal to L; shunt canal I to canal J and/or canal drainage to the drainage channel
U: Mollo Kontu monument
W-D-V-X: the drainage channel circuit around the monumental core of Tiwanaku; original depth of the canal estimated at 3–5 m at location D.
Y: drainage canal from V to the Tiwanaku River (from 1930s aerial photographic source); its inclusion in the model has a minor drainage effect compared to drainage features C, C' P, Q originating from the drainage channel Z canal below the west side of the Akapana (R) draining toward drainage canal segment D.

The CFD model is composed of a porous medium aquifer duplicating soil material properties (porosity, permeability) found at the site through which groundwater percolates. The CFD model incorporates both the east-to-west ground slope declination and a south-to-north declination observed from field measurements. The momentum resistance to flow in the porous medium representation of an aquifer (Flow Science 2018) is expressed as a vector drag term F_d **u** where F_d is the porous media drag coefficient and **u** the velocity vector $\mathbf{u} = q_x \mathbf{i} + q_y \mathbf{j} + q_z \mathbf{k}$ with q_x, q_y, q_z velocity components in the **i**, **j**, **k** (x, y, z) coordinate directions respectively (Figure 5.5). Permeability k is defined as $k = V_f \mu/\rho F_d$ where V_f is the volume fraction (open volume between soil particles /total volume), μ the water viscosity and ρ the water density. For the present analysis, k is on the order of $\sim 10^{-11}$ cm^2 based upon the site soil type (Bear 1972; Freeze and Cherry 1979) within the model area excepting monumental paved areas for which k is on the order of $\sim 10^{-5}$ cm^2. For model area soils, $0.43 < V_f < 0.54$. Based on these estimates, the average drag coefficient F_d is estimated to be ~0.80. While deviations from this value occur due to varying soil properties with depth and location, flow delivery rates from the saturated part of the aquifer to the drainage channel's seepage surface (defined as the exposed interior soil surface of the drainage channel exposed to the atmosphere) will be affected but calculations will nevertheless demonstrate qualitative conclusions regarding the drainage channel's function. In the CFD model, the deep groundwater layer is composed of saturated soil and is stabilized throughout the year at ~5–6 m below the ground surface as well probe data indicates. The saturated phreatic aquifer layer is assumed to lie above the deep aquifer for Case 1 calculations indicative of an intense, long duration rainfall period. The bottom depth of the drainage channel intersects the upper portion of the deep groundwater layer in the Figure 5.5 model and both the phreatic and deep groundwater layers provide seepage water into the drainage channel together with runoff water and canal water supplied by springs south of the city. For a less intense rainfall period, an intermediate capillary fringe zone extends upward from the deep groundwater zone to intersect the bottom reaches of a surface saturated phreatic layer. Deep aquifer recharge can occur when the phreatic layer extends sufficiently downward to penetrate the capillary fringe zone during long duration rainy periods. For minimal rainfall, surface phreatic layer is considered a small depth "evaporation zone" that vanishes in depth as the dry season continues. Again, when rainfall is intense and of long duration, the phreatic and deep groundwater layers merge – for this case, aquifer seepage into the saturated drainage channel bottom cannot occur and excess drainage water is rapidly shunted to the nearby Tiwanaku River. This effect limits the height excursion of the deep aquifer to the base depth of the drainage channel.

For Case 1 analysis, the post-rainy season phreatic layer is saturated and lies above the saturated deep groundwater layer; as no further infiltrated rainwater can be absorbed into this saturated layer, runoff occurs into the drainage channel. As the dry season progresses, surface evaporation shrinks the phreatic layer upward toward the ground surface and soil drying to depth enhanced by aquifer drainage. In times of extended drought, the phreatic and ultimately the groundwater layer contract leading to soil dry-out conditions to a large depth.

Case 1 – post-rainy season ground saturation conditions

Case 1 examines post-rainfall conditions typical of the end of the altiplano rainy season characterized by phreatic zone saturation and continuous water flow through canals O, S, N, M and J from Corocoro springs (Figure 5.5). Aquifer seepage to the bottom of the drainage channel from the saturated phreatic layer is transferred to drainage channel arms D, V and W to X-Y and then to the Tiwanaku River (A'-B', Figure 5.5) as all these canals have a down-slope toward the river. Additional seepage occurs from the top reaches of the deep groundwater layer into the drainage channel. Water from the drainage channel's east and west arms then led to the Tiwanaku River through the C canal branch and seepage from the C' area. Water arriving into inlet N was conducted by canals K and L into either (or both) canals D and then from I to J. A summary of rainy season water inflows/outflows from a representative section of the drainage channel is shown in Figure 5.6.

Numerical solutions (FLOW-3D 2018) of equations governing saturated aquifer and surface/subterranean canal flows give a picture of transient water transfers to/from the drainage channel from seepage and canal flows given estimates of flow rates based on supply spring flow rates. Given Case 1 post-rainy season conditions, surface runoff has been largely collected into the drainage channel and transferred to the river; further water transfer to the bottom of the drainage channel is from aquifer seepage and adjacent canal water flow input – this water

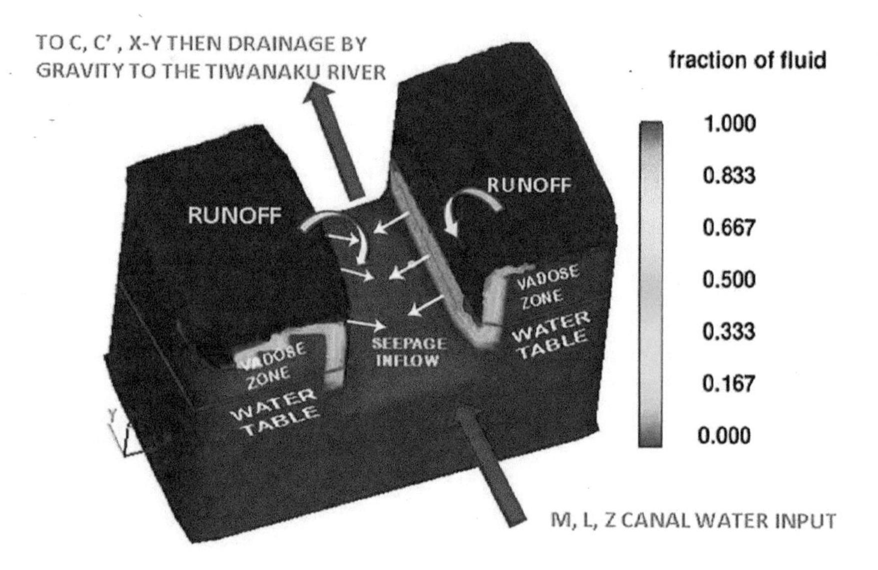

Figure 5.6 Summary diagram on water input/output flows on a typical drainage channel section near D (Figure 5.5).

is rapidly discharged into the Tiwanaku River as the drainage channel bottom intersects the saturated deep groundwater layer. Figures 5.7 and 5.8 show a time progression of water seepage from the drainage channel's open surface area and progressive surface drying as the phreatic layer deflects downward due to drainage into the drainage channel.

Rapid water removal from the drainage channel via canals C and X-Y to the Tiwanaku River (Figure 5.5) reduced water transfer from the phreatic aquifer to the deep groundwater layer promoting deep groundwater height stabilization. The Akapana monumental core experienced limited rainfall infiltration due to extensive terrace and side wall paving and compound roofing that promoted runoff into the drainage channel. Water that managed to infiltrate between paved open areas then drained into the interior of the Akapana where it reemerged from base openings to join drainage canal and subterranean channel extensions P and Q (Figures 5.4, 5.5 and 5.9) that directed water toward C' and to the Tiwanaku River through C' drainage and C, X-Y canals. As the drainage channel depth extended to the top fringe of the deep groundwater layer, it stabilized the deep groundwater layer depth below the phreatic layer.

The location of the Sunken Temple floor (F, Figure 5.5) above the stabilized deep groundwater layer and its nearness to the drainage channel helped to promote a dry floor through seasonal changes. Here rainfall accumulating on the floor infiltrated into the phreatic layer then drained out to the nearby drainage channel. This effect supplemented channel and piping drainage to the drainage channel.

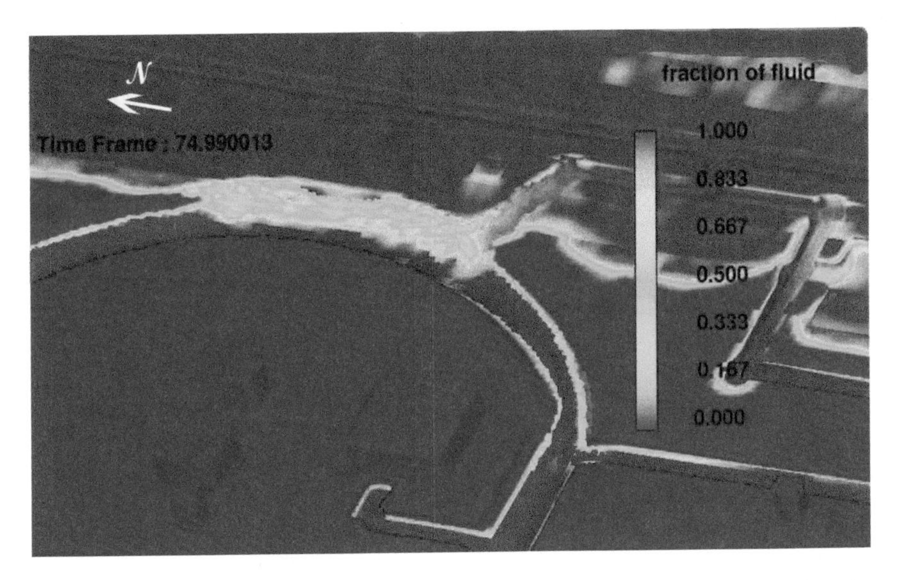

Figure 5.7 Post rainy season detail of the east arm of the drainage channel showing the seepage surface conducting water to the bottom of the canal and progressive surface drying for Case 1 conditions.

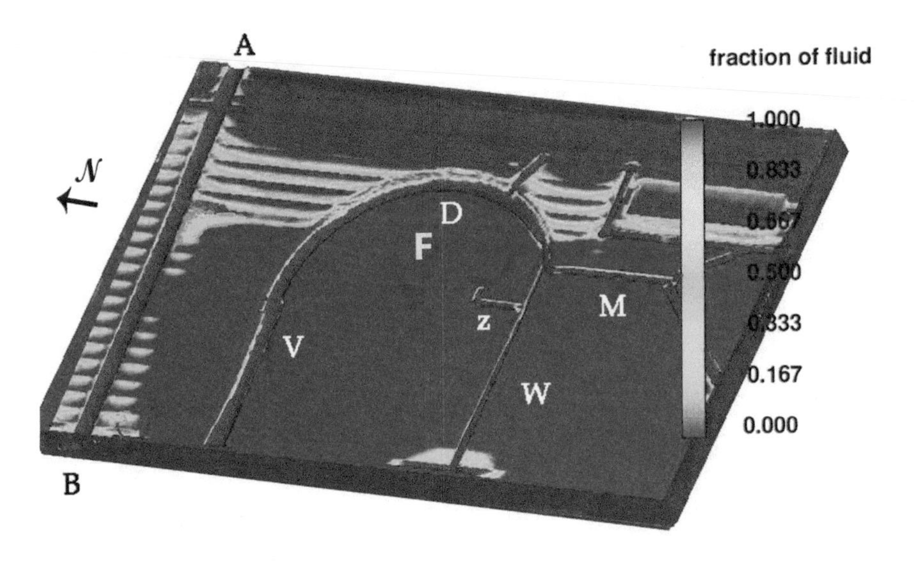

Figure 5.8 Later time surface drying achieved by aquifer seepage and surface evaporation; note low values of fluid fraction starting eastward on ground surface for Case 1 conditions.

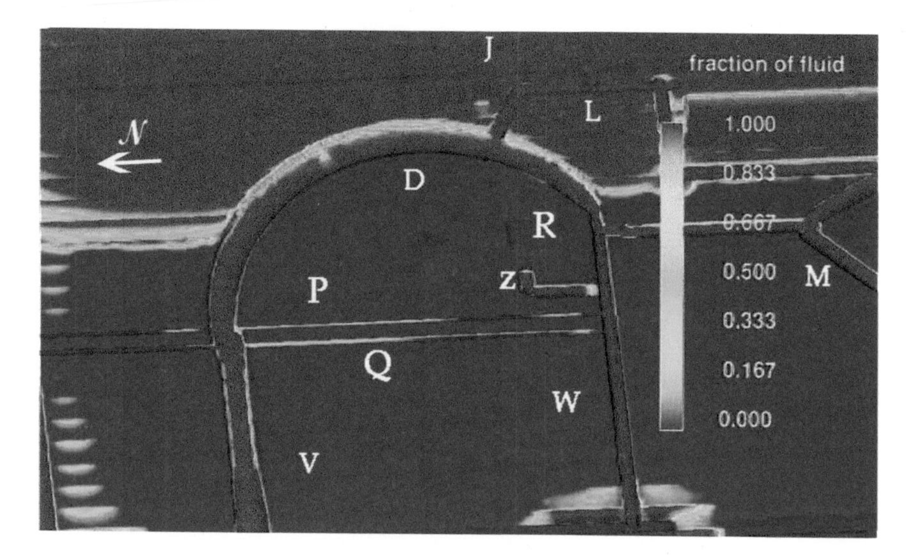

Figure 5.9 Case 1 fluid fraction results showing water input from canal M to subterranean canals P and Q flushing wastewater from the Putuni complex toward drainage canal arm V and subsequently to the Tiwanaku River A'-B'.

Figure 5.10 Dry season (Case 2) fluid fraction results on a cut plane below the ground surface; moisture levels in *cocha* region K and depressed area C' indicate sustainable pasturage and agriculture due to contact with the deep water table.

Figure 5.11 shows fluid fraction results at the inner face of the drainage channel bounding the ceremonial center. The fluid fraction (ff) is defined as the volume of water per unit volume of the porous aquifer; here ff = 1 denotes saturation conditions where water fills the spaces between soil particles and ff = 0 denotes that air fills the spaces between soil particles. Intermediate ff values denote partial water

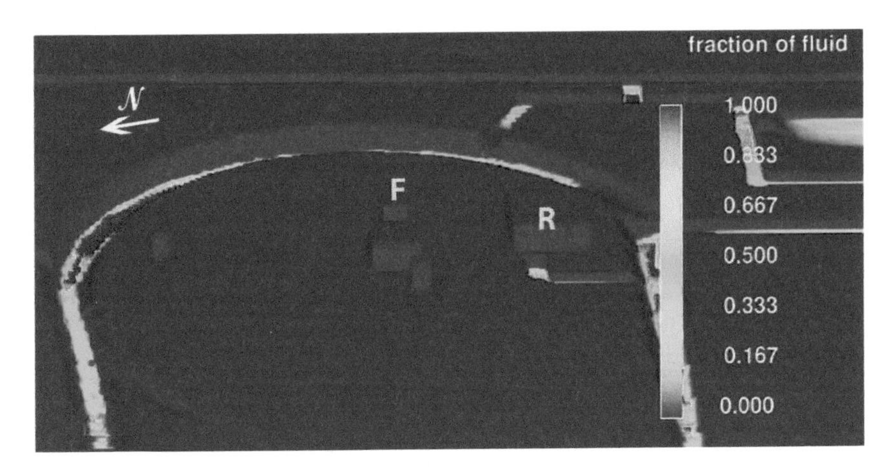

Figure 5.11 Dry season (Case 2) fluid fraction results for the east arm of the drainage canal indicating decreased seepage water into the drainage channel and extensive surface drying.

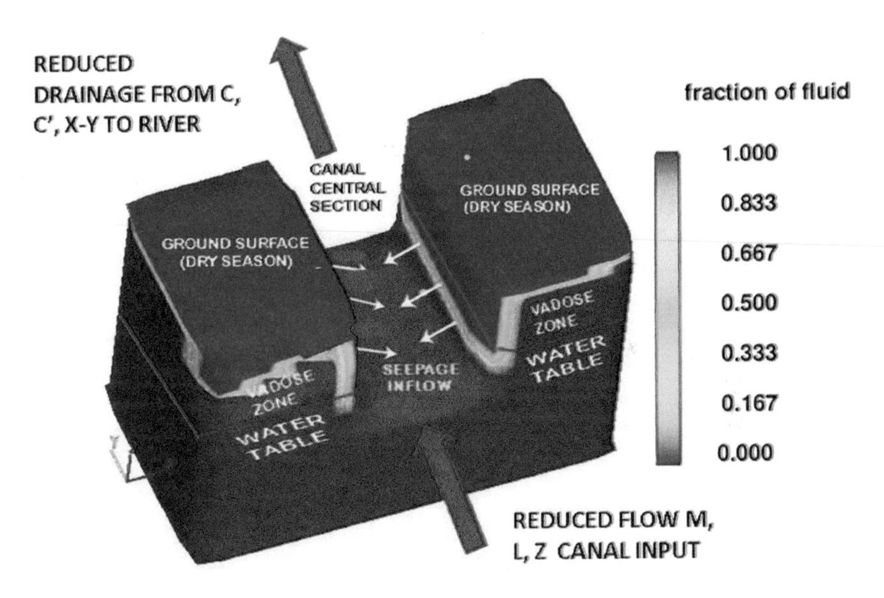

Figure 5.12 Summary of dry season water inputs/outputs into a section of the inland drainage channel near D (Figure 5.5).

filling of the spaces between soil particles denoted as damp soils. Results confirm seepage was minimal from what little infiltrated rainwater existed in this largely paved and roofed elite area as any infiltrated water was conducted into the saturated drainage channel's bottom and quickly removed by canals C and X-Y to the

Tiwanaku River. From the paved elite areas, rainfall runoff constituted the main water contribution to the drainage channel. Figure 5.9 shows the water transport in subterranean channels P and Q. Channel P lies below the floor of the Putuni; channel Q lies at the same depth as P but ~10 m west of P (Kolata 2003:115). Vertical pipes connected drainage areas in the Putuni Courtyard and Palace (Couture and Sampec 2003) to canal P with collected water directed toward the V arm of the drainage channel. The P and Q subterranean canals required a regular input of flowing water from canal M and seepage water from phreatic and deep aquifers into drainage channel arm W to maintain dry residential area hygienic conditions. As the P, Q channels, the C canal, and the drainage channel bottom all sloped downhill toward the Tiwanaku River, water flow from drainage channel arm W directed water and waste solids into the river.

Case 2 – dry season initiation

Case 2 considers the drainage channel function under dry season conditions (zero rainfall and continuous, but limited, water supply from Corocoro springs into surface canals N, O, S and M *qocha* regions K and C'. Figure 5.13 indicates that water supply from the M channel plus aquifer seepage continues to supply subterranean P and Q channels to flush human waste from the Putuni elite compound structures.

These regions remained functional due to their depth penetration into the receding phreatic zone water level indicating agriculture and pasturage were possible during the dry season. By rapid water transfer of seepage, runoff and adjacent

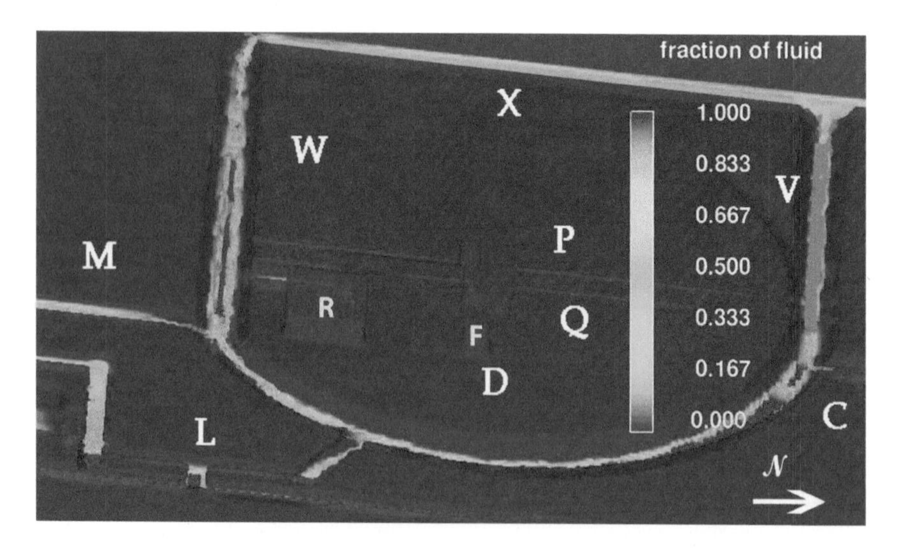

Figure 5.13 Dry season (Case 2) fluid fraction results for the monumental center with decreased water supply from spring-supplied canals; rainfall infiltration and seepage limited by large paved and roofed areas of the ceremonial center.

canal water through multiple canal paths in the rainy season to the Tiwanaku River and continued aquifer seepage into the drainage channel together with Corocoro spring-sourced canal water supply in the dry season, the deep water table remained stabilized in height year-round below and distinct from the phreatic zone aquifer. Channels P and Q indicate continued water transport and flushing activity from drainage channel segment W as the dry season progressed with canal M providing water supply during the dry season.

Additional features of the drainage canal

During the construction of the platform base of the Akapana pyramid, the phreatic aquifer layer was compressed between the heavy base and the deep groundwater layer expelling water into the nearby drainage channel. As additional heavy platforms were added (up to seven total) and the structure weight increased, further consolidation of the phreatic aquifer below the pyramid base resulted making less water intake available to the phreatic aquifer due to reduction in aquifer porosity. While some rainfall infiltration into the increasingly consolidated foundation base soil occurs, less water is available due to lowered foundation soil porosity and rapid drainage into the nearby drainage canal. As platforms were added and the compressive structural weight increased, a consolidated foundation impervious to water infiltration was created ensuring further minimal structural deflection and distortion. It is likely that city planning included creation of the drainage channel contemporary with construction of ceremonial core region structures given the planning inherent to the subterranean canals P and Q as well as anticipated hygienic benefits of rapid site drying after the rainy season together with positive consequences of monument stability. The Akapana to this date still retains its structural integrity without settling distortion as a testament to the original planning. As a side note, a similar technology involving the stabilization of groundwater levels through canal and moat water supply regulation was used to stabilize foundation soils under the temples of Cambodian Angkor (Ortloff 2009) to limit structural deflection; use of this technology is responsible for the good state of preservation of major temple structures ~900 years after their construction.

Rainy and dry season groundwater profiles

Figure 5.14 (A) shows a constant y transect through location D (Figure 5.5) that indicates a fluid fraction of unity (ff = 1) consistent with ground saturation during the rainy season. As the rainy season concluded, drainage into the drainage canal from adjacent saturated soil accelerated surface and ground drying. Figure 5.6 summarizes all drainage paths relevant to maintain the deep groundwater level during the rainy season. As the dry season progressed, Figures 5.12 and 5.13 summarize drainage and water input from canal M that maintained and stabilized the deep groundwater level.

As the drainage canal bottom (e) intersects the top of the deep groundwater layer as indicated in Figure 5.14, soil drying to the depth of the deep groundwater

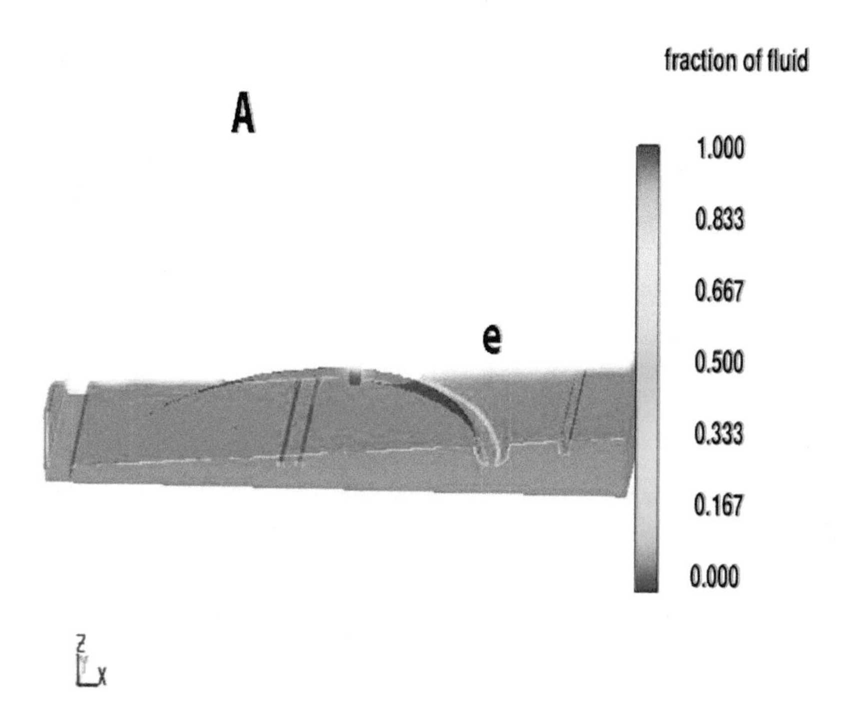

Figure 5.14 Constant y transect through D (Figure 5.5) of groundwater profile for rainy season ground saturation conditions.

layer top is indicated by lower fluid fraction values. This figure indicates that the role of water input from canal M prevents further recession of the deep groundwater layer as the dry season progresses. The intersection of the drainage canal bottom with the deep groundwater layer top provides deep groundwater stabilization that underlies the conclusions of the prior sections of this chapter. Although a case has been made for large monument foundation stability and its relation to the drainage channel, this result may have been fortuitous as knowledge of aquifer dynamics under compressive forces known to Tiwanaku engineers is as yet subjective with the present case the only known example to draw from.

Hydrologic applications exterior to city precincts: further examples of Tiwanaku mastery of groundwater science

Research conducted on Tiwanaku's raised field agricultural systems in the Pampa Koani region and systems under Tiwanaku influence on the northwest regions of Lake Titicaca has indicated use of advanced hydrological methodology underlying crop sustainability and yield improvement. Among the advances in agricultural science are use of heat transfer technology to limit crop destruction by freezing during cold altiplano nights (Kolata and Ortloff 1996; Ortloff 2009:86–92) as well

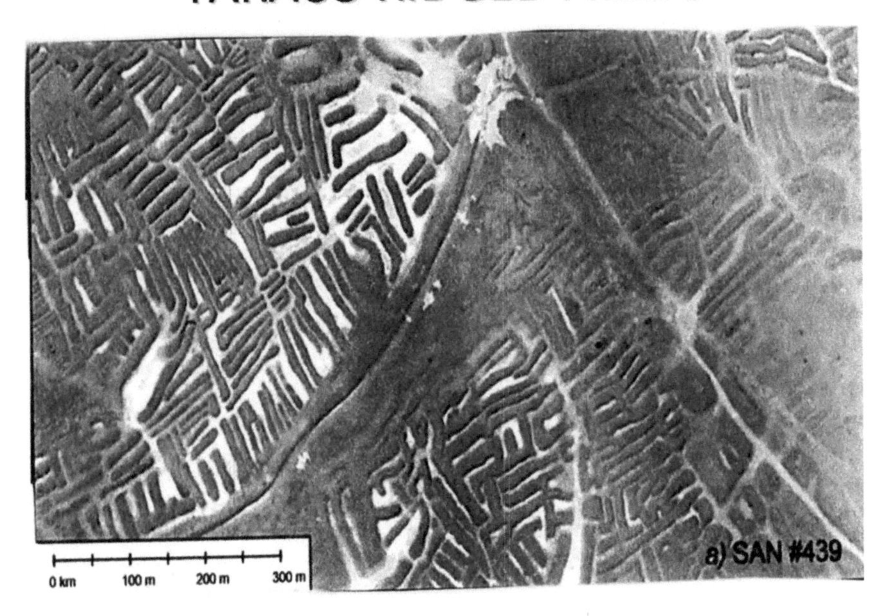

Figure 5.15 Taraco raised field aerial view.

Source: Henderson 2012

as hydrological and hydraulic control mechanisms providing groundwater height control to stabilize raised field swale water height through seasonal changes in water availability (Ortloff 1996). Additionally, raised field technology has been shown the most appropriate design choice to limit short-term drought effects on crop yield – this due to continual groundwater supply from intercepted rainfall over vast collection areas continually flowing to the Lake Titicaca basin (Kolata and Ortloff 1989b). Further analysis presented in Chapter 2 demonstrates that raised field berm design is optimum to yield the maximum agricultural output per unit land area. Analysis of groundwater control mechanisms in the Pajiri agricultural area (Ortloff and Kolata 1989a) reveals different berm heights and swale depths appropriate to the needs of different crop types. All these observations point to an advanced agricultural science used to maximize and sustain crop yields in the Tiwanaku heartland.

In the Pampa Koani and northwest Tarraco regions of Lake Titicaca, different patterns of raised field lengths, widths and orientations are frequently inserted within more regular patterns – each pattern appropriate for the water needs of different crop types (Kolata 1993; Henderson 2012). Excavation of raised field berms in the Pampa Koani area has indicated stone base lining and clay layers (Kolata 1993:115–128) to limit cold capillary water transfer from deep groundwater regions into berm interior regions; capillary water to the berm interior region is mainly provided from swale water. Due to higher swale water temperatures from solar radiation input (Ortloff 1996:129–134), the additional storage heat to berm interior regions limits convection and radiation heat withdrawal during cold altiplano nights to prevent freezing damage to crop root system. In other words, the latent heat removal for water to ice transition within berm interiors during cold altiplano nights is limited by additional heat transfer from elevated temperature swale water capillary heat transfer into berm interiors (Ortloff 1996:129–134). Examination of early north most Tarraco raised field berm patterns (Henderson 2012) reveals an average berm shape consistency. Figure 5.16 shows that swales are interconnected leading to a continuous water path surrounding berms. When a typical berm is described as an elongated ellipse with major axis **a** and minor axis

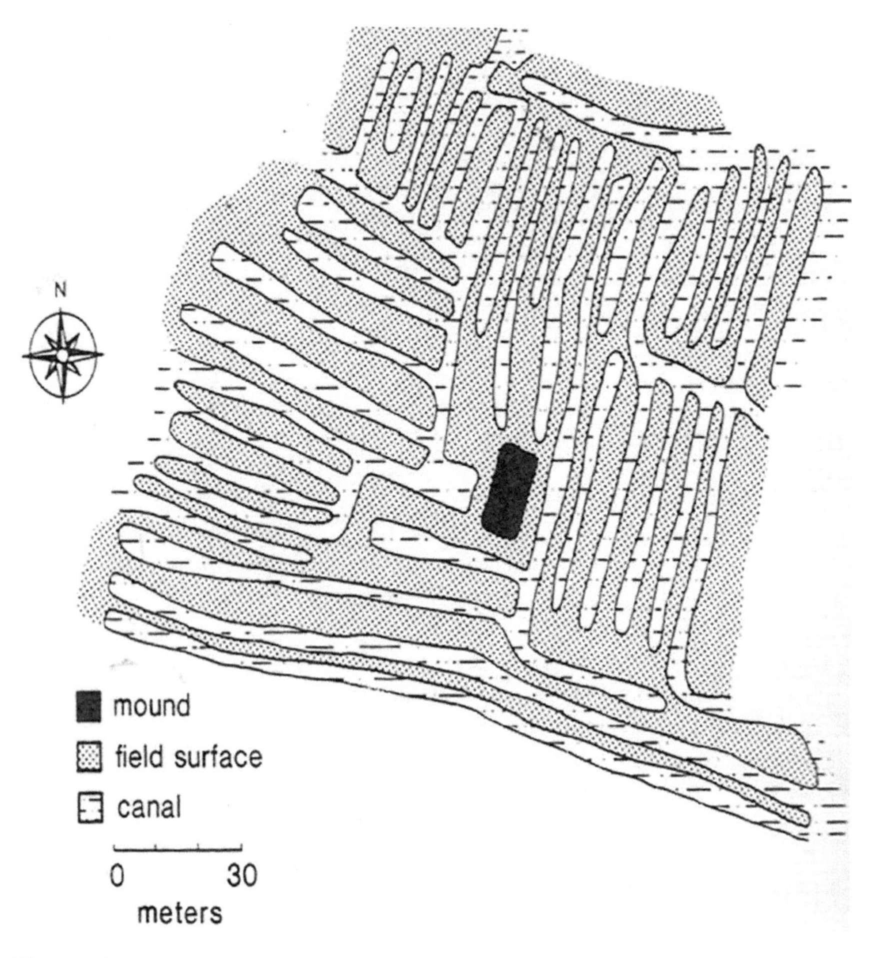

Figure 5.16 Lakaya sector raised field geometry in the Pampa Koani system

Source: Kolata 1993

b, (**a-a** and **b-b** in Figure 5.16), then **a/b** ratio from 10 to 15 appears to characterize average of berm geometries. This ratio for an elongated ellipse (**a** >> **b**) is significant in that the ellipse perimeter is a maximum for the given berm surface area (**π a b**) for this class of ellipse. This indicates that the average berm pattern configuration yields the maximum wetted berm perimeter and thus requires a minimum of interconnected swale lengths and widths and that narrower swale channels serve to provide capillary water transfer to narrow berms. Here the narrow berms (**b** << **a**) provide an easy path for elevated temperature capillary water to reach berm interiors. This, in turn, reduces the exposed water surface area of the interconnected swales reducing evaporation loss that helps to locally maintain a constant groundwater profile to maintain swale water height. The net effect is that a greater number of closely spaced berms can be watered properly per unit area to maintain the crop freezing defense while the greater area under cultivation produces more agricultural product for a given topographic area. These advantages are a key indicator of an advanced agricultural science being employed to protect and increase the yields of raised field agricultural systems. The **a/b** ratio of individual berms contains important information related to Tiwanaku engineering practice as their design incorporates a level of optimization to limit swale water area to reduce evaporation losses.

The conclusion that applies for the Tarraco raised field geometry also applies in the Pampa Koani region as Figure 5.17 indicates and makes similar claims of use of the Terraco technology to maximize agricultural land area in Tiwanako raised field systems. Although regional and time differences may exist in raised field designs, the Tarraco system design may reflect different groundwater water availability, ambient air temperatures and crop types than those for Tiwanaku raised field designs that require more efficient use of water resources.

Retrospectives on Tiwanaku societal structure

Ceramics and other objects in derived Tiwanaku style are widely distributed throughout the south central Andes from the southern coastal valleys of Peru and Chile to the lowland eastern slopes of the Andes (Goldstein 2005). The distribution of Tiwanaku artifacts exhibit stylistic variations with different types and quantities of Tiwanaku style materials occurring in different regions. The governance principles operational in the Tiwanaku polity that led to the distribution of cultural material provides insight about the social structure of the Tiwanaku polity and their expansionist policies. Current theories related to Tiwanaku territorial and agricultural expansion are broadly summarized as:

1 The distribution of artifacts results from Tiwanaku imperial expansion outside the Titicaca Basin with colonies and conquest aimed at lowland, highland and tropical base resource extraction.
2 The widespread distribution of cultural materials results from the growth of an archipelago system where discontinuous territorial and ecological niches were exploited through placed colonies.

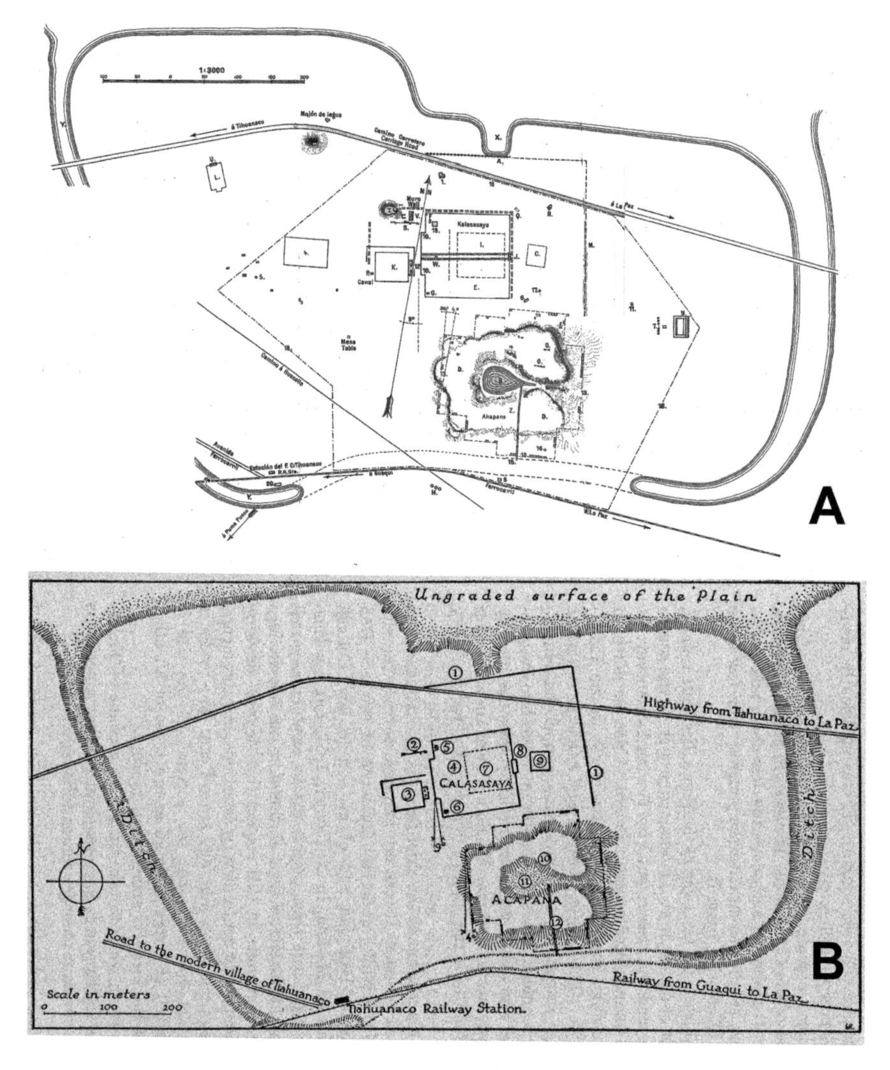

Figure 5.17 Early explorer drawings of the drainage channel.

3 Tiwanaku style materials spread spatially through trading networks headed by the Tiwanaku polity.
4 Tiwanaku expansion was ideological and/or ritual in nature devoid of political control or colonization. Tiwanaku expansion may occur as a combination of some of the above four paths and was used by Tiwanaku populations in distinct regions depending upon local conditions.

From the analysis thus far detailed and its relevance to the four versions of proposed Tiwanaku expansion based in part on the increase and diversity of agricultural resources available to the Tiwanaku urban center, it is apparent that different regions of expansion were associated with different ecologies. For the Tiwanaku urban center, agricultural systems incorporated groundwater based raised fields, rainfall supplied terraces, combination spring/groundwater supplied raised field areas, cochas and urban canal-supplied agricultural basins (Figure 5.5, K). For sites distant from the Tiwanaku heartland that exploited different ecological conditions for farming, typical agricultural systems incorporated lowland valleys with river supplied canal irrigation systems, channeled spring-supplied agricultural fields and a variety of field systems located at different altitudes with different soil, water supply and temperature conditions that permitted a different range of crop types not possible to cultivate at the high altiplano elevation of urban Tiwanaku. Individual satellite sites required mastery of different hydrological regimes for irrigation. From the analysis of Tiwanaku urban and raised field hydraulic/hydrological knowledge (Kolata and Ortloff 1989b, 1996; Ortloff 1996, 2009, 2016a; Ortloff and Janusek 2014), aspects of this technical base were likely exported to remote sites to maximize food production to economically justify the effort to maintain extensive trade and food import supply networks. As some site areas were already occupied by different societies, some cooperative and others not accommodating an intrusion from a dominant competing society for use of limited water and land resources, the existing agricultural technology at distant areas outside of direct influence from the Tiwanaku urban core may not have initially generated sufficient surplus to interest incorporation into the Tiwanaku archipelago; however, by incorporation of advanced agrotechnical knowledge from urban Tiwanaku's hydraulic/hydrology experts, the agricultural output could be increased to justify import/export status. The establishment and economic success of satellite archipelago sites appears directed from a central authority based in the Tiwanaku urban center that included advice and council of technical experts versed in hydraulic and hydrological matters. This observation is best stated (Dillehay 2011:208) as the challenge of new technologies – "certain individuals were probably empowered by technological or knowledge status and decisions made regarding the adoption, invention and use of certain technologies (e.g., plants, canals, . . .) must have been made by technocratic and expert-centered individuals." Here in ancient times engineering knowledge was a valuable export commodity just as it is today. While trade in sumptuary goods (Goldstein 2005) to outlying societies was prevalent in the Tiwanaku sphere of influence and served to expand Tiwanaku influence and cultural traits into outlying areas, the exportation of agricultural knowledge had value to outlying societies that promoted benefits of association with the Tiwanaku urban core. While export of technical groups from the Tiwanaku urban core experienced in raised field agricultural production to Tiwanaku colonies in the Moquegua Valley occurred as urban Tiwanaku declined in the post 1100 AD time period (Goldstein 2005), given different coastal Moquegua Valley ecological conditions, new agricultural technologies involving canal irrigation from river and spring sources were invented to sustain limited agriculture for small village sites with different ecologies than those existing at the Tiwanaku urban core heartland. In this respect, the many surface and subterranean canals associated with Tiwanaku city drainage and water control canals and the

Pampa Koani raised field heartland contain a comprehensive water engineering technology base that would be of limited use in the coastal Moquegua Valley canal irrigated areas of Omo and Chen Chen. While some researchers claim that that individual farming communities could invent optimum agricultural field systems in a bottom-up manner without the need of a central controlling administration overview group, they underestimate the technical complexity involved in surface and groundwater control related to agricultural production over vast raised field areas and the city water control engineering necessary to regulate the groundwater level through seasonal changes to produce hygienic benefits for city dwellers as well as the water control complexity needed for floodwater and human waste elimination from the Tiwanaku city environment. The subterranean channels P and Q (Figure 5.5) are examples of advanced technology applied to this end. The ~75,000 hectares of Pampa Koani raised fields required expertise in local groundwater height control by means of a canal supply and drainage network operational over vast areas (Ortloff 1996, 2014b, 1997, 2009) to provide tailored agricultural berm moisture levels and different berm geometries for different crop types. The agricultural system at the outlying site of Pajchiri (Kolata and Ortloff 1996) is a prime example of water control mechanisms of this type developed for specialty crops. Thus extrapolating from bottom-up success at the local level by small *allyu* groups exploiting small farming areas to what is required to reliably support a Tiwanaaku urban population of 20,000–40,000 through seasonal variations in groundwater swale water levels and the necessary moisture content of agricultural berms to maintain an adequate food supply must incorporate a top-down management overview structure capable of assigning and relocating agricultural zones as well as fallow area use timing over a vast area. This capability was manifest during the last stages of 10th- to 12th-century drought as the groundwater interface with declining lake levels (Kolata and Ortloff 1996:181–201) forced abandonment of near lake edge raised field agriculture as groundwater levels declined below swale bottom levels. This caused agriculture to move to outer fringes of Pampa Koani, as noted by Graffam (1992), where the water table remained high in swales far from Lake Titicaca due to incoming intercepted groundwater from distant sources and earlier rainfall events. Summarizing, an overview of vast agricultural land and water management to ensure successful agricultural yields required knowledge of optimum berm designs (Figure 5.16) and groundwater height control as is only possible to create from a top-down overview perspective of land and water management for vast areas under their control. While other outlying sites had value for imports of non-agricultural resources (Goldstein 2005), sites with the potential for optimization of agricultural resources could be improved by optimization technologies demonstrated in the previous sections of this chapter for raised fields and other coastal valley sites to improve the economic basis for agricultural exports. Clearly optimization of river/spring source irrigation, raised field agriculture and control of urban water supplies for hygienic advantage to the city population demonstrates that exported Tiwanaku oversight to apply engineering methodology to optimize food production and city living benefits would be of definite advantage for candidate sites to associate with the Tiwanaku hierarchy and share the mutual benefits of association. It would be expected that this oversight activity was applied to rate potential archipelago satellite sites given that the economic burden of long distance transport of perishable goods to urban Tiwanaku would be an important consideration in evaluating the economic potential and incorporation of remote sites. Within the Tiwanaku governmental structure were religious rites, rituals and ceremonies elaborated with elaborate ceremonial, royal and administrative architecture to provide the religious accompaniment to the worldly success of their agroscience both local and distant from the urban core of Tiwanaku. Thus aspects of all the preceding categories likely provided the basis and rationale for Tiwanaku expansion from its heartland – this was only made technically and economically possible with the underlying centrally planned agrotechnical base provided by a top-down corporate management structure at urban Tiwanaku. The Tiwanaku corporate structure provided the success basis for satellite trade networks in agricultural goods together with the export of cultural traits and artifacts from the urban center of Tiwanaku to cement cultural ties back to the homeland source as observed from the archaeological record. A further argument for a top-down Tiwanaku management structure can be posited on an economic advantage basis. From similitude analysis methods (Chapter 2), a mathematical model of two competing *allyu* groups is considered: the first group sets up localized, near lake raised field agriculture supplied with a local spring-fed canal system; the second group arriving at a later time sets up an outlying raised field system with long canal lengths to irrigate their distant fields. The first group clearly has a benefit due to better water access (shorter canals) and resulting less labor required to tend agricultural land (Figure 5.11, (1) and (2) field system designs). Analysis shows a computable economic advantage to both groups to combine land and water resources by use of a newly designed canal irrigation network [the (2) option] that more effectively irrigates both land areas, reduces labor input from both groups to maintain the combined agricultural land area while raising the agricultural output of the combined land area. The advantages to both groups to combine their resources under a collective management that demonstrates economic advantages to both groups serves to promote top-down oversight of the vast raised field area to the advantage of all participating *allyu* groups through state managed oversight. Here the formalism of the similitude methodology permits a calculation of the increase in food production through collective top-down management oversight.

While certain researchers suggest less centralization and more local autonomy in the Tiwanaku core region (Albarracin-Jordan 1996; Berman 1997; McAndrews et al. 1997; Janusek 2003, 2004, 2008) as opposed to Kolata's reconstruction of a highly centralized state-directed agrarian production (Kolata 1986, 1991, 1993, 1996a), the present work demonstrates that the profound agricultural science base together with remarkable knowledge of urban Tiwanaku water control design and function essentially largely defines the success of the Tiwanaku society. Thus the massive public reclamation and construction projects requiring a large and coordinated labor force envisioned by Kolata appears supported by an advanced technology base much in the same way that modern progressive societies function. As to the demise of Tiwanaku colonies located in the Moquegua Valley (Ryan 2002), collapse dates are consistent with, or follow somewhat, the final collapse dates of Tiwanaku urban complexes. As detailed by Sharratt (2019), evidence of marginal Moquegua colonies persisted into Ilo-Tumilaca-Cabuza coastal phases and highland Tumilaca Phases l past AD 1000 times indicating in many cases extension of some of the Tiwanaku city traditions. As slow development of altiplano drought initiates in the 10th century, the rainfall runoff-based canal agriculture of the valley colonies invariably responded to rainfall runoff decrease in valley rivers leading to ultimate sustainable population contraction of these societies. The establishment of Estuquiña highland valley society at higher altitudes with higher rainfall levels is a natural survival consequence. As groundwater decline is a slow process due to recharge from distant infiltrated rainwater sources continually flowing through the aquifer toward Lake Titicaca, slow groundwater level decline permits longer continuation of elements of raised field agriculture in outlying regions of the Pampa Koani area (Graffam 1992) well past that of rainfall runoff supplied agricultural system of the Tiwanaku Moquegua colonies. It is expected that the colonies ultimately diminish in size due to drought but at different rate than hardier raised field systems of Tiwanaku proper due to their different agricultural water supply means. To assign Tiwanaku societal collapse to sociopolitical mechanisms would likely reflect the catalytic effects of drought-induced agricultural contraction on the sustainability of a society. The Tiwanaku collapse appears to be a slow process over centuries as the near lake raised fields appear to decline first as lake level subsides leaving agriculture to more distant raised fields where groundwater swale water decline lags that of near lake fields. This is consistent with Graffam's (1992) observations of agriculture still being practiced at later drought stages in areas remote from the Titicaca lake edge.

Visions on the last days of the city of Tiwanaku

The final stages of urban Tiwanaku due to extended drought are described in Ortloff and Kolata (1993). Essentially the drought slowly lowered the water table supporting raised field agriculture for the city's 20,000–40,000 inhabitants; additionally, absent water levels in raised field swales promoted loss of crops to freezing events (Ortloff 2009:88–92). Given the vast area devoted to raised field agriculture, restoration of the fields by excavating swale depths to penetrate the declining groundwater level together with lowering field system berm heights to accommodate crops with root system depths necessary for plant growth proved to be an impossible task given the vast labor requirements to perform these tasks. Evidence of use of raised field systems remote from the edge of Lake Titicaca where the groundwater height remained high exist requiring relocation of elements of city population to distant areas from the city. The presence of scattered *cocha* farming pits excavated to groundwater phreatic levels located away from the city center indicated population fragmentation to conduct localized survival farming. New information pertinent to the last days of Tiwanaku city life (Miller et al. 2014) is available from use of multiple stable isotope methods involving analysis of skeletal remains dating from the AD ~1100 time frame which correspond to city abandonment dates. Noted are dietary changes from previous norms experienced by city population as drought intensifies: these changes include absence of fish from the diet and no reported instances of child remains incorporating nutrients from fish or marine sources. This later observation may represent partitioning high nutrient food types to the most productive society members capable of generating food resources in emergency situations. As population decline and migration continued in this time period, specialized industries randomly lost key members necessary to sustain the group's function effectively; hence the loss of skilled fisherman can diminish the amount of fish available from the lake source. From modern observation of villages' use of lake resources, small minnows can be gathered from the near shoreline by nets. As this practice likely continued from ancient origins, and as drought conditions lowered availability of this food resource due to lowered lake levels that limited access to shallow shoreline depths together with increased salinity that affected fish stocks, marine resources by lake fishing and shoreline collection means was likely reduced from previous norms. Results from Miller et al. (2014) indicate the substantial presence of maize as a food source in the AD ~1000 time period – this indicates a likely increase in importation from different satellite areas (Logan et al. 2012) where this crop could be successively raised. Throughout the existence of Tiwanaku, maize importation constituted a large fraction of the population's diet and source of *chicha* for celebratory, social binding rituals (Logan et al. 2012). Although the totality of effects on population decrease and dispersion are yet to be brought forward under extended drought conditions, the use of stable isotope methods opens new paths to understand the final days of Tiwanaku city closure.

Conclusions

CFD results suggest that the drainage channel accelerated Tiwanaku's ceremonial center and the surrounding urban areas dryness to greater depth throughout the year promoting the city's hygienic benefits. For example, reduction of dampness in indoor habitable structures limits the occurrence of many respiratory and mould borne diseases (Clark and Ammann 2015). In the rainy season, the deep groundwater table upper boundary was stabilized by runoff and aquifer drainage into the drainage channel; in the dry season, additional seepage from the phreatic layer aquifer and canal M flow kept the deep groundwater upper boundary from subsiding. The resulting stabilization of the deep groundwater boundary prevented settling of monumental structures in the ceremonial core that originated from the design feature of the drainage channel's depth intersection with the top fringe of the deep groundwater layer. Additionally, stabilization of the groundwater level through seasonal changes in rainfall permitted interior-city agricultural and pastural areas integral to the urban city center to continue to provides specialty crops and continuous supply of pasturage. This engineering innovation originated from knowledge of groundwater manipulation previously shown as vital to sustain Tiwanaku's raised field agriculture. Channels P and Q largely served the hygienic requirements of the Putuni and Kerikala structures by providing continuous water flow from drainage channel seepage water and canal M water arriving into canal arm W. Each major monumental structure maintained an intricate drainage system that simultaneously served practical and symbolic purposes as exemplified by the Akapana's elaborate drainage network that limited rainfall infiltration into its compartmentalized earth-fill interior to preserve its structural integrity. Canals O, N south of the drainage canal directed water to the *qocha* complex K for inter-city agricultural and pasturage purposes. Rainy season runoff water that washed into canals N, O and S exceeding their carrying capacity was diverted into canals L, I and J leading to the river thus protecting urban regions from canal overflow flooding. In totality, the drainage channel was the linchpin of an intricate hydraulic network that controlled surface and aquifer flows as rainfall amounts varied from rainy to dry seasons. Analysis of the canal's hydrological function indicates that *qocha* systems were an integral feature of urban Tiwanaku. Interlinked by canals fed by Corocoro springs, *qocha* clusters K and C' occupied massive portions of the city and likely supported camelid herds and caravans brought to the center during key social gatherings (Janusek 2004, 2008). Recent excavations in adjacent Mollo Kontu residential compounds support the hypothesis that llama and alpaca herds were important in this part of Tiwanaku (Vallières 2012) and were served by K and C' *qocha* pasturage areas. Raised-field and *qocha* systems that occupied the edges of some of the city's canals are evident from aerial photos of the edges of canals I, J, L and C to support localized in-city agriculture and pasturage. The floodplain at the south edge of, and several meters below, the main portion of the city area of Tiwanaku supported an extensive cluster of integrated raised-field networks and *qochas* to support additional intra-city agriculture and pasturage.

Prior studies focused on Tiwanaku's hinterland demonstrated an understanding of hydrologic principles to develop intensive raised-field farming systems (Kolata and Ortloff 1996; Ortloff 1996, 2009). Present research indicates that the urban center of Tiwanaku incorporated an intricate hydrological network focused on the drainage channel that effectively managed seasonal water variations through surface canals, subsurface canals and aquifer drainage manipulation. CFD results detail many practical hydrological features of the drainage canal related to environmental and population livability concerns: these include rapid drying of subsurface soils underlying elite ceremonial and secular housing districts to limit soil dampness and its negative health effects on the city's population. The drainage canal further supplied water to flush the subterranean canal network underlying the elite ceremonial core region to transfer human waste material to a nearby river. These health-related features and remarkable plumbing features are the first reported for any Andean pre-Columbian city. Tiwanaku city planners demonstrated an extraordinary level of knowledge regarding hydrologic and structural maintenance principles based upon surface and groundwater manipulation to maintain high livability standards under harsh altiplano environmental conditions. Building on prior studies of the groundwater-based raised field systems that supported agriculture for the large population of Tiwanaku, analysis results demonstrate that knowledge of surface and groundwater flows within urban Tiwanaku merit further consideration in assessing New World engineering science.

6 Inka hydraulic engineering at Tipon

The site of Tipon, located in the proximity of Cuzco, provides an example of Inka hydraulic engineering knowledge and civil engineering practice as demonstrated by the design and operation of the site's complex water system. The water engineering knowledge base is revealed by analysis of the site's use of river and spring sourced surface and subterranean channels that transport, distribute and drain water to/from multiple agricultural platforms, reservoirs and urban ceremonial centers. Complex intersecting surface and subterranean channel systems that regulate water flows from different sources provide water to Tipon's thirteen agricultural platforms to maintain different ground moisture levels to sustain specialty crops. Additionally, within the site are fountains and multiple water display features requiring sophisticated hydraulic engineering to provide aesthetic displays. To understand the water technology used by the Inka to design some of the site's water system, use of modern Computational Fluid Dynamics (CFD) methodology and hydraulic theory is employed. Computer models of key elements of the Principal Fountain and the Main Aqueduct are made to reproduce water flow patterns in these features matching those intended by Inka engineer's design and calculations. From observation of these water flow patterns, the design intent and civil engineering knowledge base of Inka engineers is revealed. Results of the analysis show an Inka hydraulic technology utilizing complex engineering principles similar to those used in modern civil engineering practice centuries ahead of their formal discovery in western hydraulic science.

Tipon is a demonstration of Inka hydraulic engineering expertise. This is illustrated by the site's complex water supply and distribution system comprised of surface and subterranean water channels supplying multiple agricultural platforms and elaborated water display fountains. As with all hydraulic engineering projects, both ancient and modern, application of an engineering knowledge base underlies the design and function of complex water supply and delivery systems to ensure their successful operation. To uncover aspects of the knowledge base available to Inka hydraulic engineers involved in the design and operation of Tipon's water systems, modern hydraulic engineering analysis is applied to Tipons's Principal Fountain and Main Aqueduct to extract the engineering principles underlying their design and function. This reverse engineering approach reveals aspects of the hydraulic engineering base available to Inka engineers although their usage of hydraulic principles resides in a prescientific format as yet unknown.

Site description

The site of Tipon, located in Peru approximately 13 km east of Cuzco along the Huatanay River at south latitude 13°34' and longitude 71°47' at 3700–4000 masl is known for its many unique hydraulic features coordinated in a practical and aesthetic manner to demonstrate Inka mastery of water control principles. The site area indicates early Middle Horizon (AD 500–1200) presence evidenced by an encircling 6.4 km long outer wall attributed to Wari control of the enclosed area (McEwan 1991). The ~2 km^2 site area was under Inka control past AD ~1200 and was later converted into the royal estate of Inka Wiracocha in the early 15th century (Bray 2015a:170; Bauer 2004:87; Mithin 2012) evidenced by the nearby royal residence compound of Pukara. Figure 6.1 (Wright et al. 2006:6) shows the site composed of 13 major agricultural platforms, several of which are shown in Figures 6.2 and 6.3. These platforms were irrigated by a combination of natural spring source channels, multiple aqueduct and river source canals and groundwater drainage channels through a series of interconnected surface and subsurface channels that provided water to each of the platforms as well as domestic, elite residential and ceremonial areas (Figures 6.2 and 6.3).

The multiplicity of canals and channels and their water sources serving different agricultural platforms and display fountains is described by Wright et al. (2006:34, 35) and Bray (2013, 2015b) to show the complexity of Tipon's water supply and distribution systems. Figures 6.4 and 6.5a illustrate several of the overfall channels providing channeled spring and river source water to different level agricultural platforms; these structures are indicative of a water control design distributing surplus water from higher to lower elevation agricultural platforms. Figure 6.5b illustrates the continuance of water flows from elevated platforms shown in Figure 6.5a to continuation channel leading water to lower elevation destinations.

Several of the agricultural platforms contain an open channel collecting aquifer drainage from higher elevation platforms. Water from these sources was designed to combine with independent channeled water sources originating from spring and river sourced channels to provide the correct moisture level for different crops on individual platforms. Water from several sources was led to a series of intra-platform surface and subsurface channels (Figure 6.6) to further distribute water to other agricultural platforms (Mithin 2012:274–277).

Platform numbering convention starts with platform 1 at the lowermost south altitude and sequentially proceeds upward to platform 13 at the highest altitude (Wright et al. 2006:34, 35). In one case, water was led from multiple sources to combine with a natural spring source on platform 11 (Wright et al. 2006:34–35) to provide water to the Principal Fountain (Figure 6.8) designed for aesthetic as well as for drainage control purposes. According to Wright et al. 2006:39, Rio Pukara was the river water source for the uppermost altitude platforms 11-NW, 12 and 13 through the Principal Aqueduct (Figures 6.1, 6.12 and 6.13) while the spring on platform 11 was the water source for platforms 1 to 10 and part of 11 as well as several side platforms (Figure 6.1). Among the more prominent hydraulic features

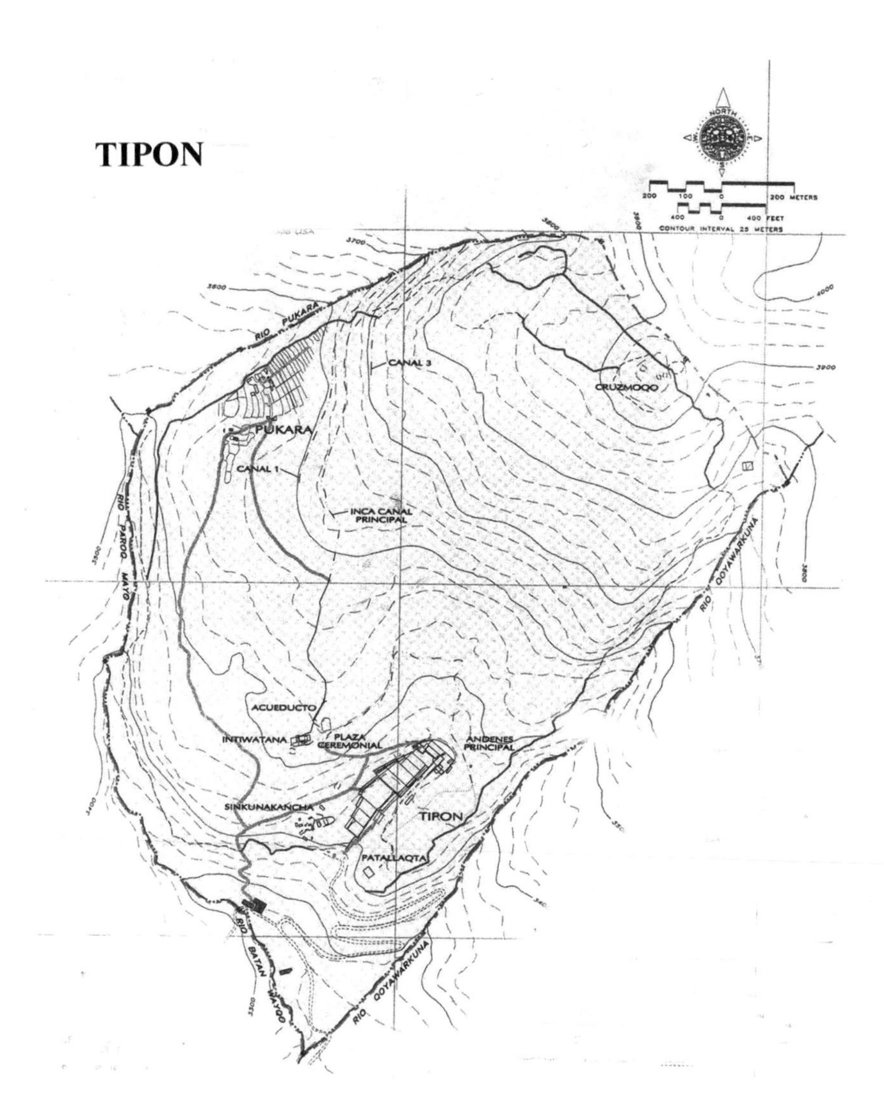

Figure 6.1 Map of the Tipon site area.

Source: Wright et al. (2006) by permission

of Tipon, the Principal Fountain on platform 11 had an elaborate spring supplied, multichannel branched water delivery system (Figures 6.8 and 6.9) leading to four independent waterfall streams shown in Figure 6.7.

A further hydraulic feature is the Main Aqueduct (Figures 6.12 and 6.13) vital to supply water from the Rio Pukara river source to central Tipon through different branch channels. Inherent to the water regulation system are drainage channels (A) and (C) (Wright et al. 2006:34, 35; Bray 2015b) of which (A), shown in

Figure 6.2 View of the upper level agricultural platforms.

Figure 6.10a, is located at the eastern edge of several of the platforms to convey excess water past that required for different crop types on individual platforms to a lower drainage area. Figure 6.10b shows a redistribution basin derived from the Figure 6.10a channel with a water overfall directed to provide water to a lower elevation destination. Encoded within the complex multi-channel water supply and distribution system of Tipon are examples and demonstrations of the hydraulic science base available to Inka hydraulic engineers. This chapter explores, by use of modern hydraulic engineering methods, the hydraulic knowledge base available to Inka engineers and used to construct the elaborate water system of Tipon. The analysis serves to yield new information about Inka water technology not previously reported in the literature and provide refined estimates of important flow parameters in two main hydraulic structures at Tipon.

As the terminal Late Horizon society (AD 1400–1530) familiar with water systems of conquered and occupied territories, the Inka had access to the hydraulic knowledge base of contemporary and earlier societies appropriate for use in their royal estates, cities and agricultural systems. As surveys of several contemporary and earlier Andean societies' water system technologies are available in the

Figure 6.3 View of the walled Sinkunakancha ceremonial plaza location shown in Figure 6.1.

literature (Ortloff 2009) as well as descriptive expositions of water systems at other Inka sites (Bray 2013, 2015b; Kendall 1985; Brundage 1967), it is instructive to determine if water technologies used by the Inka, particularly at Tipon, had borrowings from earlier predecessors. Given that application of different water technologies from different predecessor and contemporary societies were specific to their environmental and water resource types, only a limited number of these technologies would be applicable given the mountainous terrain and water resource constraints available to Inka hydraulic engineers. A further question, beyond observation and analysis of Tipon water systems and the water technologies used in the design of these systems, relates to indigenous innovation not previously noted from precedents derived from other societies. Given the multiple sources of water engineering knowledge available to the Inka, a further question arises as to how close to modern hydraulic engineering practice Inka water systems were. As codified observations of hydraulic phenomena provide a common basis for hydraulic engineering construction in both ancient and modern practice,

Figure 6.4 Drop structure conveying water from a higher to lower agricultural platform; water is channeled laterally into a central canal network at platform base.

analysis of the Inka hydraulic constructions at Tipon reveal the early application of hydraulic principles predating their later discovery in western science.

While appropriated usage of hydraulic engineering knowledge was available to Inka engineers, native innovation and inventiveness would have played a role given the mountainous terrain of the Inka homeland that would require special technologies to irrigate and productively farm. As many different land area types were available for farming by *mit'a* labor extracted from Inka conquered populations transferred to different regions together with the reciprocity and gift giving strategy of the Inka to assimilate different Andean societies into their multiethnic state, transferred water control and distribution technologies were an important part of the Inka strategy to expand their multiethnic state structure. Additionally, control of water systems for urban and agricultural use demonstrate aspects of political power exercised by Inka elites (D'Altroy 2003; Moseley 2008; Patterson 1991; Earle et al. 1989; Rowe 1946; Morris 1982; Zuidema 1990; Quilter 2014;

Figure 6.5a Water drop structure collecting agricultural platform drainage from an upper platform and channeling it to a lower platform's water supply network.

Bauer and Covey 2004; Brundage 1967) and symbolic manipulation of water symbols of the sacred (Bauer 1998; Urton 1999; Rowe 1979; Moore 2005; Moseley 2008). The importance of understanding Inka water control technologies lies in its importance to maximize agricultural production through elaborate irrigation systems both within the Cuzco area and Inka conquered territories; here surplus production was vital for Inka storage facilities (Morris 1992; LeVine 1992) that served as a defensive measure against extended drought periods. As Inka royalty controlled portions of agricultural lands for state governance functions involving

Figure 6.5b Top view of continuation channel of Figure 6.5a.

ritual bonding ceremonies that included social participation of all classes of Inka society (Murra 1960, 1980a; D'Altroy 2003:268–276), agricultural success through water irrigation technology was vital to demonstrate the management intelligence and reliance on the Inka elite class. While agricultural success based upon water control technology was a key concern of the Inka state, the provision of potable water to Cuzco inhabitants (and at other Inka sites) through display fountains and water basins (Kendall 1985) involved aqueduct design technologies to provide water from distant spring and river sources. Investigation and analysis of water technologies used for the Principal Fountain and the Main Aqueduct in the discussion to follow therefore are vital to expose aspects of Inka water technology.

The Principal Fountain

Wright et al. 2006:42 channel measurement data is used for the Principal Fountain analysis and calculations related to the water supply; this data precedes later repair

Figure 6.6 Bifurcation water distribution network element – water flows from two side channels that lead to a trans-platform channel at the base of a platform.

and reconstruction modifications. In conformance with modern civil engineering practice, English unit convention is used for hydraulic engineering calculations.

A first example used to determine the scope of Inka water technologies used at Tipon derives from examination of the channel system supporting the Principal Fountain shown in Figures 6.7, 6.8 and 6.9. A channel contraction occurs from a 0.9 m width, ~2.5 m long channel to a 0.4 m width, ~10.5 m long channel upstream of the Principal Fountain area on platform 11 as illustrated by Figures 6.8 and 6.9. Both channel sections have a rectangular cross section and have the same mild slope (Wright et al. 2006:47). The water source to the wider channel section derives from eight separate water supply conduits (Wright et al. 2006:47) and a major spring. Measured flow rates into the wide channel from two different tests yielded 0.68 ft^3/sec and 0.58 ft^3/sec leading to an average 0.63 ft^3/sec flow rate

Figure 6.7 Principal waterfall – spring flow into a wide channel, then diverted into channels directed into four waterfall channels.

(Wright et al. 2006:47). The question arises as to the water engineering design intent of the abrupt width change of the channel section shown in Figures 6.8 and 6.9.

To understand Inka hydraulic engineering in terms of modern hydraulics technology, use of the Froude number (Fr) is convenient to explain water behavior (Woodward and Posey 1941; Morris and Wiggert 1972; Ven Te Chow 1959; Henderson 1966). For shallow depth (D) flows, the Froude number definition is $Fr = V/(g\ D)^{1/2}$ where V is the water velocity and g is the gravitational constant (32.2 ft/sec^2). Physically, Fr is the ratio of water velocity V to the gravitational wave velocity $(g\ D)^{1/2}$ – when $Fr > 1$, water velocity exceeds the signaling gravitational wave velocity so that water has no advance warning of an obstacle – this leads to the creation of a sudden hydraulic jump at an obstacle as illustrated in Figure 6.16. In this figure, a shallow, high velocity water flow ($Fr >> 1$) encounters a

Figure 6.8 Channel from spring reservoir showing the width contraction zone.

plate obstacle at the leftmost exit region of a hydraulic flume causing an elevated, highly turbulent hydraulic jump. In physical terms, for Fr >1, there is no upstream awareness of the presence of an obstacle until the obstacle is encountered by the water flow as the gravitational wave signaling velocity that informs the flow that obstacle exists $(gD)^{1/2}$ is much less than the V flow velocity. For Fr < 1, the gravitational wave signaling velocity travels upstream of the obstacle faster than the water velocity V to inform the incoming water flow that an obstacle lies ahead. This causes the water flow to adjust in height and velocity far upstream of an obstacle to produce a smooth flow over the obstacle. Here the obstacle for the present application is the large contraction in channel width. In the discussion to follow, Fr > 1 flows are denoted as supercritical, Fr < 1 flows are denoted as subcritical and Fr = 1 flows are denoted as critical. While the presence of subcritical, critical and supercritical flows can be calculated from modern hydraulic theory,

Figure 6.9 Multiple channel branching from the single supply spring supplied channel.

a simpler method exists to determine flow types that were perhaps used by Inka engineers. Insertion of a thin rod into a flow that produces a downstream surface V-wave pattern is indicative of supercritical flow; if the surface pattern shows upstream influence from the rod, then subcritical flow is indicated. If only a local surface disturbance upstream or downstream of the rod is noted, then critical flow is indicated. Thus this simple test can determine flow regime types and promote the various usages associated with the different flow regimes.

Figure 6.11 (Woodward and Posey 1941) is derived from the Euler fluid mechanics continuity and momentum equations and is useful to describe the flow transition from sub- to supercritical flow in the Principal Fountain supply channel. As all water motion is governed by the mass and momentum conservation equations, Figure 6.11 indicates flow transitions based on Froude number change due to channel geometry changes. While modern conservation equations

Figure 6.10a South side drainage channel.

are a shorthand method of summarizing governing principles, similar results are obtainable by codified and recorded observation methods as likely employed by ancient water engineers. Using the subscript notation (1) for flow conditions in the wide channel section and (2) for contracted width conditions, the width contraction ratio is W2/W1= 0.44. The (1) flow entry value using the average flow rate is based on a ~0.3 ft water depth for which V_1= 0.72 ft/sec and Froude number is $Fr_1 = V/(g\ D)^{1/2} = 0.23$, ($Fr_1^2/2 = 0.03$ in Figure 6.11). The contracted channel (2) Froude number is $Fr_2 \approx 1.14$, ($Fr_2^2/2) = 0.65$ in Figure 6.11), based on ~0.3 ft depth. The channel contraction shown in Figures 6.8 and 6.9 takes the flow from subcritical flow ($Fr < 1$) in the wide channel to a near critical ($Fr \approx 1$) in the narrowed channel. From Woodward and Posey 1941:140, the flow rate per unit width in the contracted rectangular cross-section channel is $q_2 = 0.68/W2 = 0.52$ ft^3/sec ft and the critical water depth (Henderson 1966) is $y_c = (q_2^2/g)^{1/3} = 0.2$ ft in agreement with the previous critical water depth value calculated for near critical $Fr \approx 1$ flow in the contracted width (2) channel. The special case for which $Fr = 1$ is

Figure 6.10b Continuation and redistribution chamber with free overfall stream connected to the waterfall area.

accompanied by flow surface wave instability derived from the presence of translating large scale vortex motion within the body of the water. Figure 6.15 (Wright et al. 2006) illustrates a surface ripple pattern (Figure 6.15) consistent with near critical $Fr \approx 1$ flow in the contracted channel region as well as a slight decrease in water depth resulting from the (1) to (2) width change transition.

The parallel ripple wave structure normal to the flow direction is consistent with $\sin \theta = 1/Fr$, where θ is the half angle of the surface wave so that when $Fr \approx 1$, $\theta \approx 90°$ verifying the near normal surface ripple wave structure shown in Figure 6.16 (Morris and Wiggert 1972:190; Ven Te Chow 1959:451). Surface wave instabilities associated with $Fr = 1$ flow are to be avoided in the supply channel as a uniform, stable flow is ultimately required to produce the proper aesthetic display in the downstream fountain waterfall channels.

Figure 6.11 (Woodward and Posey 1941) illustrates the flow Froude number transition arrow (K to J) resulting from the wide (1) to narrow channel (2) shape

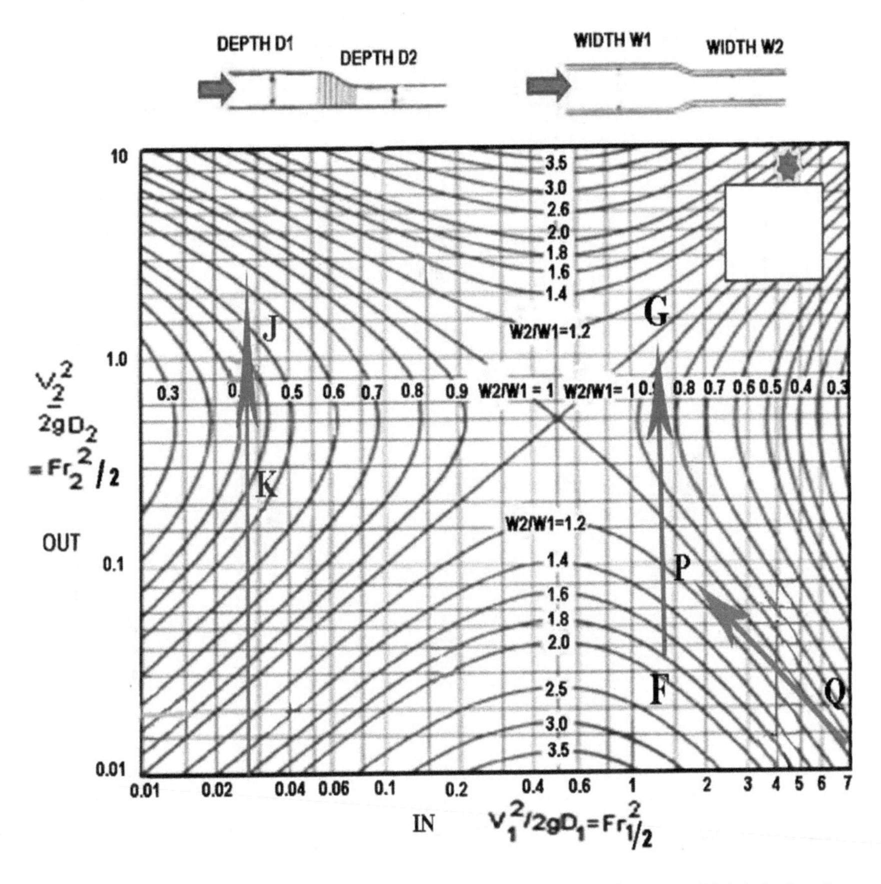

Figure 6.11 Width and depth change diagram (Woodward and Posey 1941) indicating channel entry and exit Froude number effects on flow characteristics for the Principal Fountain and the Main Aqueduct.

change that incorporates transition from sub- to near critical flow. This indicates that Inka engineers deliberately designed the (2) contracted channel section to support near critical $Fr \approx 1$ flow, but not exact critical $Fr = 1$ flow. In modern channel design practice, $\sim 0.8 < Fr < \sim 1.2$ is the prescribed Fr range to obtain high flow rates that accompany $Fr \approx 1$ flows to avoid surface waves associated with translating large scale vortex motion below the water surface associated with exact $Fr = 1$ flows. The Fr_2 Froude number range in the contracted (2) channel lies between $\sim 0.8 < Fr < \sim 1.2$ and the computed $Fr_2 \sim 1.14$ value is consistent with near stable flow according to modern hydraulic engineering standards. The width reduction construction shown in Figures 6.8 and 6.9 thus yields the narrowest supply channel (2) at the maximum flow rate per unit channel width without significant internal surface wave structures causing transient flow instabilities. Observing the narrow channel leading to the stilling reservoir ahead of the waterfall, it appears that Inka hydraulic engineers wanted to limit disturbances and currents within the reservoir to help promote equal, stable flows into the four waterfall channels. The contracted channel (2) then serves this purpose as no upstream influence exists from flow disturbances in the downstream direction. From Figure 6.8 (Wright et al. 2006:2, 47), the contracted width channel led to transecting channels A-B, which when blocked, served as a reservoir with a further channel to the platform immediately ahead of the four channel waterfall (Figure 6.7). This channel design promoted near symmetrical flow conditions in the reservoir so that waterfall channels on each side of the inlet channel have symmetrical input flows from the reservoir. As flow into the A-B channel is directed into a smaller channel almost directly across from the supply channel (Figure 6.8), there is an increase in the Froude number but this value is still close to critical according to the Figure 6.12 process path connecting F to G. With $Fr > 1$ flow in contracted channels (Figure 6.8) ahead of the waterfall, no upstream disturbances can affect the stability and aesthetics of the waterfall.

Further, near critical flow is associated with the maximum flow rate per unit channel width that a channel can support (Morris and Wiggert 1972); this is closely achieved for $Fr_2 = 1.14$ conditions. Flow stability is achieved when Fr is either slightly less than or more than the critical $Fr = 1$ condition – knowledge and use of this hydraulic engineering practice is evident in the Principal Fountain design and is vital to produce a constant, stable water delivery flow to the waterfall area. The presence of near critical flow in both channel (2) and its continuance channel has the advantage of preventing upstream influence of any downstream channel flow resistance element (channel bends, wall roughness effects, non-symmetric flow into reservoirs creating surface waves and vortices, disturbances from the A-B channel) from creating flow instabilities in the subcritical (1) wide channel that would translate and translate instabilities into downstream flow patterns. For $Fr > 1$ flows in all downstream channels, uniform flow to the waterfall is guaranteed to preserve its aesthetic display.

The contracted channel flow design is important as: (1), the flow from the spring source channeled into the wide channel (1) to a downstream narrow width channel (2) is associated with the maximum flow rate per unit channel width; (2) the near critical flow values in the two channels downstream of the wide section channel (1) eliminate flow disturbances propagating upstream derived from flow into the A-B channel and other downstream resistance sources that would alter the stability of flow to the waterfall; (3) channel A-B, when blocked, serves as a stilling basin to help eliminate any water motion that would challenge symmetrical flow delivery to the fountain area and (4) any disturbances from flow into the downstream stilling basin are not propagated upstream into the wide (1) channel to destabilize its flow to downstream channels. Why is limiting upstream influence important? If Inka engineers chose a channel design that had a wider width throughout than those shown in Figure 6.8, then the value of $Fr < 1$ would exist in all channels so that unstable disturbances from flow width transitions to the waterfall display would propagate in both upstream and downstream directions. An example disturbance associated with subcritical flows into sequentially wider channels would be

Figure 6.12 Main aqueduct canal section showing the steep channel section leading to the lower slope region.

a sudden water velocity decrease that would be felt as an "obstacle" and have an upstream influence the subcritical flow in the upstream wider channel section. The low velocity, unstable upstream water height changes would then interact with the incoming spring flow producing transient surface wave instabilities that propagate downstream in all wider channels and ultimately lead to an erratic, non-aesthetic waterfall display. This negative design was anticipated by Inka engineers and avoided by their design shown in Figures 6.8 and 6.10 with its many benefits.

An overview of the Inka technology used for the Principal Fountain reflects modern hydraulic design principles to preserve fountain aesthetics during seasonal water supply changes. Knowledge of channel width change effects on flow regime change from sub- to supercritical lows to achieve stable flow at the maximum flow rate to the Principal Fountain is apparent in the channel and is consistent with modern practice.

As for Andean precedents for channel width change to regulate Froude number, examples from the Late Intermediate Period (AD ~1000–1300) Chimu Intervalley Canal (Ortloff 2009:44–53; Ortloff et al. 1982, 1985) show a similar hydraulic technology designed to regulate Froude number within the $\sim 0.8 < Fr < \sim 1.2$ range to ensure stable flow conditions at a prescribed maximum flow rate of 4.6 ft^3/sec to reactivate drought desiccated irrigation canals in the Moche Valley. This was achieved by changing the channel cross-section geometry consistent with channel slope and wall roughness variations to make the required flow rate close to maximum flow rate the channel can transport with $Fr \approx 1$ flows (Ortloff 2009:91). In this sense, the Inka use of channel width changes to regulate Froude number may derive from the Chimu precedent as both technologies involve creation of stable, near critical flows maintained by channel cross-sectional shape changes. Unstable, pulsing flow in the Chimu Intervalley Canal would have the effect of accelerating erosion of unlined canal banks as well as causing oscillatory forces on canal stone lined banks that would cause leakage into unconsolidated foundation soils. The Inka use of subsidiary channel A-B (Bray 2013:171) to maintain the design flow rate to the fountain during seasonal changes in spring flow rate is further acknowledgment that flow addition/subtraction is necessary to preserve the fountain's aesthetics.

Noting that the Figure 6.8 waterfall streams are close to the back wall, the x distance of impact of the overfall stream from the vertical wall on to the flat bottom base platform is $x = V_0 (2z/g)^{1/2}$ where V_0 is the water velocity just before the fountain overfall edge and z is the height difference between the overfall edge and the flat bottom base here given as ~4.0 ft. As x is approximately ~0.25 ft, then $V_0 \sim 0.5$ ft/sec. As the input flow rate (assuming canal A-B extraction/supplement is zero) is ~0.63 ft^3/sec and, as four channels supply the fountain, then the water height in each of the four channels is ~0.3 ft which is consistent with the four waterfall input channel depth measurements. The point is that in order to achieve an aesthetic fountain display, a precise and equal water velocity into all four channels was required. This was achieved by the Figure 6.9 channel geometry change to transition $Fr < 1$ flow in the wide channel to a near critical $Fr > 1$ flow in subsequent contracted channels to eliminate downstream flow resistance effects on upstream flow conditions. Additionally, as seasonal rainfall varies affecting supply/drainage canal flow rates and the platform 11 spring flow rate, channel A-B served as flow additions/subtractions to maintain a constant flow rate to the fountain.

The realization that the channel contraction effect (W2/W1 in Figure 6.12) leads to an $Fr \approx 1$ near maximum flow rate in the contracted (2) channel was a necessary first step in the Inka design. This feature was necessary to deliver a stable water flow to the downstream delivery system to largely eliminate instabilities that would compromise equal water delivery to all four waterfall channels. The attention paid to flow stability and elimination of upstream influence effects that would compromise downstream flow delivery stability demonstrates Inka water management techniques to achieve waterfall aesthetics. In modern hydraulic engineering terms, the Froude number of the incoming water stream determines the Froude number in an expansion W2/W1 > 1 or contraction W2/W1 < 1 stream (Figure 6.11). Here Inka water engineers demonstrated mastery of this concept to design channel expansion/contraction control of water flows as being Froude number dependent. Together these design features indicate a sophisticated hydraulic technology in use to preserve fountain characteristics during water seasonal supply changes.

The main Tipon aqueduct

The Main Canal's Section 4 aqueduct (Wright et al. 2006:6, Plate 1) upstream of the site of Intiwatana (Figure 6.1) demonstrates several additional facets of Inka hydraulic engineering. A 40 by 25 m reservoir above the Intiwatana sector (Hyslop 1990:137) conducted water into the Main Aqueduct from the Pukara River (Figure 6.1) to supply fountains in the adjoining main plaza. Section 4 of the Main Aqueduct includes a long, steep channel of ~16 degrees measured slope followed downstream by a ~1.7 degree slope (Figures 6.12 and 6.13). Channel width dimensions range from ~0.65 to ~0.8 ft while channel depths range from ~0.8 to ~1.0 ft indicating near constant dimensions of the rectangular cross-section

Figure 6.13 Main aqueduct canal steep slope section.

aqueduct channel (Wright et al. 2006:53–63). Upstream of the aqueduct is a built-in water diversion channel (Wright et al. 2006:56) activated by a movable sluice plate to control the delivery flow rate to the aqueduct. As flow in the steep section is supercritical (Fr > 1) and approaches normal depth asymptotically on a steep slope (Morris and Wiggert 1972; Henderson 1966), when the flow approaches the mild slope, nearly flat channel section, a hydraulic jump occurs in the flat channel section (Bakhmeteff 1932:254, his Figure 193b) as no channel width and shape change is present to lower or cancel the hydraulic jump. Figure 6.14 schematically shows the formation of the hydraulic jump (HJ) at the lower slope segment of the aqueduct.

In terms of the Froude number, supercritical Fr > 1 flow on the steep section of the aqueduct is converted to subcritical Fr < 1 flow in the aqueduct flat section through creation of a hydraulic jump. For flow downstream of the hydraulic jump on the shallow, mild slope, Fr < 1, and as the aqueduct slope continues downhill to the final destination and V increases, the post-hydraulic jump flow has no significant upstream effect on the location and height change of the hydraulic jump. The hydraulic jump feature provides a means to calculate the flow rate in the aqueduct channel given that the hydraulic jump flow height produced by the supercritical to subcritical flow transition is to be contained within the channel to eliminate spillage from the channel. As spillage represents a waste of precious water intended for both agricultural and urban use purposes, the canal design reflects Inka engineer's concern to eliminate water wastage. To determine the correct canal flow rate to eliminate spillage, modern hydraulic engineering analysis methods are

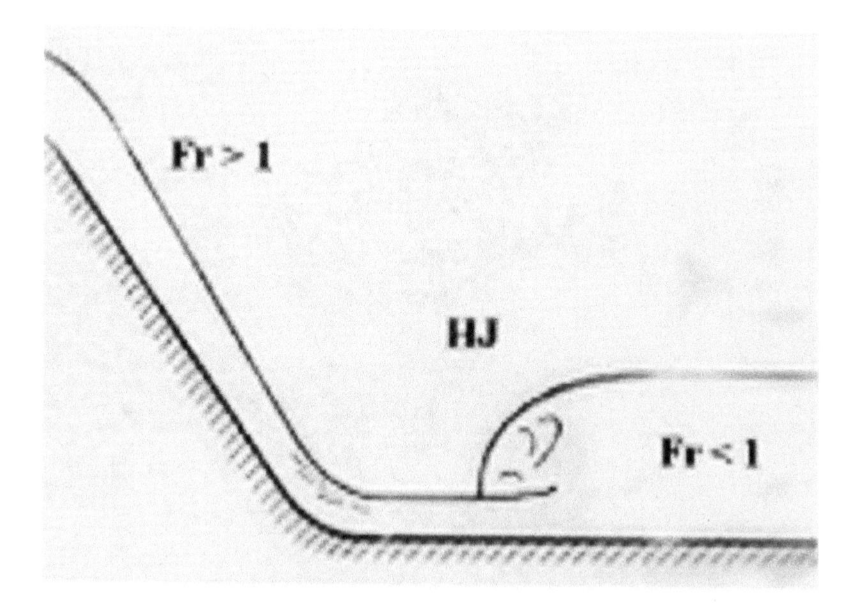

Figure 6.14 Schematic representation of the formation of a hydraulic jump (HJ) due to the aqueduct slope change from steep to mild slope.

Figure 6.15 Water ripple patterns in the Figure 6.9 contracted channel section indicating critical flow.

employed. Here trial water depths and the associated flow rate calculated from the Manning equation (Morris and Wiggert 1972) are used to determine if the hydraulic jump height produced at the base of the steep aqueduct section is contained within the channel wall height dimensions. As the correct flow rate in the Main Aqueduct is as yet unknown to eliminate spillage, trial water flow heights (h_1) and associated trial flow rates for different steep slopes (as indicated in the Wright et al. 2006:60 Table) together with the water velocity consistent with trial heights and slopes, permits trial Fr_1 values can be calculated at the base of the steep slope. From the trial Fr_1 and h_1 values, the hydraulic jump height relation shown as Equation (1) (Rouse 1946:145; Henderson 1966:69) is used to determine the hydraulic

Figure 6.16 Hydraulic jump caused by a high Froude number flow encountering a submerged plate obstacle at the leftmost exit of a hydraulic flume.

jump height h_2 and the post hydraulic jump Froude number Fr_2 on the mild slope. Based on calculations of the trial hydraulic jump water height change from Equation (1), the correct hydraulic jump water height and the correct aqueduct flow rate is determined when the h_2 water height is fully contained within the channel to eliminate channel spillage. The height change h_2 from the original trial water height h_1 due to a hydraulic jump is given as:

$$h_2 / h_1 = (1/2)\left[\left(1 + 8\,Fr_1^2\right)^{1/2} - 1\right] \tag{1}$$

and the post hydraulic jump Froude number Fr_2 is:

$$Fr_2 = \left\{(1/8)\left[1 + 2\,(h_1/h_2)\right]^2 - 1/8\right\}^{1/2} \tag{2}$$

Using elements of the Wright et al. 2006:60 Table for different trial water depths at different trial slopes where the volumetric flow rate is in ft^3/sec where and H is the post hydraulic jump total water height in feet, Fr_1 is the supercritical Froude number characterizing flow at the base of the steep slope channel. Fr_2 is the subcritical Froude number characterizing the post hydraulic jump flow, and h_2/h_1 is the hydraulic jump height change ratio. Depth is the trial water depth in the channel in inches. From Equations (1) and (2), Table 6.1 results follow.

Table 6.1 Determination of main aqueduct hydraulic jump properties

Depth	*Slope (degrees)*	*Velocity (ft/sec)*	*Flow Rate*	Fr_1	h_2/h_1	*H (ft)*	Fr_2
2	8.53	6.6	0.80	2.85	3.65	0.61	0.42
2	11.3	7.6	0.92	3.29	4.18	0.70	0.24
2	16.7	9.3	1.12	4.01	5.19	0.87	0.34
4	8.53	8.6	2.08	2.63	3.25	1.08	0.45
4	11.3	9.9	2.40	3.03	3.81	1.27	0.41
4	16.7	12.2	2.93	3.72	4.78	1.59	0.36

Based on Table 6.1 results, the maximum flow rate (assuming the flow diversion channel plate is completely raised and flow in the aqueduct is from the Rio Pukara river source only, a flow rate on the order of ~1.12 ft^3/sec is sufficient for water to be safely contained in the channel without spillage given the ~1.0 ft depth and ~1.0 ft width of the channel (Wright et al. 2006:62). For this case, the 16.7° slope case closely matches the measured aqueduct steep slope. For higher aqueduct flow rates on the same steep slope, Table 6.1 indicates a large degree of spillage as the hydraulic jump water height exceeds the channel wall height dimension. An upstream Section 1 part of the channel has a water diversion structure (Wright et al. 2006:56) used in conjunction with an adjustable height sluice plate used to divert a fraction of the Main Aqueduct's water into a side turnout irrigation canal. This diversion structure likely was used in times of heavy rainfall periods to divert water from the aqueduct channel to prevent spillage in downstream portions of the Main Aqueduct. Some evidence of a slight widening of the channel is apparent from Figure 6.12 in the near post hydraulic jump region – this effect, if deliberate and more substantial in width, would reduce spillage and permit a higher flow rate on the order of ~2.0 ft^3/sec. This feature is not evident in the aqueduct design. The Figure 6.11 Q-P arrow indicates the Froude number transition from supercritical Fr_1 to subcritical Fr_2 due to the hydraulic jump located at the steep-mild slope transition point for contained flow within the channel; results shown indicate consistency with the governing equations of fluid motion. For a flow rate higher than ~1.12 ft^3/sec, only the lower amount would be deliverable to destination sites and the rest subject to channel bank overflow. This consideration defines the intent of Inka water engineers to design the aqueduct for the maximum flow rate fully contained within in the aqueduct channel. Based upon Figure 6.12, the upstream channel ahead of the steep slope section appears to have a mild slope supporting subcritical flow; therefore only the steep to mild slope segment of the aqueduct shown in Figures 6.12 and 6.13 proves useful in determining the aqueduct maximum flow rate. Since the Main Aqueduct canal was to supplement site water supplies during the drier parts of the year when the Rio Pukara river source was at a low flow rate, the water contribution to the urban Intiwatana area (Figure 6.1) and its storage reservoir would be vital to maintain its potable water source as well as to supply adjacent canal systems for agricultural use throughout seasonal changes in the river water supply. Here the sluice plate placed far upstream in the aqueduct channel served the purpose to regulate the channel flow rate to a design value to eliminate flow spillage.

The question arises as to why the Main Aqueduct flow rate is important to determine. As the aqueduct was the water source for the ceremonial plaza fountains and reservoir and had connection to several of the agricultural platforms, knowledge of its maximum flow rate determined the geometry of several downstream channels designed to accommodate its water flow without overflows or spillage. Additionally, water delivery amounts (ft^3/sec-unit land area) for specialty crops on different agricultural platforms with different soil compositions were important to provide the correct moisture levels for plant growth and determine water supply structure dimensions within the plaza to maintain their functionality and aesthetic presentation. Thus Inka designers apparently had the means to determine channel width, depth, channel slope and wall roughness to tailor flow characteristics dependent upon Froude number to support different delivery flow rates to different agricultural platforms, aesthetic waterfall displays and ceremonial plaza usages much in the same way that modern hydraulic engineering practice dictates.

As for precedents observed from earlier Andean societies, in canals particular attention was paid to limiting canal flow rate beyond design values to prevent canal over-bank spillage. In a (possibly late Moche or Chimu) Jequepeque Valley canal (Ortloff 2009:111–122), a choke consisting of opposed stones a given distance apart was installed in a water conveyance channel to limit canal flow rate to a design value. Excessive runoff from El Niño rains flowing into the canal causing a low rate excess over the design flow rate (as determined by the choke geometry) backed up ahead of the choke and was shunted into an elevated side weir ahead of the choke that emptied water into a lower drainage area thus limiting damaging erosive over bank flow spillage. Elsewhere along the Chimu Chicama-Moche Valley Intervalley Canal, drainage chutes were placed high up along canal walls to convey excessive water from El Niño flood events away from the main canal (Ortloff 2009:91; Ortloff et al. 1982). In this canal, for excessive flow rates over the design flow rate (4.6 ft^3/sec), sophisticated channel shaping was used to create vortex regions in concave channel side pockets that effectively narrowed the streamline path of the channel flow to convert a sub- to supercritical flow inducing a hydraulic jump water height change (Ortloff 2009:47–50). The amplified water height from the hydraulic jump was then diverted into an elevated side weir and then conducted to a diversion channel that led to a lower farming area (Ortloff 2009:49). From known cases observed from Chimu (and possibly late Moche) hydraulic engineers' work, special attention was paid to preserving water transport canals from erosive damage due to water spillage over canal banks. As the Inka incorporated the Chimu Empire by conquest, water technology experts were likely exported along with craft specialists back to the Inka capital to serve in advisory roles for hydraulic engineering projects. The preoccupation of Chimu (and hydraulic engineers from different societies) to preserve their agricultural systems by innovative adaptive strategies under climate change duress (principally flood and drought) is documented in Ortloff 2009:197–205) and in Chapter 3. Given the same preoccupation of the Inka society to preserve their agricultural and urban water supply systems, importation of hydraulic technology was a vital concern. As a further example of technology available to Inka engineers, the complex channel shaping noted at the exit of a steep sloped channel at Tiwanaku's Lukurmata area essentially created a hydraulic jump of zero height (Ortloff 2009:100–105) necessary to limit erosion damage of an unlined canal. This technology to raise the channel's flow rate, although available to Inka engineers, was not evident to limit over bank spillage as Figure 6.12 indicates. In modern hydraulic practice, a hydraulic jump can be eliminated when the aqueduct mild slope section is made equal to the critical slope (Bakhmeteff 1932:243) together with channel widening to produce a *neutralizing reach* condition.

A further precedent involving control of groundwater for agricultural purposes involves Tiwanaku raised field systems. Water supplied to raised fields through a channel was used to regulate groundwater height for different crop growth requirements. This provides a distant analog to Tipon's controlled moisture content agricultural terrace systems. An elevated channel side weir cut into in the Tiwanaku Pampa Koani water supply channel directed excess water to an adjacent channel that led water directly to Lake Titicaca. This control regulated the permissible water height, and thus the flow rate, in the main channel that supplied the correct amount of water to the field system swales (Ortloff 2009:84–86). This control system was vital to maintain the productivity of the raised field systems distant from Lake Titicaca during high seasonal rainfall periods as it limited field system saturation destructive to agriculture. In the seasonal dry period, reduced main channel water height continued directly to raised field systems to maintain the required groundwater and swale water height to support crops. Given that differences in distance, centuries between the Tiwanaku Middle and Late Horizon Inka times, water supply types and other environmental conditions were vastly different, it remains questionable if the Inka understood and utilized aspects of this technology.

Conclusions

The utilization of multiple spring water sources together with a river sourced aqueduct used to supply surface and subsurface channel networks controlling the water supply and drainage systems of Tipon was part of an intricate Inka design to control the moisture content of the agricultural platforms for specialty crops. The use of subsidiary channels with independent water sources intersecting main channels was part of a complex water control system designed to supplement (or drain) water to achieve design flow rates to key site areas. For the Principal Fountain, the water flow rate was carefully controlled by an intersection A-B canal to supplement (or drain) the spring flow rate to ensure fountain aesthetics during seasonal changes in spring water supply. For the Main Aqueduct, the flow rate into the aqueduct from the Pukara river source was regulated by a movable sluice gate to ensure aqueduct spillage was eliminated. Here the maximum aqueduct delivery flow rate (for the sluice gate in full open position) was determined by the maximum contained water height in the post hydraulic jump region of the aqueduct. The realization that a control sluice gate to divert excess flow from the aqueduct beyond its design flow rate was a necessary part of the aqueduct system design to eliminate spillage was acknowledgment of the thought process behind Inka hydraulic engineering practice. In summary, the presence of intersecting canals as part of the design for both the Principal Fountain and the Main Aqueduct indicate supplemental (or drainage) water controls designed to achieve precise flow rates necessary for the intended destination purpose of these hydraulic features. As revealed by modern hydraulic theory illustrated by Figure 6.11, the relation between input channel Froude number (Fr_1) and its relation to the Froude number (Fr_2) in an expanded (or contracted) channel is a complex hydrodynamic procedure. As observed from Figures 6.8 and 6.9, Inka engineers understood, likely by trial and error observations, that the channel contraction geometry they chose for the Principal Fountain converted the low speed subcritical flow in the wide channel to high speed, near critical flow in the contracted channel section to give the Principal Fountain its proper function and aesthetics. The use of near critical flows in supply channels (Figures 6.8 and 6.9) to the waterfall eliminated upstream resistance influence that would influence flow stability. As with all hydraulic engineering projects, both ancient and modern, application of an engineering knowledge base underlies the design and function of complex water supply and delivery systems to ensure successful operation. As in modern hydraulic engineering practice, test work involving models in a hydraulic flume (for example, Ortloff 2009:47) provide visual confirmation of theoretical flow predictions or, in many cases, empirical correlation equations derived from test observations to describe flow phenomena. It may be assumed that some similar form of observational data was available to Inka engineers to know in advance how to design channel geometry changes to achieve a desired effect.

Both the Principal Fountain and the Main Aqueduct exhibit hydraulic design features that show a consistent high level of hydraulic knowledge to fully utilize local on site and distant water resources efficiently. Inherent to the design of the Tipon water network is an as yet unknown Inka format for analyzing and recording observations of water engineering experiments and tests necessary to achieve the Tipon system's success. This process would involve pre-scientific notations for water velocity, flow rate, water height change, flow stability and farming aquifer moisture levels, among other parameters, and their mutual interaction to predict water flow patterns and efficient farming productivity results. Of interest are the many words in the Quecha language used by the Inka related to water (Brundage 1967:379–380). Many relate to the hydraulic technology that was prevalent in Late Horizon Inka times such as *pincha*, a water pipe, *rarca*, an irrigation ditch, *patqui*, a channel, and *chakan*, a water tank, while other words describe water motion such as *pakcha*, water falling into a basin and *huncolpi*, a water jet. While these words were commonplace among the general public given their dependence upon water for agricultural and urban use, there must have been more complex vocabulary to describe more complex hydraulic phenomena used by Inka hydraulic engineers involved in the design Tipon and other water systems (Brundage 1967:379–380; Kendall 1985; Wright et al. 2006; Bray 2013, 2015b) given the complex water technology used in those systems.

Given the complex water supply system of Cuzco (Brundage 1967:89–91) that included multiple spring supplied channels as well as pressurized pipe systems supplying potable water to city inhabitants, water technology was in an advanced state of knowledge. It may be speculated that the complex terrace system at Moray (D'Altroy 2003; Wright et al. 2011) played a role as an experimental facility to help Inka engineers design the many water transport facilities used in agriculture in the Inka realm. The engineering base used to design complex water facilities would likely involve use of *yupanas* (and *quipus*) for analysis and recording of flow phenomena much in the same way that a canal surveying problem can be represented by *yupana* calculations used by Chimu water engineers (Ortloff 2009:80). The use of multiple canal supply and drainage systems for agricultural, living compound and ceremonial center functions, as well as for the aesthetic fountain display at Tipon, are prime examples of Inka mastery of hydraulic engineering principles and constitute a notable contribution to the history of hydraulic engineering. The water system designs exhibited at Tipon are notable in their coincidence to modern water engineering practice and, as such, anticipate discovery of modern hydraulic principles by western science by several centuries.

It is of interest to compare the state of hydraulic knowledge in ancient Andean societies with that of continental Europe in similar AD centuries. The water engineering technology exhibited by AD 300–1100 Tiwanaku, AD 800–1480 Chimu and the AD 1480–1532 Inka societies as described in Ortloff (2019) and earlier chapters indicate a reservoir of advanced hydraulic technologies comparable in many ways to modern technology. Roman water technology as demonstrated in Chapter 2 appears to exhibit comparable level advanced hydraulic knowledge. For both ancient Old and New World societies thus far investigated, accumulated hydraulic knowledge likely relied upon test, experiment and nature observations put into codified formats and recording procedures that are unfortunately lost due to lack of surviving written records or other forms of recording hydraulic knowledge. To recover ancient societies' versions of hydraulic science, use of modern hydraulic engineering methodologies involving computer simulations, theoretically derived equations and laboratory test procedures applied to analyze ancient water structures provide one way to uncover lost the knowledge used in their design and operation to bring forward the design intent of ancient engineers – although their techniques of obtaining and applying this knowledge and the format and data recording and data storage methods used to produce sophisticated water system designs is as yet unknown. Only in the 17th and 18th centuries in Europe did the invention of mathematical descriptions of physical phenomena usingcalculus methodologies together with concepts of mass, momentum and energy conservation theorems finally lead to basic calculation methods applied to predict fluid motion. In this regard Bernoulli's 1738 publication *Hydrodynamica* initiated and advanced hydraulic calculation methodology to levels that continued in complexity to the present day. That ancient societies of both Old and New Worlds utilized advanced hydraulic technologies remains a subject for future research to uncover – most fascinating is that ancient science demonstrates alternate ways to describe and utilize hydraulic knowledge.

7 2600–1800 BC Caral

Environmental change at a Late Archaic Period site in north-central coastal Peru

Introduction

The presence of ENSO climate variations in the form of long-term drought, flooding and sediment deposition/erosion transfer events and their effect upon the agricultural sustainability of Andean civilizations is of vital importance to understand the influences that affected Andean historical development. While the timing and intensity of flood related deposition and erosion events is manifest from geophysical analysis of deposition layers and erosion profiles, the soil transfer and deposition physics causing landscape change from such events remain elusive. To understand the physics of such events, Computational Fluid Dynamics (CFD) analysis is of use where a computational model composed of El Niño derived, high velocity flood water containing silt, gravel, rocks, soil particles form a non-Newtonian viscous slurry that then proceeds to erode landscape rain-saturated soil to settle and deposit slurry mixed with captured landscape soil sediments to form a deposition layer. The present chapter investigates phenomena of this type for the particular case where the formation of beach ridges from flood soil/slurry sediment deposition occurs as a result of interaction with ocean currents. While linear beach ridge structures (Figure 7.16) are well noted in the literature (Moseley et al. 1992a; Knighton 1998; Rogers et al. 2004; Sandweiss 1986; Sandweiss et al. 1996; Wells 1992, 1996), the underlying physics of their formation is basically a problem in fluid dynamics amenable to solution by use of computational fluid dynamics methodology. As landscape change brought about by ENSO events altered the agricultural base of a society through erosion/deposition events and caused damage to the marine resource base through disturbance of offshore fisheries and shellfish gathering beds, societal continuity and sustainability were adversely affected. These events are subsequently shown to influence and affect the sustainability of Peruvian north central coast (Norte Chico) societies in the Preceramic, Late Archaic period (2600–1800 BC).

While field system modifications and defensive technologies against flood events play a vital role in societal sustainability, in a worst case condition, flood damage can be irreversible and abandonment of preexisting field systems occurs leading to societal dispersal and termination. Changes derived from ENSO events affecting both the agricultural landscape and the ocean littoral affecting the marine resource base of a society are key elements to understand to interpret societal structural change events. The present discussion provides an example of the use of CFD methods to provide information as to modification of the agricultural and marine resource base of the Late Archaic society based at Caral centered in the Supe Valley of Peru due to multiple major ENSO events – such events present a case for the ultimate collapse and abandonment of Supe Valley and other Norte Chico sites in the ~1800 BC time frame.

Evolution of late archaic sites in the Peruvian Supe Valley

The Late Archaic Period of north central coast Peru witnessed increased El Niño ENSO events that transferred flood sediments from coastal valleys into ocean currents forming extensive beach ridges. Typical landscape sediment layers resulting from multiple deposition/erosion events, when datable, yield timing of ENSO events and intermediate stable weather patterns characterized by aeolian sand deposit layers. As a result of the formation of multiple barrier beach ridges formed from a sequence of ENSO flood events, geological history of coastal littoral zones reveal that river drainage to ocean currents was impeded, bay infilling and coastal marshes appeared behind the beach ridges, aeolian sand deposits infilled lands behind and in front of beach ridges and sediment deposits covered coastal agricultural lands compromising their productivity. Sand dune incursion from exposed beach flats induced by constant northwesterly winds penetrated inner valley margins – evidence of this transfer process in Late Archaic times is evident from datable sand layers noted in excavation test pits. Datable beach ridge formations and interior sand and flood debris deposition layers provide the basis for a series of ENSO events that played a role in the collapse of coastal and valley societies in the Late Archaic period as further discussion details.

Many sites within the Supe Valley, with its ceremonial center at Caral, were based upon trade of marine resources from coastal sites exchanged with agricultural products from valley interior sites (Shady 2000, 2001, 2004, 2007). Major ENSO – landscape disturbances with no possibility of return to previous norms for the agricultural and marine resource base compromising the Supe Valley society posed a possible reason for valley site abandonment past ~1800 BC and thus indicates that dynamic landscape change likely played a role in the collapse of Late Archaic Period sites in the Norte Chico region of Peru.

The Norte Chico region of Peru is characterized by many preceramic sites (Figure 7.1) with different existence dates (Figure 7.2); within the Supe Valley are many individual sites (Figures 7.3 and 7.4) characterized by complex social organization and urban centers with monumental architecture dominated by truncated pyramid structures of which Huaca Major is typical. (Figure 7.5).

The earliest preceramic societies of north central coast Peru developed a cooperative economic model based upon trade, irrigation agriculture and exploitation of marine resources to sustain large populations in the Late Archaic Period. The nearby sites associated with Caral in the Supe Valley form a collective integrated societal complex (Figure 7.4) participating in the exchange network. Coastal sites

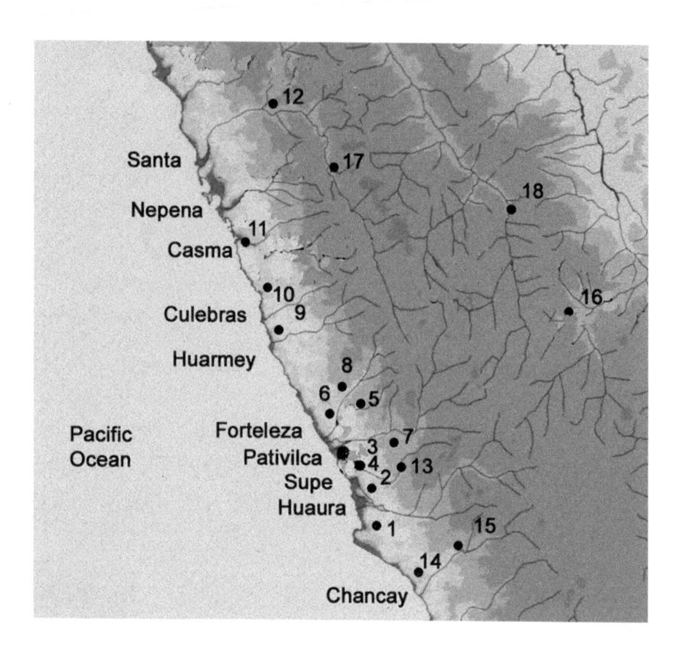

Figure 7.1 Peru map with preceramic sites: (1) Bandurria, (2) Vichama, (3) Áspero, (4) Upaca, (5) Pampa San Jose, (6) Caballete, (7) Vinto Alto, (8) Haricanga, (9) Galivantes, (10) Culebras, (11) Las Aldas, (12) La Galgada, (13) Caral, (14) Rio Seco, (15) Las Shicras, (16) Kotosh, (17) Huarico, (18) Piruru.

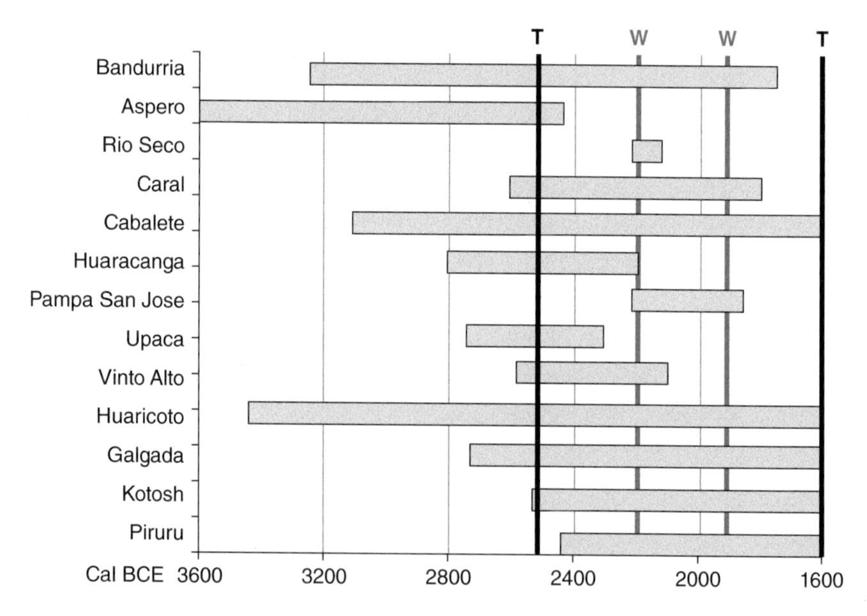

Figure 7.2 ^{14}C date range of major North Central Coast preceramic sites.

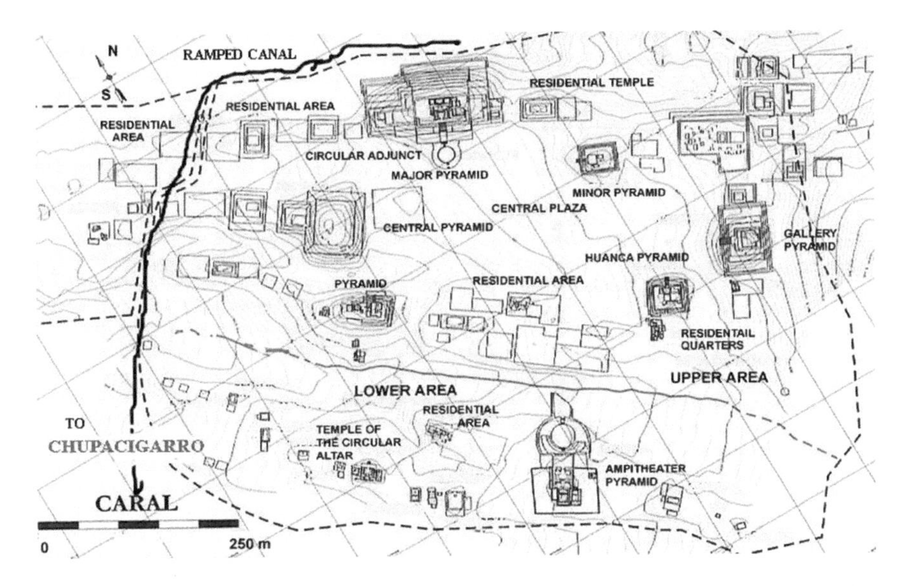

Figure 7.3 Caral site map.

Figure 7.4 Major Supe Valley sites: (1) Áspero, (2) El Molino, (3) Liman, (4) Era el Pando, (5) Pando, (6) Pueblo Nuevo, (7) Cerro Colorado, (8) Allpacoto, (9) Huacache, (10) Piedra Parada, (11) Lurihuasi, (12) Miraya, (13) Chupacigarro, (14) Caral, (15) Penico, (16) Cerro Blanco, (17) Capilla, (18) Jaliva.

Figure 7.5 The major pyramid (Huaca Major) shown in Figure 7.3.

exploiting marine resources traded with inland sites for agricultural and industrial crops (cotton for fishing nets and lines as well as gourds for net floats. In the Supe Valley alone, 18 sites evidenced the success of this economic exchange system (Shady 2000, 2001, 2007; Shady and Leyva 2003; Shady et al. 2004) over many centuries starting from early site origins.

The Late Archaic Period experienced major changes in environmental conditions brought about by Holocene sea level stabilization, Peru Current establishment and increased frequency of El Niño flood events (Sandweiss et al. 2001, 2009; Sandweiss and Quilter 2009). The flood events were the basis for sedimentary beach ridge deposits inducing river drainage blockage, creation of coastal marshlands, valley water table changes and aggraded sand sea formation behind and in front of beach ridges subject to aeolian transport – all of which influenced the agricultural and marine resource base of valley and coastal sites. Beach ridge dates and locations at given in Figure 7.6.

Strong coastal winds were the source of Aeolian sand dune transfer to interior Supe Valley farming areas from exposed beach flat area as noted by datable sand layers from excavation pits. Additional aeolian sand transfer to interior Supe Valley areas originated from vast sand seas in the Huara Valley south of the Supe

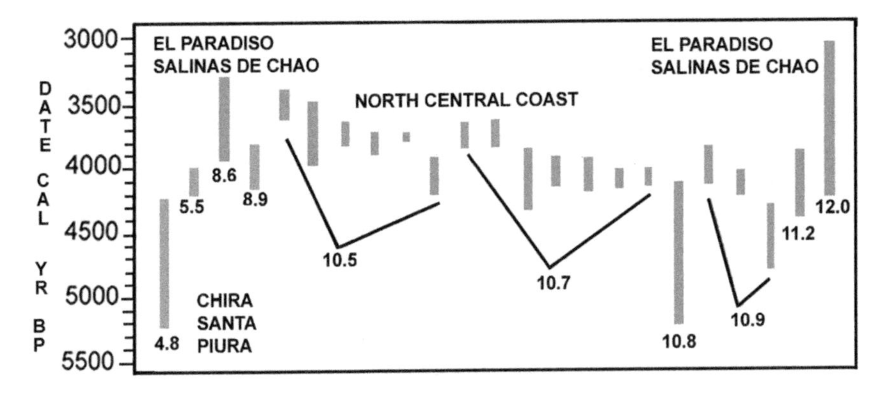

Figure 7.6 North-central coast beach ridge dates and locations.

Figure 7.7 Sand accumulations in the Supe Valley from aeolian transport across the Huara Valley.

Valley. Here strong northwesterly winds carry sand across the low mountains separating the two valleys to deposit on the southern slopes and valley margins of the Supe Valley (Figure 7.7). As the formation of a major beach ridge from a major ENSO event noted in the Late Archaic Period record starts a chain of events that threatens both the agricultural and marine resource base of both coastal and interior valley sites, the origins of beach ridge formation are next discussed in detail.

As flood-induced sediment transfer into ocean currents and the formation of beach ridge deposits are fluid mechanics phenomena, recourse to Computational Fluid, Dynamics (CFD) techniques provide insight into the sediment formation and deposition processes involved during major ENSO events. To substantiate CFD predictions of beach ridge formation, recourse to Google Earth satellite photographs of actual beach ridge formations in the Supe Valley area was made in areas under study to verify computer predictions. As beach ridge formation dates are contemporary with large flood sequences occurring in the late Archaic Period, CFD analysis provides the methodology to show both how ridges were formed and the geometry behind their shape. Creation of a new beach ridge on the north Peruvian coast within years after the massive 1982 El Niño flood event is

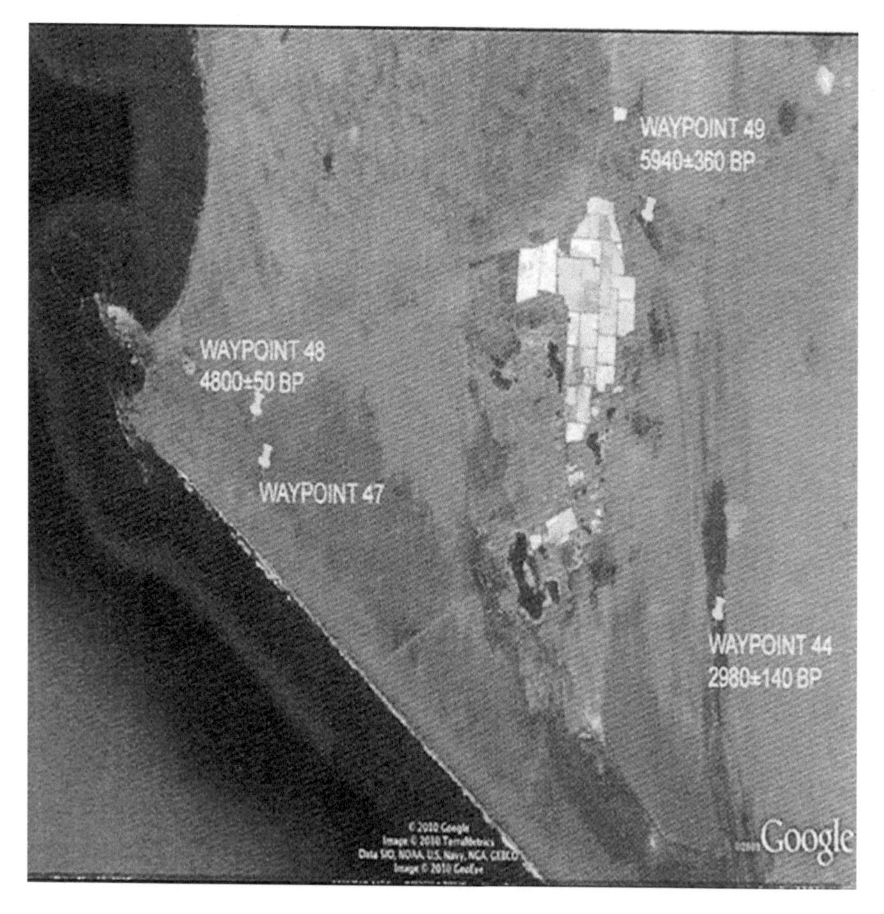

Figure 7.8 ^{14}C BP shell dates at Salinas de Huacho Bay waypoints interior to the present-day coastline indicating points close to the ancient shoreline at earlier times caused by sediment infilling over millennia (Daniel Sandweiss, personal communication) – note Figure 7.13 prediction of bay infilling for the same region.

well documented (Moseley et al. 1992b). This event verifies that El Niño floods were the origin of beach ridge formation over a relatively short time period given geomorphic conditions typical of the Peruvian north coast. Coastal progradation stemming from sediment accumulation behind barrier beach ridges is demonstrated by ^{14}C dating of mollusk species known to occupy shallow coastal waters but later found far inland. Figure 7.8 provides an example of shoreline change over time and sand infilling related to ENSO induced events as indicated by dates associated with test pits given in Figure 7.8.

The geomorphic landscape changes affecting the agricultural and marine resource base of Late Archaic societies played a catalytic role in the fate of many Late Archaic Period sites at ~1800 BC as subsequent discussion details. While research continues into the social, political and economic structure of Late Archaic Period sites and their response to climate change environmental stress to determine details of societal structure modification (Renfrew 1979; Paulson 1996; Moore 1988a; Manners et al. 2003; Contreras 2000; Bawden and Reycraft 2000, 2009), the present chapter focuses on the underlying fluid dynamics of beach ridge formation and the consequences of the induced geomorphic changes by flood and aeolian sand transfer events that affected the agricultural and marine resource base of Late Archaic Period sites. As such, major environmental changes are expected to be a significant contributory factor in the cultural trajectory of Norte Chico sites.

The Supe Valley Caral site

Following investigations by Lanning (1967), Moseley (1975), Moseley and Willey (1973), it was recognized that complex societies based on fishing and irrigation agriculture arose during the Late Archaic Period on the desert coast of Peru (Figures 7.1 and 7.4). There labor groups built monumental structures of increasing size and complexity indicating development of societal structural change and evolution of a managerial class to direct construction projects involving consensual communal labor participation and organization. The earliest Norte Chico region platform constructions contained restricted access rooms (Feldman 1985) indicating some degree of social differentiation (Moseley 1975, Moseley et al. 1992). Recent research by Shady (2000, 2001, 2004, 2007; Shady et al. 2002, 2004) in the Supe Valley and in adjacent valleys (Haas and Creamer 2004, 2006; Haas et al. 2009) have demonsrated that this early cultural florescence took place in other North Central Coast valleys and grew to a size and complexity not previously recognized by earlier researchers. Recent research indicates those Late Archaic Period temples of the North Central coast were abandoned by ~1800 cal BC (Shady 2001, 2004, 2007; Shady et al. 2003, 2014; Sandweiss et al. 2009; Barrera 2008) as indicated in Figure 7.2. The Norte Chico region was never again a center for cultural florescence although a small number of Formative and Initial Period agricultural sites in mid-Peruvian valleys temporarily reoccupied a few of previously abandoned sites (Chen 2018). A number of sites at the margins of the North Central Coast originated toward the end of Late Archaic Period (sites at El Paraíso in the Chillón Valley to the south and Las Salinas de Chao to the north) and survived hundreds of years well into the Formative/Initial Period as preceramic sites.

For the present analysis, discussion is focused on Caral in the Supe Valley. The site is located 182 km north of Lima and 24 km inland from the city of Haucho on the central Peruvian coast. Figure 7.3 indicates major building complexes at Caral. A foundation element for the concentration of Supe Valley sites was an abundance of water for agriculture to support the valley population.

Figure 7.9 Water pond penetration of the high Supe Valley groundwater level.

Since coastal rainfall is limited to a few centimeters per year, valley water for agriculture was mainly supplied by springs originating from valley bottomland areas. The water source for the valley springs originated from high altitude sierra lakes supplied by seasonal rainfall; high Supe Valley groundwater levels were maintained throughout the year by seepage water transferred from sierra lakes through geologic faults sourced from the sierra watershed augmented by aquifer seepage from the Supe River. Such systems are noted as *amunas* (sierra runoff capture-pits to augment the valley groundwater supply). An additional water source amplification of the valley groundwater level originated from canalized lagoons and water reservoirs that formed in low valley land areas that penetrated the groundwater level (Figures 7.9, 7.10 and 7.11.). These reservoirs served to maintain valley groundwater levels constant during seasonal water availability change and permit multiple-cropping to sustain valley population increases. As the near-surface water table varies about a meter from the wet to dry season, many

Figure 7.10 Canalized reservoir supplied by groundwater level penetration.

Figure 7.11 Large-scale reservoir supplied by artificially deepened bottom to penetrate the groundwater level.

springs were canalized to irrigate specific bottomland agricultural areas devoted to specialty crops including varieties of beans, squash and maize types as well as many fruit varieties (Shady 2000, 2001, 2004, 2007; Shady et al. 2004) and industrial crops such as cotton and gourds.

As the water table height was close to a permanent level, valley bottom field systems permitted multi-cropping to occur throughout the year. Other archaic Norte Chico valley sites (in the Nepeña valley in particular) were likewise associated with functional springs and large dams traversing upland valley gullies to trap rainfall runoff water to provide off-season irrigation water for crops supplied only by canals emanating from intermittent river water sources. Typical reservoirs shown in Figures 7.9, 7.10 and 7.11 are of ancient origin and are still in use today to support extensive valley agriculture.

Early canal development in the Supe Valley

Figure 7.12 indicates the Supe Valley bottomlands north of the Supe River and irrigated areas on the elevated plateau south of the Supe River.

Canals originating from springs and reservoirs located on valley bottomlands served agricultural field systems; a ramped canal constructed on the plateau sidewall (Figure 7.13) took water from a Supe River inlet (Figure 7.14) to the western part of Caral and further on to the inland site of Chupacigarro on a channel built on top of a mounded aqueduct structure. The careful surveying associated with the ramped portion of the canal (Figure 7.3) and its continuance over many kilometers to the Chupacigarro site is notable given its early provenience of ~2500 BC; this canal system is likely one of the oldest canals yet discovered in Peru and is notable for its size and complexity. Figure 7.15 indicates the totality of canal systems in the Caral plateau area.

Figure 7.15 indicates that the ramped canal plateau extension was the water source of canals on the Caral plateau as well as the water source for the Chupacigarro site (Figure 7.4). Figure 7.16 shows the canal path to the inland Chupacigarro site now largely buried by drifting sand. Figure 7.15 shows canal cross-section profiles located near the Supe Valley excavation house – although the source of water for the different canal cross sections remains obscure (but likely the water source is a branch canal from the plateau ramped canal extension (Figure 7.15), its continuance may point to an early transverse canal running laterally across Caral as indicated by the red line in Figure 7.15. Presently a. large erosion gully divides the east and west sides of Caral so traces of an

Figure 7.12 Valley bottomlands observed from the Caral center plateau.

Figure 7.13 Plateau sidewall holding a contour canal from the Supe Valley River inlet.

earlier transverse canal segment are no longer present to extend data taken from the cross-sectional profiles shown in Figure 7.15. Since no dating is currently available for the ramped canal and its Chupacigarro extension, the time sequence of canal construction components remain unknown.

The Supe Valley had (and still has) the advantage of a continuous water supply to source agriculture throughout the year while adjacent Late Archaic Forteleza, Pativilca and Huara valley sites had access only to intermittent rainy season runoff

Figure 7.14 Supe River inlet.

for canal irrigation. The agricultural base and water supply systems of sites in these adjacent valleys remain a topic of future research.

Late Archaic Period climate change evolution

Excavation data from Caral (Shady et al. 2002, 2003, 2004; Shady 2000, 2001, 2007) indicates that marine products transferred from the coast were plentiful

Figure 7.15 Major canal systems in the Caral area.

Figure 7.16 Sand incursion over the ramped canal extension to the Chupacigarro site.

at interior valley sites as evidenced by large marine shell deposits at locations within Caral while gourds and cotton grown at interior Supe valley sites were traded and used for fishing nets and lines at coastal sites indicating cooperative trade underwrote the valley's economic base. Irrigated farming products within coastal valleys included *guayaba* (Psidium guajava), *pacae* (Inga feuillei), *achira* (Canna edulis) as well as avocado, beans, squash, sweet potato, maize varieties and peanuts attesting to the wide variety of comestibles available for coastal trade.

Key to the importance of Caral and subsidiary Supe Valley sites is that they comprise the earliest New World example of an integrated valley economic unit deriving benefits from valley bottomland spring sourced canalized irrigation systems and plateau agricultural systems (Figure 7.3) supplied by the tramped canal. As such, the multiple Late Archaic Supe Valley inland and coastal sites initiated early complex civilization development in Peru that, were it not interrupted by large-scale ENSO environmental change effects in later phases of its existence at ~1800 BC would have accelerated added complexity at Norte Chico sites to follow.

Sea level stabilization, El Niño floods and beach ridges

Sea level stabilization between 6000 and 7000 cal BP set the stage for Late Archaic Period developments. The onset of El Niño rains about 5800 cal BP, after a mid-Holocene hiatus, had implications for the social processes that found expression in the temple centers of Supe and surrounding valleys (Sandweiss 1986; Sandweiss et al. 2001; Sandweiss, Shady-Solis et al. 2009). By ~1800 BC, preceramic North Central Coast sites were \ abandoned (Figure 7.2) suggesting a common influence of a large-scale environmental change affecting all areas of the Peruvian north central coast. Prior to ~1800 BC the agricultural and marine resource base apparently developed over time in a relatively stable climate period. Accretion processes influencing the geomorphology of the Peruvian north central coast littoral and in land valleys were determined by deposition of large sediment loads originating from El Niño flood events interacting with oceanic and wave-induced near-shoreline northward flowing currents together with major aeolian sand incursion events from beach flat area and wind-borne sand from the southern Huanca Valley. Apparently in late phases of Late Archaic Period sites around ~1800 BC significant changes in the geophysical environment brought about by major ENSO flood event and amplified aeolian sand transfer into interior valley lands challenged the continuity of the valley agricultural and marine resource base. Sediment transport and offshore sediment deposition patterns into the Pacific offshore seabed depend upon El Niño flood magnitude, rainfall duration and spatial distribution, valley landscape geometry, landscape soil types, sediment and slurry physical properties, seabed shelf angle, coastal uplift/subsidence and geometric details of river channels and watershed collection areas. Flood sediment load is influenced by earthquake activity that produces large quantities of loose surface material available for runoff transport. Large rainfall events cause changes in the equilibrium profiles of drainages affecting sedimentation and drainage patterns

(Moseley 1993, 1999; Moseley et al. 1992; Sandweiss 1986) that influence the amount of flood transported sediment and the formation of offshore the sediment deposits in form of beach ridges. Sediment transfer processes result not only d from El Niño flood events but also from rivers that flood during rainy seasons and carry sediment into ocean currents and/or deposit sediments behind beach ridges that serve as barriers to river drainage into ocean currents.

North of 9° S, the continental shelf widens abruptly and extensive beach ridge plains exist formed from sediment deposits that trail north from the largest rivers of Peru: the Santa, Piura, Chira and Tumbes rivers. Beach ridges formed from sediment deposits are present at Colán, where El Niño rains have eroded an uplifted marine terrace (Ortlieb et al. 1993; Rogers et al. 2004). All of these northern beach ridge plains consist of eight to nine separate ridges and each plain formed well after late Holocene sea level stabilization. To the south of 9° S latitude, where the continental shelf is narrower and the seabed angle steeper (Barrera 2008), Figure 7.6 shows dated beach ridge formations from 4.8° to 12.0° south latitude on coastal Peru. In Figure 7.6, bar heights represent one standard deviation from the central mean. Earliest beach ridge dates are from Chira, Piura and Santa coastlines while remaining dates are for beach ridges at Salinas de Huacho and El Paraíso and are shown with their latitude positions. Earliest beach ridges appear at far southern latitudes at 4.8° S with later ridges occurring to the north. This trend implies that early El Niño activity as a source of deposition material was prevalent past ~5500 BC (but not earlier) consistent with late Holocene sea level stabilization.

As the shoreline transgressed during post-glacial sea level rise, prograding beach ridge plains could not form. Subsequent sediment deposits from aeolian sand transfer, intermittent flood and oceanic current sources and silt entrained in farming drainage runoff over millennia promoted bay infilling to constitute the present-day shoreline (as a later discussion details). Only with the relatively stable sea level of the last 6000 years could beach ridge plains form and alter landscape geomorphology and littoral resource suites along the coast. Sea level stabilization is linked to climate change in the Pacific Basin with El Niño events starting after a hiatus of about three millennia (Sandweiss et al. 1996, 2001) during which northern Peru was characterized by annual warming and a fishery base absent of small schooling fish such as anchovy and sardine. With ecological changes accompanying the northward extension of the Peru Current at the start of the Late Archaic Period, small fish species began to dominate the fishery, calling for different capture strategies that required intensive production of cotton nets, lines and gourd floats consistent with cotton dominant among domesticated plant assemblages in all the Late Archaic coastal centers (Moseley 1993, 2008; Shady et al. 2003, 2004; Shady 2000, 2001, 2004, 2007). The founding dates of Áspero, Vichama and Bandurria on the shoreline of north central coast valleys preceded dates of the inland Late Preceramic temple centers (Feldman 1985; Moseley and Willey 1973) suggesting that access to irrigated agricultural lands was linked to intensified production of cotton and gourds and was key to the integrated economic model existing in the north central coast area.

Geophysical origins of beach ridge formation

Various types of beach ridges observed at different locations along the Peruvian coastline (Figures 7.17, 7.18 and 7.19) result from complex interaction of river-borne El Niño flood sediments with oceanic and wave action currents to produce beach ridge sequences.

Post sea level stabilization, sediments and aeolian sand transfer began to accumulate west of the Quaternary sea cliff that marks the back of the original Supe Bay and the smaller Albufero and Medio Mundo inlets to the south and the Paraiso Bay south of Huacho. In time, narrow beach ridges developed from a series of ENSO events inducing sand accumulation behind each ridge; this period was followed by stable progradation that largely buried each minor ridge by aeolian sand transfer and dune formation. At about ~1800 BC, a major ENSO event created the massive Medio Mundo beach ridge along ~114 km of coastline sealing off former fishing and shellfish gathering bays (Moseley 1992). This event created large scale sand flats that accumulated behind beach ridges and promoted large scale aeolian sand dune inundation of coastal plains and inland valley areas thus compromising a significant part of the agricultural base of Supe Valley society. Coastline geomorphic change affecting the marine resource base of Norte Chico societies by a combination of flood sediment accumulations amplified by aeolian sand transfer processes that infill previously established fishing and shellfish

Figure 7.17 Inland beach ridge deposit along the coastal Supe Valley.

Figure 7.18 Further beach ridge type. Degree of exposure and inland position depends upon time of ENSO event.

Figure 7.19 Supe Valley inland sediment deposit.

Figure 7.20 Sediment layer deposits from ENSO flood events at Aspero.

gathering areas. Although river-borne sediment constitutes a major source of sediment transport during flood events, additional opportunistic drainage paths develop to provide a multiplicity of drainage channels to coastal areas. Again, many north central coast preceramic sites terminated occupation by ~1800 BC suggesting that a major geophysical change over a wide coastal area compromised their economic base – thus the importance of understanding the geophysics of beach ridge formation and its consequence on the food base of archaic societies. The abandonment of major preceramic sites, with limited Formative Period reoccupation of a limited number of former sites, motivates the discussion to follow as to what these changes were to the economic base of archaic societies and their relation to beach ridge formation.

Tectonic coastal uplift rate differences from north to south latitudes influence beach ridge typology. A significant uplift rate may strand and separate sequent ridges from individual flood events while in the absence of uplift, single subsea ridges form and sediment deposits accreting landward from the ridge appear to strand a ridge above the land/water interface. Additionally, offshore ridge formation may alter the deposition history of subsequent ridges by altering seabed shape, river mouth geometry and river positional shifts. The relation between flood events, beach ridge formation and agricultural landscape change, as demonstrated by computer modeling results, is next detailed to illustrate their connectivity to provide the basis for conclusions supporting observed geomorphic transformations of Supe Valley agricultural and littoral areas in Late Archaic times.

To illustrate the fluid mechanics basis of ridge formation, a three-dimensional computer model of the Peruvian coast from Santa to Virú Valleys (Figure 7.21) was made using FLOW-3D software. The minor intermediate Chao Valley drainage between the Santa and Viru Valleys is omitted from the CFD model due to its small effect on beach ridge formation compared to the major Santa and Viru

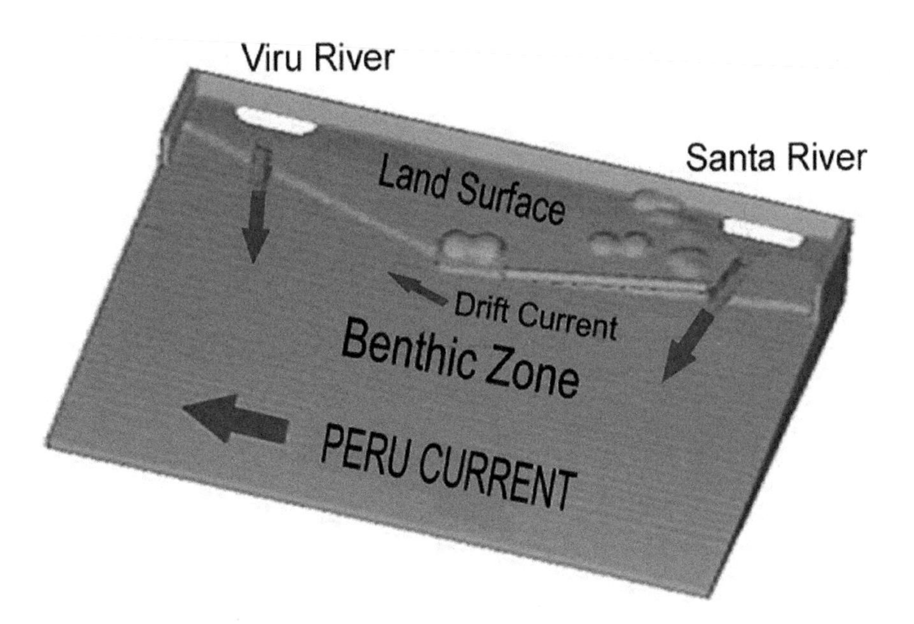

Figure 7.21 FLOW-3D computer model of the Santa-Virú coastal area (Santa River located south, Virú River located north).

drainages. The CFD model area was selected due the availability of data (Sandweiss 1986; Wells 1992, 1996) that allows for qualitative verification of modeling results compared to field observations. Shown in the Figure 7.21 CFD model are the sloped offshore seabed and land areas with southern Santa and northern Virú river Valleys providing El Niño flood sediment conduits to ocean current areas. The river basins downstream from sierra watershed collection zones funnel sediment-laden flood water into individual rivers. The composition of flood-transported material is drawn from the size, gradation, cohesivity and stratification of erodible bank sediments over the coast-to-mountain watershed area and surface material washed into the streambed. Profiling a selection of beach ridge cores provided indication of the percentage by weight of different size sediment particles.

High concentrations of large size particles (gravel and small boulders) in the wash load damp slurry turbulence, increase the apparent viscosity of the slurry flow and reduce settling slurry velocities enabling the transport of coarser grains and a larger bed-material load that would otherwise be the case. While only river-borne sediment transfer is considered for CFD model purposes, additional opportunistic drainage channels and ephemeral streams develop during flood events leading to sediment transfer to lower coastal areas that contribute additional sediment to ridge formation/accretion processes.

A two-fluid model represents sediment-ocean mixing interactions. Fluid 1 is ocean water characterized by kinematic viscosity ν_1 and density ρ_1; Fluid 2 is sediment-laden flood water slurry characterized by large values of kinematic viscosity ν_2 and density ρ_2 compared to ocean water. Here $\nu = \mu/\rho$ where μ is the absolute viscosity. Depending upon the sediment load of floodwater, ν_2/ν_1 can range from 1 (no river-borne flood sediment) to 10^3 (heavy river-borne sediment loads with very high absolute viscosity).

For purposes of demonstrating the fluid dynamics phenomena involved, a selection of input properties typical of observed sediment composition is given. Due to local variations in sediment composition, only a generic, illustrative solution is given typical of conditions in the area of study. Model river current velocity is set to a value to induce riverbed erosion and sediment transport mobility. For silt (< 0.001 mm diameter, ~15% by weight), sand to 0.1 mm diameter, ~30% by weight), 0.01 gravel (0.1 to 5 mm diameter, ~35% by weight) and an assortment of rock sizes from 5 mm to 100 mm diameter, ~20% by weight) composing the sediment solids, estimates of both properties and critical mobility stress are provided by Knighton (1998:107–112), Bain and Bonnington (1970) and Abulnagy (2002) to substantiate typical υ_2/υ_1 values used in the simulation. The northward offshore current velocity is set low to represent near cessation during El Niño events. A near shoreline, northward drift current is induced from the difference between incoming wave vector angle and a normal vector to the shoreline. The model sea level is set to the stabilized ~5000 BC level. By using model length scales (model area is ~1035 km^2) and velocity ranges approximating actual values, Reynolds and Froude numbers are duplicated and computer time is equal to real beach ridge formation time. Results provide a lower-bound time to determine the flood duration required to deposit beach ridges of known size and volume.

Note that the early Quaternary version of the shoreline consisted of deep river-downcut bays consistent with the lower sea level and that wave-cut bluffs (now inland from the present-day accreted shoreline) resulted from rising sea level; subsequent deposition of sediment over millennia from flood, river and aeolian sand transport that served to infill this landscape. The present computer model is representative of an intermediate stage in this landscape transition process. Results of the computations then represent the early stage of beach ridge formation; subsequent millennia of accretion and erosive effects then serve to represent current day shoreline patterns.

Results of Santa-Virú coastal zone simulation are next summarized. This zone has a well documented ridge sequence (Sandweiss et al. 1996; Wells 1992, 1996) and is used to test computer predictions with observed geomorphic features that have survived and persisted from early creation stages to present-day stages. Figure 7.21 shows the initial offshore sediment deposition density distribution from a single large El Niño flood pulse concentrated in the Santa-Virú Valley area; the density scale ranges from 1.94 slugs/ft^3 (1000 kg/m^3) for seawater to 5.40 slugs/ft^3 (2783 kg/m^3) for heavy flood slurry. For a lower limit slurry grain size of 0.01 mm, erosion, transport and entrainment is maintained from 0.001 to >10 m/sec velocity; for a grain size of 10–100 mm, erosion, transport and entrainment is maintained for >1.0 m/sec velocity (Knighton 1998:110). On this basis, a river near-surface velocity is assumed to be ~1.0 m/s for purposes of demonstrating the sediment-ocean current mixing and deposition process by CFD simulation. Since slurries of the type encountered in flood debris are highly non-Newtonian power law fluids, shear thickening with increasing shear rate is expected. Here a higher apparent viscosity applies and the kinematic viscosity proposed is used for purposes of the demonstration problem. Although the velocity profile for Newtonian, viscous channel flow can vary substantially from that determined for non-Newtonian slurries (Skelland 1967), for the present demonstration problem, the slurry is assumed to have a constant absolute viscosity rather than a shear rate dependent value. The offshore Peru Current velocity is assumed to be ~5 cm/s and coastal drift current is ~3–5 cm/s at the lower model boundary with local drift velocity values computed based upon the geometry of the coastline. Elapsed time is ~35 hr of continuous El Niño flood activity with $v_2/v_1 = 10^3$ indicating a high absolute viscosity, heavy sediment load carried by flood currents. Offshore average seabed slope is ~0.15° from horizontal for the Santa-Virú north central coast area; offshore seabed slopes after 3800 BC are estimated from Barrera (2008, Figure 11) and Pulgar-Vidal (2012) as well as values for oceanic current velocity. Figure 7.7 indicates formation of a long sediment deposit (beach ridge) nearly parallel to the coastline; the scale shades represent the mixture density of seawater and the initial flood slurry density emanating from the rivers. The northward current together with the near-shore drift current enhances the northward deposition of sediment. Predicted high density, larger size sediment compositions appear along the ridge length as observed from field studies (Sandweiss 1986) while transport and sorting of sediment fines continues in time from wave-induced drift currents. For steeper seabed angles, the formation of a close-in ridge deposit to the shoreline occurs – here a steeper seabed angle results in a greater sea depth closer into the shoreline and, as resistance to a sediment particle's forward motion is related to the hydrostatic pressure encountered on its front projected area (in addition to viscous drag and dynamic pressure effects), sediment deposition occurs more rapidly for steeper benthic seabed angles.

A velocity vector plot (Figure 7.22) shows that the out-rushing sediment stream creates a flow reversal pattern leading to deposits of lighter sediments back toward the shoreline where low drift velocities prevail. The "U-turn" of the sediment stream is consistent with the path of least resistance of small and intermediate size, low-inertia sediment particles that alter direction away from increased

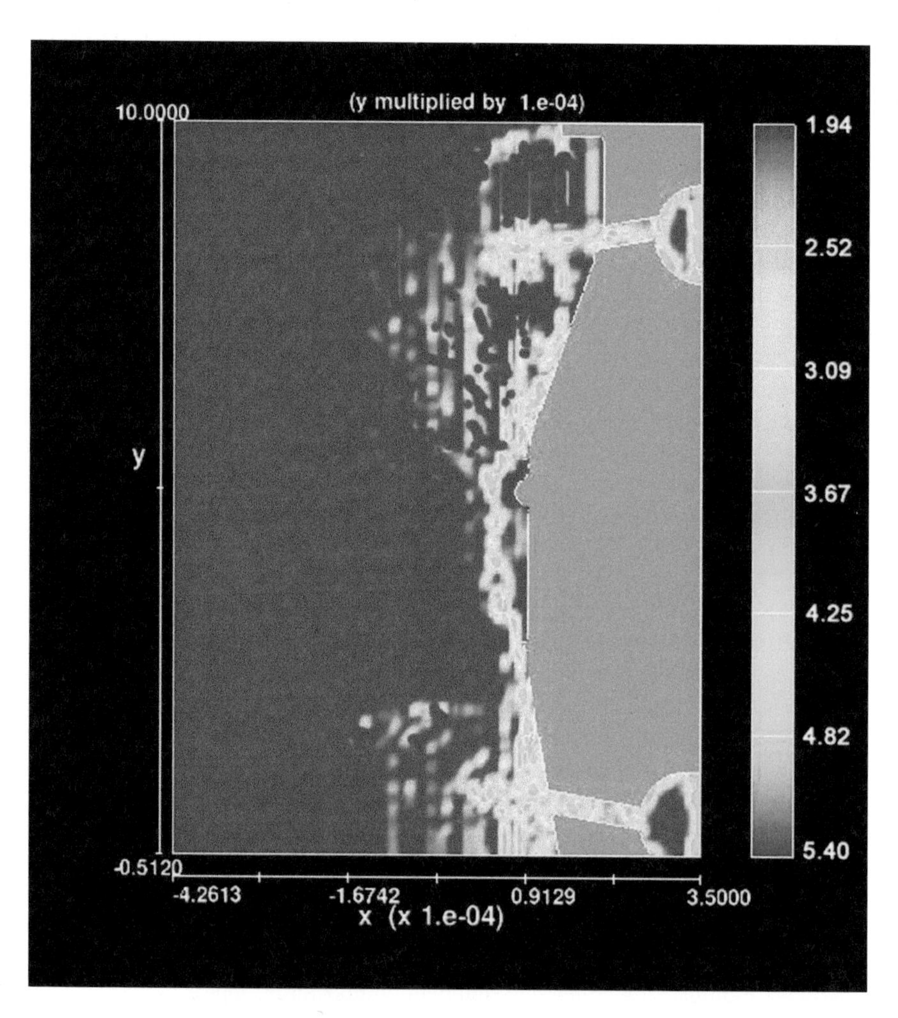

Figure 7.22 FLOW-3D oceanic sediment density distribution for an El Niño event the Santa-Virú coastal area of northern Peru; sediment density 5.43 slugs/ft³ (1937 kg/m³).

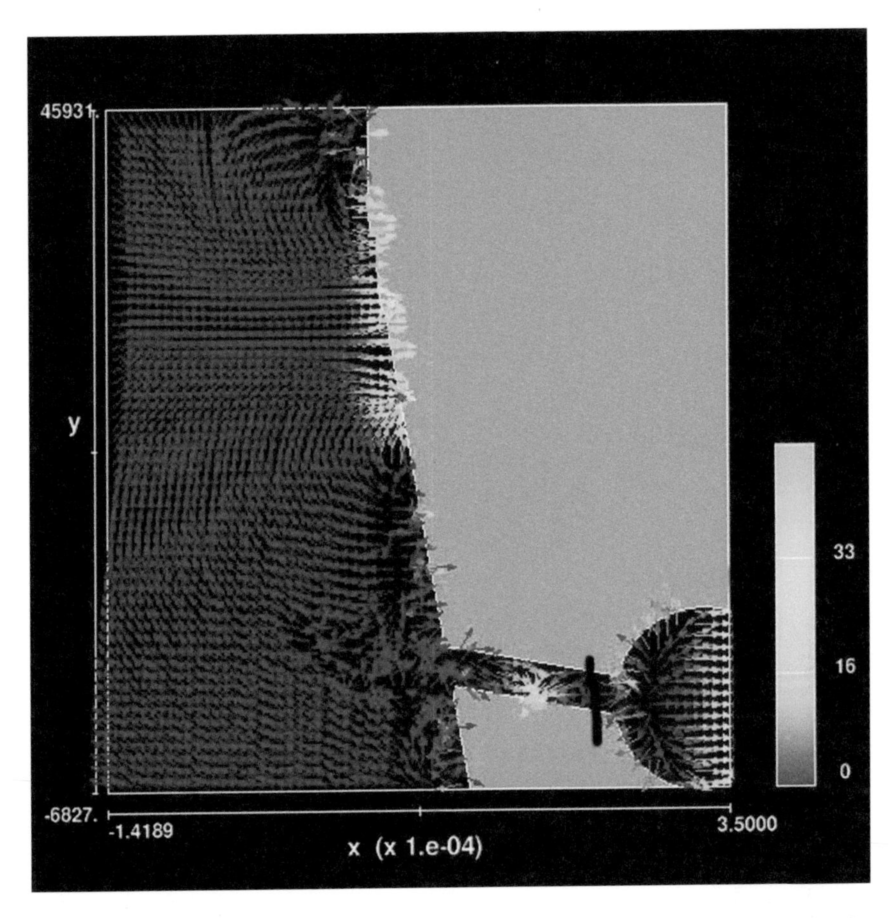

Figure 7.23 Velocity vector patterns for sediment flows into oceanic currents from the Santa River.

hydrostatic pressure resistance encountered further from the shoreline. As offshore Peru and drift current directions are not aligned, agitation and transport of deposited fines cause gradual northward ridge extension over time. Calculation results provide a qualitative relation between an El Niño flood event and details of ridge formation: floods with large sediment loads are capable of producing extensive near-shoreline subsea deposits whose size depends upon flood duration and amount/type of sediment available for transport by ocean and drift currents.

Once barrier ridge deposits are in place, accretion of sediment from later flood, river and aeolian sand transport events leads to gradual infilling of the shoreline littoral. That aeolian sand transfer was a continual threat to Caral's city environment was demonstrated by the multiple stone wall capture sand barriers constructed in open areas east Caral. Figure 7.26 details the effect of a subsequent

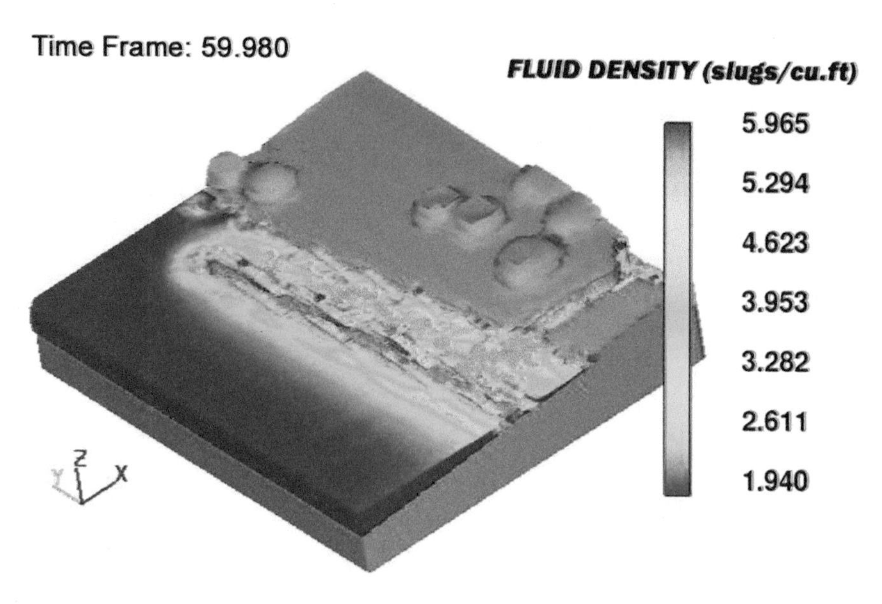

Figure 7.24 Computer results for sediment deposition behind an established ridge from a flood event. Sediment density is 5.43 slugs/ft³ (1937 kg/m³).

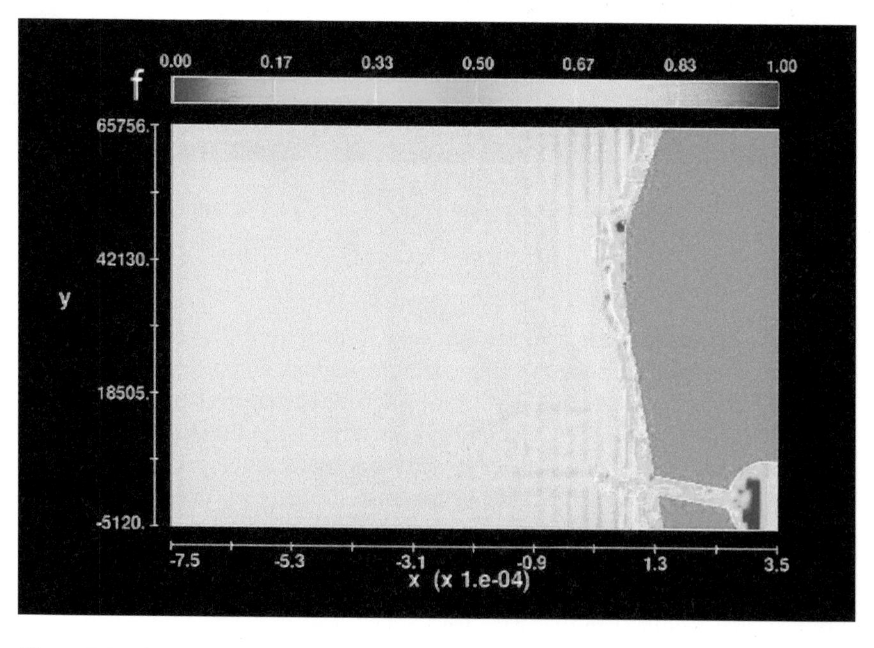

Figure 7.25 Computer results showing accumulated sediment deposits and fraction of fluid (f) for a flood event on the Santa-Virú coastline – light sediment loading case.

Figure 7.26 Sand infilling of the Salinas de Huacho Bay from Chancay River sand sources from a flood event.

Figure 7.27 Medio Mundo beach ridge south of the Supe Valley showing brackish marsh areas created by blocked drainage inland of the ridge.

flood event superimposing sediment deposits upon a previously established subsea ridge; results are specific to shallow seabed angles. Shown in Figure 7.25 are fluid fraction (f) results: f = 1.0 represents ocean water, f = 0.0 represents the sediment slurry stream and intermediate f values indicate mixture states. Sediment deposits on top of, and on each side of the original subsea ridge provide the barrier mechanism for accretion of inland sediment deposits. A trace ridge accumulates seaward of the main ridge and affects subsequent ridge development as the sea bottom geometry has been altered together with the river outlet orifice geometry. This result, when repeated for multiple flood events, qualitatively demonstrates how sediment accumulates inland behind (and to a lesser degree, in front of) a beach ridge by multiple flood, river, canal and aeolian sand/fines transport. Such changes in seabed geometry from multiple ENSO events alter the composition of fish species available for catch as well as the availability of shellfish types that can only exist at certain water depths. Current research (Chen 2018) demonstrates the time change in marine resource dietary patterns derived from beach ridge formations originating from weaker ENSO events occurring over a long time period on the Peruvian shoreline. Here the cumulative effect of ridge formations proceeding seaward on the seabed from sequential ENSO events gradually infill's bays and thus creates a changed marine environment accommodating different fish species and shellfish types over time. For a major ENSO event such as occurred in the Supe Valley area in late Archaic Period times, adaptation to a drastically changed marine environment coupled with induced changes in the valley agricultural environment proved difficult to overcome and eventually led to site abandonments. As each coastal valley has different soil types and landscape geometry, it is expected that the effect of a major ENSO event will have somewhat different effects on separate valleys with those valleys with a concentrated population showing the greatest response to a major threat to continuance to their food supply. As aeolian sand transfer across the northern mountain barrier and exposed coastal plains was a continuous problem to cover agricultural field systems, use of stone wall barriers (Figures 7.28, 7.29 and 7.30).

For Figure 7.10, the density of the slurry stream is 5.2X that of ocean water. Many beach ridges observed in the Santa-Virú sequence are composite indicating accumulations from multiple, closely spaced-in-time flood events – where aggraded sediment material separates ridges, a longer time interval had occurred between flood events; use of ^{14}C dating on organic material within the ridges then gives dating of major ENSO flood events. Ridges often contain mollusk shell material indicative of sea bed sediments being agitated and entrained during flood events. Once ridges are stranded on land by surrounding aggraded material, new flood events create new subsea deposits that accretes material to create the ridge sequence noted in the Santa-Virú coastal area where shallow seabed angles prevail. South of this zone where steeper seabed angles prevail, a composite, unitary ridge type is predicted. Situations occur where loose surface sediment is minimal and/or surface runoff water velocity too low to carry or erode sediment so that ridge formation is minimal or absent during lower magnitude flood events.

Figure 7.28 Several sequential stone wall barriers across open regions on the south side of the Caral plateau served to limit sand incursion on to the urban center of Caral.

Figure 7.29 Further use of sand barriers in the same area as that of Figure 7.28.

Figure 7.30 Last of he exposed aeolian sand barriers in the northern part of the Caral plateau.

Often trace sediment deposits occurs seaward and inward of the main ridge – this trace ridge alters the river outlet shape and influences (Figures 7.22 and 7.25) river discharge patterns as well as altering the local seabed slope from settled sediment – all of which alter conditions under which subsequent ridges form. In Figure 7.22 a trace subsea ridge is formed close to the shoreline from a flood event and the sediment load from a subsequent flood event was deflected outward as a result. A sequence of distinct ridges form that gradually and accrete sediment between them to form a ridge sequence typical of that observed in the Santa-Virú area. Based on observations of the Santa-Virú ridge sequence formed from datable multiple flood events, computer predictions provide the underlying fluid mechanics mechanism to explain ridge formation and their observed orientation, shape, width and composite nature.

Figure 7.2 represents a case where seabed slope in the Supe Valley area from Huara to Forteleza Valleys is steeper than that for Santa-Virú calculations. Results indicate ridge deposition close to the coast owing to the higher hydrostatic pressure close into the shoreline encountered by low-mass, low-inertia sediment particles; here higher pressure drag resistance leads to rapid particle settling closer into the shoreline. Subsequent flood events combine to produce a unitary, composite, multi-layered ridge which is continually reworked and redistributed as currents shift deposits and extend the deposition length.

When a mega El Niño flood simultaneously affects multiple river valleys with a steep offshore seabed slope, sediment fields coalesce to form a massive unitary ridge spanning the coastline typical of the ~100 km long Medio Mundo ridge observed to span the littoral of five north central coast river valleys. A result of prograding processes behind ridge barriers is formation of brackish water lagoons and marshes such as observed at Bandurria, El Paraíso and Albufero represent the physical reality of the computer prediction of the Medio Mundo ridge creating an inland marsh region; Figures 7.24 and 7.26 show ridge formation in a beach area south of the Supe Valley where aeolian sand deposition and later flood sediment deposition behind the original beach ridge gradually infill this area (Figure 7.27 shows an aerial photograph of the reality of this type of event).

To date, direct dating of the full extent of the Medio Mundo ridge is not available. However, an estimate of its formation date can be made by examining the broader history of beach ridges along the northern coast of Peru (Figure 7.6). The Medio Mundo beach ridge could not have been deposited prior to sea level stabilization, so it is younger than ~6000 cal BP. Given ridge-forming processes identified in the region, El Niño floods were active for Medio Mundo to form – this provides a maximum limiting date of 5800 cal BP. Further, rains associated with El Niño events are generally attenuated to the south. Given these formation date limits, the northernmost dated beach ridge plain (Chira and Colán) began forming ridges earlier than ridge plains further south. Available dates place the origin these ridges between about 5000 and 5200 cal BP (Ortlieb et al. 2003; Richardson 1983). To the south, the earliest Piura date is around 4100 cal BP (Ortleib et al. 1993) and the earliest Santa ridge date is around 4000 cal BP. Following this trend, the Medio Mundo ridge should date between ~3900 and 3700 cal BP. This time span overlaps the latest dates for most of the North Central Coast Late Archaic centers and exists in a time frame necessary to influence the marine resource base of Norte Chico societies.

A further computer result relates to the Salinas de Huacho area where vast sand accumulation infilled bays. During El Niño events occurring in southern reaches of the north central coast, flood sediment is mainly composed of sand transferred predominately from the Chancay and adjacent southern rivers valleys – a calculation of sand-rich sediment emanating from the Chancay River (slurry density ~2.34 slugs/ft^3 = 1206 kg/m^3) indicates sand transfer to the Salinas de Huacho area ~25 km north of the Chancay River. Multiple flood events would continue the infilling processes to create the observed vast beach flat area. Again, individual velocity vector patterns according to geophysical values (Figure 7.23) prevail to create different types of near-shore deposits. ^{14}C dates of shallow-water mollusks that lived 0.5–1.0 m below shoreline sands indicate the 4000–4500 BC shoreline was about 3–4 km from the present-day shoreline (Figure 7.27) indicating large scale sediment accretion continuously altered both the marine and inland farming area resource base through numerous flood and sand transfer processes. Similar shoreline and inland infilling processes behind the Medio Mundo ridge characterizes north central coastal valleys bounded by the ridge.

A further example illustrates the ridge formation process in the presence of the irregular coastline of the Sanú Peninsula which forms the southern boundary of the bay on which Bandurria and Áspero are located (Figure 7.1). The far offshore Peru Current velocity is ~4 cm/s while shoreline drift currents are ~3–7 cm/s but vary northward due to coastal geometry effects. Figure 7.27 indicates the coastal current caused shifting and deposit of sediment creating a curvilinear bay ridge and an inland marsh area as sediment drainage was blocked by the ridge. Figure 7.31 shows the (dot) trajectories of light sediment particles from flood runoff entering the coastal current from southern reaches. The sediment trail loops around the Sanú promontory to deposit sediments along the shoreline adding to the expansion of the original pre-ENSO event shoreline area. This process explains the date sequence shown in Figure 7.27. The northward dot sequence shows vortical currents depositing sediment to form a further coastal shoreline extension consistent with Figure 7.27 aerial photo imagery. CFD results indicate the qualitative nature of the prograding process; quantitative determination involves detailed knowledge of rainfall intensity and duration, geographic extent, event time duration and the amount/type of surface material available for transport by the eroding action of floods.

Sediment transport from upper-valley areas from flood events, river and aeolian transport gradually shifted sediment to coastal areas to accelerate infilling behind the Medio Mundo beach ridge extending from the Huara to the Fortelaza Valley. Based on survey of the lower Supe Valley, ~3 to 5 km of accreted sand, fines and clays deposited to a depth of 3 to 5m over the Holocene beach littoral inland of the present shoreline since El Niño floods began at the time of sea level stabilization. The sand seas in the Huara area south of the Supe Valley subject to strong onshore winds sourced aeolian transport of vast quantities of sand over the southernmost mountain chain that bounded the Supe Valley to inundate inland valley farming

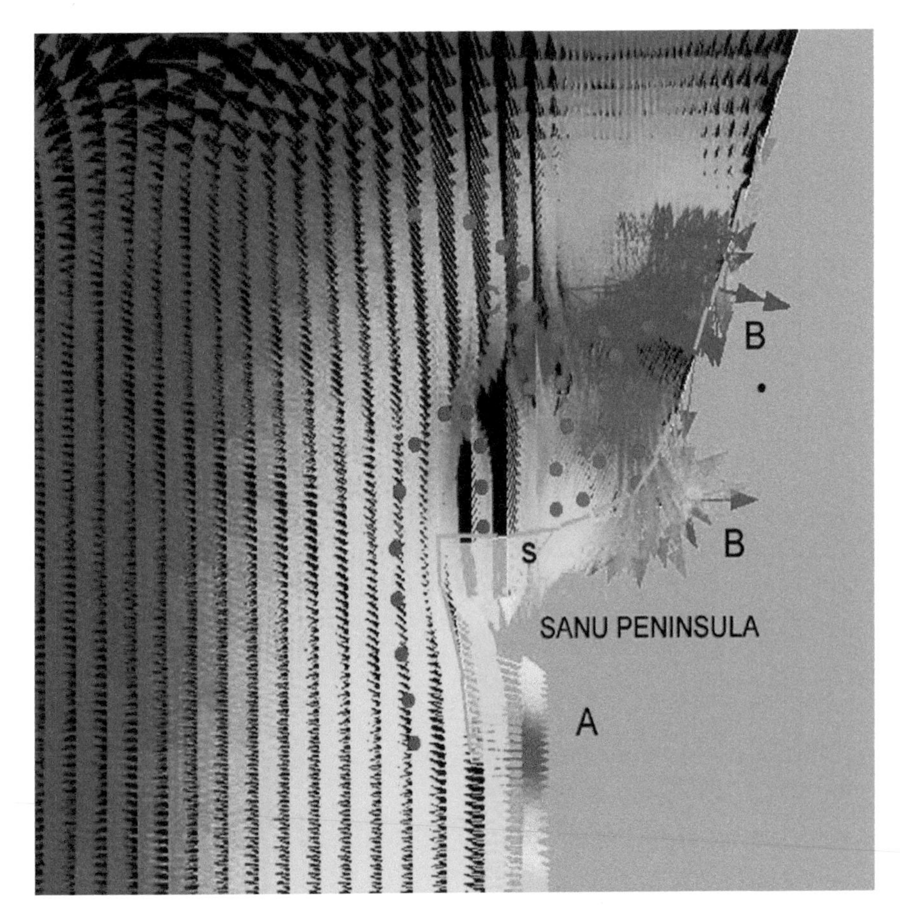

Figure 7.31 Velocity path plot of ocean current influenced by coastal shape that influence flood sediment deposition.

areas. Sand accumulations appear on north and south sides of the Supe Valley as the small, intermittent discharge Supe River presented no barrier to cross-valley aeolian sand transport. Buried sand layers covered over by later farming surfaces are present throughout Supe Valley profiles (Figure 7.20) indicating continuous aeolian sand and flood sediment transfer events over millennia. As a consequence of infilled bays and lagoon formation landward of ridges, coastal lagoons dominated by reeds were created under brackish water conditions; this environment exists in lower valley plains landward of Áspero.

Beach ridge formation

Figures 7.22 and 7.25 show the beach ridge formation from a single large scale El Niño event to produce the Medio Mundo ridge. This large scale event (or closely spaced in time events) sealed off the bay at Supe, the shallower Albufero and Medio Mundo inlets and the Pariso bay below Huacho and to the South created the Salinas de Huacho sand flats. To the north bays and inlets through Bermejo were closed off. With this ridge in place, northward ocean currents narrowed the width of the ridge in time and deposited material westward of the ridge to form a new beach area. In time, another El Niño event followed with river borne material cutting through the earlier ridge and depositing new material further westward of the earlier ridge on the beach deposited by the prior El Niño deposit. This process leads to sequentially spaced ridges propagating westward into the ocean – due to northwesterly winds carrying sand, the land between ridges gradually are buried leading to beach areas shown in Figure 7.27. Sequences of ridges formed in this manner are clearly seen in the infilled Supe bay as well as in the Santa Valley areas among others. With exposed beach flats covered with aeolian sand plus sand transport over the northern mountains of the Supe Valley, large sand dunes formed in the lower Supe Valley limiting agricultural land areas, as indicated in Figure 7.22. The high water table in the Supe Valley with moist soils provides for increased erosion transport during El Niño flood events further reducing valley bottom arable land – this in combination with sand incursion and loss of marine resources as a result of a sequence of major El Niño events clearly limited Caral's survival. Limited agriculture on the plateau adjacent to Caral provided the only option left to provide sustain the small population remnant left in Caral as Figure 7.19 indicates.

Groundwater amplification processes affecting Supe Valley agriculture

Limitations on groundwater drainage due to hydraulic conductivity resistance from accreted sediment and clays behind beach ridge affected areas led to gradual elevation of the up-valley groundwater profile causing numerous springs to appear in upper reaches of the Supe Valley. Since lower valley, near coastal delta areas originally comprised most of the agricultural lands, farming land loss due to sand accumulation overlays and deposited flood sediments consisting of eroded sierra gravels and stones gradually led to agriculture being transferred further inland to narrow bottomland and plateau locations nearer Caral (Figure 7.4). In near coastal areas, sediment buildup caused the water table to appear lower with respect to the ground surface limiting spring formation. To provide surface water to coastal field systems, river or spring flow would have to be channeled from far upriver locations to achieve elevation over coastal plains. The rough mountain corridor topography incised by erosion gulleys and sand deposits covering Supe Valley margins prevented long canals originating from valley neck areas to be extended along valley sidewalls to provide water to lower elevation lands.

Reconnaissance of the southern Supe Valley mountain corridor areas yielded no trace of long, high elevation canals. As a result of flood sediment accretion over coastal farming zones and sediment infilling of coastal zones behind coastal beach ridges forming the Medio Mundo ridge, only narrow, valley bottomland farming areas irrigated by spring-sourced, short canals remained to replace extensive coastal zone agricultural areas. As coastal rainfall is on the order of a few centimeters per year producing an intermittent Supe River flow, springs resulting from inland groundwater elevation and sierra *amuna* sources supported valley agriculture throughout the year albeit in narrow inland valley bottom areas. As testament to the high volume of groundwater underlying the Supe Valley, a drainage channel adjacent to the access road to Caral from the Pan-American Highway flows continuously throughout the year with a deep, high velocity drainage flow indicating that water abundance, rather than shortage, supports multi-cropping throughout the year now as in the past. This drainage channel, presumed to have an ancient counterpart, was vital to drain fields of excess irrigation water; this in turn limited the salt deposits in agricultural fields that, over time, would limit field system productivity.

Survey of the Supe River choke point revealed Canal A with a river inlet (Figure 7.1) to support the ramped canal to Chuapacigarro. Due to riverbed meander And braiding characteristics of low slope rivers, rainy season canalized flow to valley bottomland farm area proved unreliable as the river channel frequently deviated from established canal inlets. As springs developed in valley bottomland areas distant from the coastline from *amuna* based groundwater elevation as shoreline prograding progressed, the changeover to spring-supplied canals provided reliable, year-round irrigation systems that additionally maintained the high valley groundwater level.

The long, low slope, ramped canal (whose entrance and path is now obscured by by dense plant growth (Figures 7.12 and 7.13) supported the spring-sourced canal B-C to provide water to plateau field areas. The embankment ramped canal brought water to Caral and Chupacigarro and was the remedy to add plateau agricultural land area to supplement limited narrow inland bottomland areas subject to flood erosion and/or coverage by flood sediment. As an anecdotal note, one local farmer employing valley bottomland for agriculture reported that as a result of the latest 1989 El Niño event, 150 hectares of his farmland were washed away – this observation also held in ancient times so that devastating floods reduced valley bottomland agriculture irreversibly requiring new lands to be developed on the Caral plateau (Figure 7.15) to avoid land loss.

In the Supe Valley, the cross-section of Canal E (Figure 7.15) located near the Caral excavation house indicates that a canal provided water integral to Caral city precincts although its path remains unexcavated and perhaps no longer available. Canals E, G and C served the site of Chupacigarro; canal D serves modern field systems and an early Canal B-C (Figure 7.1) provided irrigation water from a spring at the origin point of the ramped embankment canal.

Sand incursions affecting the Supe Valley agricultural base

With the formation of the Medio Mundo beach ridge, beaches and sand flats formed near the Supe Valley mouth near the modern shoreline. Runoff sediments from later ENSO events together with later occupation irrigation sediment transfer transformed former Medio Mundo and Paradiso embayments into sand flats and lagoon wetlands. Constant northwest directed winds transferred sands into valley margins from the exposed coastal sand flats and buried the windward face of the low mountains separating the north side of the Huara Valley from the south side of the Supe Valley resulting in huge sand dunes accumulating within the Supe Valley burying lower valley agricultural fields and former drainage paths (Figure 7.7). Although defensive measures were taken by means of sand barriers to block sand transfer from the adjacent southern valley, the volume of aeolian sand proved insurmountable. Remnants of archaic dunes added to by years of intravalley sand transfer persist today on both sides of the Supe River.

This transfer process continues to the present day and it is surmised that in Late Archaic times sand incursion extended up to the site of Caral compromising agricultural fields as far down river to the coast. This sand incursion deposited a ~3 cm sand layer atop final archaic occupational deposits in northeastern areas of the site. In certain areas the sands are stratigraphicly overlaid by early ceramic bearing midden dating to the Initial Period (1600–800 BC). Excavation pits reveal this sand layer is now overlain by soil deposits from extensive modern-day farming. Eventually, the Rio Supe carved a path to the sea and agriculture was returned to near river margins in present-day times but both sides of the valley still show remains of the early inundation of archaic sand seas. As sand inundation in the post-Medio Mundo epoch compromised valley agricultural lands and further compromised the marine resource base, an argument for the demise of the Supe Valley society may be proposed resulting from the creation of the Medio Mundo beach ridge in Late Archaic period times.

Termination of a Late Archaic Period site

Based upon the estimated formation date of the Medio Mundo ridge, a major El Niño event (or sequence of events) started a progression of geophysical changes that compromised both the Supe Valley agricultural and coastal marine resource base through bay sediment infilling. Additionally, flood erosion of fertile valley bottomland topsoil reduced available farming land areas and led to slurry deposition areas whose soil composition was unsuitable for agriculture. Despite use of drainage channels, groundwater levels remained too high after a major flood event to enable crop growth for an extended time period. ENSO flood activity combined with aeolian sand incursion into wide expanses of lower valley bottomlands thus led to abandonment of near coastal agricultural lands and use of narrow upper valley land that limited agricultural production to lower levels necessary to maintain high population levels.

Effects of the Medio Mundo beach ridge barrier thus reduced the agricultural and marine resources to the extent that large valley populations dependent upon the preexisting food resource base experienced a rapid collapse of the coastal-inland trade network established during earlier periods absent of major flood events. As further research may show, the large number of sites in the upper valley region (Figure 7.4) may have been an attempt to redistribute valley population around transitory functioning land and water sources as water and farmland availability for agriculture rapidly decreased in the broad lower valley over a short time period.

Ancient civilizations of Peru experienced episodic climate change patterns inducing floods, drought, geomorphic landscape change through inflation/deflation cycles that affected their cities and agricultural base. Despite these challenges, several of these societies demonstrated continuity throughout time by relocating population to areas with more land and water resources and/or instituting large scale inter- and intravalley water transfer projects. When such changes provided a form of societal continuity, several societies vanished from the archaeological record when their agricultural systems did not permit modification or alternate water sources were not available for transfer to their agricultural fields. While some societies managed to overcome environmental challenges by technological innovation applied to modify agricultural landscapes to maintain food productivity, other societies unable to implement modifications due to irreversible damage done to agricultural and marine resource areas and unable to shift the resource base, disappear from the archaeological record. For Late Archaic Period Norte Chico societies, landscape change induced by establishment of large beach ridges from a major ENSO flood event severely altered the agricultural and marine resource base to the extent that the intravalley trade network no longer functioned. As food resources diminished from reduced farming area and decreased marine resource availability changes in social structure to accommodate a population out of balance with the available food supply was a likely source of population decline or resettlement to life sustaining areas although details of this transformation of Caral society structure is now only in its early stages.

El Niño flood deposition events formed subsea ridges that initiated progradation processes infilling coastal zones trapped by the ridges. Comparison of duration dates of preceramic coastal societies (Figure 7.1) to beach ridge dates (Figure 7.2) indicates an overlap period accompanied by intense El Niño activity. By ~1800 BC, most local sites were depopulated (Figure 7.2) indicating a likely common cause for abandonment of the central north coast area. While not associating coincidence with causality, effects of gradual loss of the marine and agricultural base of north central coast societies are likely contributing factors in the abandonment of major sites. Evidence of flood events from depositional silt layers at Áspero (contemporary with Caral) verifies major flood events in the Late Archaic Period. Within the Supe Valley, excavation profiles reveal sedimentary layers indispersed with sand layers indicative of major erosion and deposition events from ENSO events. Recent research (Chen 2018) in the Norte Chico region related to subsistence changes in the Preceramic and Initial periods indicates that the presence of littoral changes brought about by ENSO events caused a shift in dietary composition of site inhabitants. Bay infilling in the Huaca Negra area (close to the present-day town of Barranca) apparently was slower due to different landscape and valley geophysical conditions than in bay areas to the north permitting longer term shellfish gathering in shallow bay areas as well as a shift in netting small schooling fish to catching fish species found in deeper offshore waters. Thus gradual changes in landscape and offshore bed geometry particular to different coastal valleys from sequential ENSO events permit, in certain cases, continued but limited availability of a modified food resource base sufficient to reinstitute previous food supply norms. Only when such transformations are possible, can societal continuance occur – but in a limited condition compared to previous norms.

The T-T and W-W date band notations (Figure 7.2) refer to a climate anomaly period (Weiss 2000) influencing worldwide oceanic current shifts with probable influence on El Niño frequency and intensity. As these changes occurred toward the end of the Late Archaic Period, some effect on study areas may be inferred but further research is needed to track their specificity to Pacific coastal areas.

Conclusions

Aeolian sand transfer from the Huara Valley to Supe Valley margins and coastal areas provided overlying sediment to coastal and valley agricultural areas and continues to the present day with dune formation compromising present-day agricultural areas. Figure 7.32 summarizes these processes: localized El Niño flood drainage paths (1) combined with river fluvial sediment from Fortelaza, Patavilca, Supe and Huara rivers delivered sediment to coastal areas; (2) flood sediment coalesced into the Medio Mundo ridge with ridge geometry determined

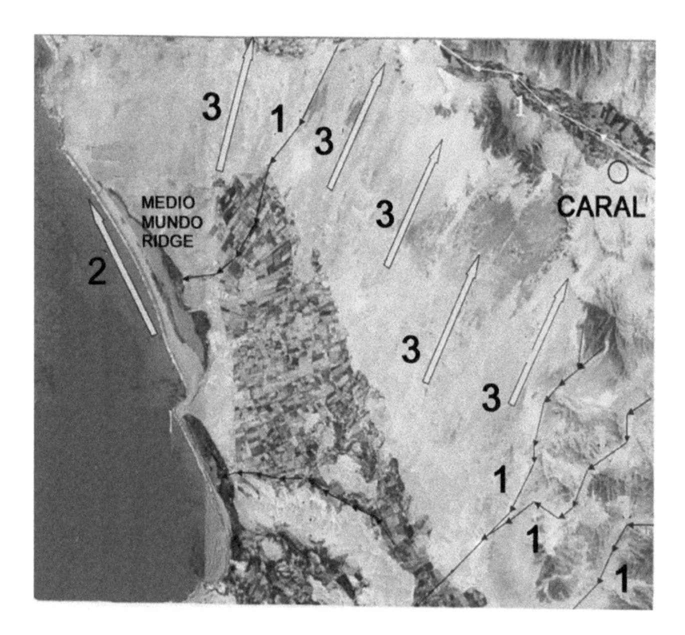

Figure 7.32 Agricultural collapse origins: drainage paths (1) carry flood sediment to coastal areas forming the Medio Mundo Ridge; (2) coastal currents extend the initial ridge length over time and alter coastal geomorphy; (3) accumulated sands from prograded beach areas behind the ridge are transported by aeolian sand transfer processes into the Supe Valley covering coastal agricultural lands.

by sequential sediment transport amounts and the oceanic/drift current magnitude; (3) prograded sand areas trapped behind the Medio Mundo ridge subject to onshore winds further compromised inland agricultural land areas through inland dune transport. Remnant sand accumulations on the Supe Valley northern side limited river flow reducing the agricultural potential of the coastal delta area. Increased hydraulic resistance to groundwater drainage from sediment deposits and clay formation in saturated coastal soils backed up groundwater height and led to increased numbers of springs appearing in valley bottom areas far inland from the coast. Since inland valley bottomlands and sierra foothill areas were the source of most springs, agriculture was limited to up-valley narrow bottomlands and limited ramped canal plateau areas (Figure 7.19) as a result of geophysical landscape changes and sand incursion from exposed beach flats. As marine resource extraction was the purview of coastal communities and inland sites that supported farming, reciprocal product trade diminished between communities as a result of a major ENSO event that altered previous trade basis norms. To sustain large-scale agriculture in the gradually infilling coastal environment, river or spring water would have to be channeled onto land surfaces lower than the riverbed choke point – this would require canal inlets originating far upriver to achieve elevation over the near coastal land surface and canal construction on the steep and erosion – incised mountainside corridor topography on the upper reaches of the Supe Valley to revitalize lower valley agricultural lands.

Sand accumulation on mountain slopes limited ambitions for canal extension to lower valley areas. Extensive surveys of the southern mountain corridor flanks of the Supe Valley revealed no high level canal construction. The coastal area was progressively removed from agricultural exploitation and could not be irrigated by canalized river sources. Since coastal areas decreased in agricultural productivity over time from erosion of topsoil, overlays of flood sediments and aeolian sand deposition, transfer of agriculture to narrow valley-bottom farming areas in inland valley locations could not support a large population. Disruption of the marine resource base from bay infilling and sediment deposits over mollusk shell beds accompanied the loss of farmlands and the viability of the economic model upon which Supe Valley society was based. A later Formative/Initial Period occupation occurred at some valley sites with limited construction overlay over earlier temple sites – sand layers between construction phases attest to massive sand incursions during the hiatus period.

Results presented relate to investigations conducted in the Supe Valley and reconnaissance of coastal areas of the adjacent Fortaleza, Patavilca and Huara valleys. These valleys contain Late Archaic Period sites and yield terminal ^{14}C dates for these sites consistent with those sites in the Supe Valley (Haas and Creamer 2004, 2006; Haas et al. 2009). The present analysis extends investigations (Sandweiss et al. 2009) detailing reasons for collapse of the agricultural and marine resource base of the Supe Valley society in Late Archaic Period times. The geomorphic changes thus far described are likely causative elements contributing to societal disruption from previously established societal norms established over centuries based upon stable environmental conditions.

Several climate-driven events that altered ecological conditions beyond recovery have influenced Andean prehistory. Notable are the collapse/transformation of the southern Moche V society in the 6th to 7th centuries AD by cycles of high rainfall, drought and sand incursion (Moseley 1993) in their Moche-Chicama Valley homeland; the collapse of the altiplano Tiwanaku society in the 12th century AD due to extended drought (Ortloff and Kolata 1993); the collapse of the Lambeyeque Valley Sican and Wari societies in the 12th century AD due to extended drought; the collapse of the Chimú intravalley (Moche Valley) canal systems in the 11th century AD (Ortloff 2009) and El Niño flood catastrophes experienced by the Chirabaya (Reycraft 2000) in far south Peru. To this list, Caral is a further example based on the rapid decline of the agricultural and marine resource base which exerted a profound influence on the continuance of the economic model of Supe Valley sites. Given the abandonment of major Late Archaic sites at ~1800 BC in nearby valleys, the same environmental change based upon formation of the Medio Mundo beach ridge likely was responsible for the similar fate experienced by the Supe Valley society.

II

Hydraulic engineering in the ancient Mediterranean world

8 Hydraulic engineering at 100 BC–AD 300 Nabataean Petra (Jordan)

Introduction

Use of hydraulic analysis methodologies involving Computational Fluid Dynamics and modern hydraulic engineering theory is applied to discover the hydraulic science knowledge base of several ancient Mediterranean societies. Analysis examples include analysis of the Roman Pont du Garde aqueduct/*castellum* (France), Nabataean Petra's (Jordan) urban water supply system and the Minoan Knossos pipeline water supply system (Crete). Water system design and use at these sites demonstrates knowledge of modern engineering principles only developed later in 18th- to 19th-century western hydraulic science. The use of port openings on pipeline top surfaces to allow air entry to eliminate partial vacuum regions associated with pipeline slope change water transfer is but one of many indications of sophisticated hydraulic engineering used by Petra's water engineers. The Roman Pont du Garde *castellum* indicates use of pipeline critical water flow technology governing flows from its 13 *castellum* water distribution pipelines – this technology serves to match the *castellum* 13 pipeline outlet flow rate *exactly* to the 40,000 m^3/day input aqueduct flow rate. This design feature explains the reason for the zero slope aqueduct terminal section requirement for water flows to the *castellum* and provides new insight to Roman water engineering. Internal pipeline shaping in the water supply system at Knossos demonstrates a methodology to produce a potable water supply to palace rooms. Roman, Nabataean and Minoan hydraulic new science discoveries demonstrate many aspects of modern hydraulic engineering methodologies.

The history of Petra's monumental architecture and historical development has been described by many authors (Browning 1982; Glueck 1965; Guzzo and Schneider 2002; Hammond 1973; McKenzie 1990; Ossorio and Porter 2009). Other scholars concentrated on technical and location aspects of water supply and distribution systems within Petra (Akasheh 2002; Bedal 2002, 2018; Bedal et al. 2007; Bellwald 2007; Bellwald et al. 2002; Oleson 1995, 2002, 2007; Ortloff 2003, 2014a, 2014b; Ortloff and Crouch 2002; Schmid 2007; Joukowsky 1998, 2004) while other surveys (Hodge 1992; Laureano 2002; Mays 2007) concentrated on the water control and distribution technology available from Roman and other eastern/western civilizations through trade and information transfer contacts during Petra's expansion period (100 BC–AD 300). Petra's openness to foreign influence is demonstrated in the city's monumental architecture that reflects elements of Greek, Persian, Roman and Egyptian styles integrated into Nabataean monuments. Later Roman occupation of Petra past AD 106 exhibits Roman pipeline technologies employed to expand the Paradeisos Pool Complex (Bedal 2002; Bedal et al. 2007) and city precincts responding to increased water demands for an expanding population as the city's status advanced as a key trade and emporium center. Petra's ability to manage scarce water resources to provide a constant potable water supply for its permanent population with reserves for large caravan arrivals and drought periods was key to its centuries of prominence as the nexus of a trade network between African, Asian and European cities for luxury goods. While knowledge of hydraulic technologies from foreign sources was available, the rugged mountainous terrain, distant spring water sources and brief rainy periods posed unique water supply challenges that required technical innovations consistent with the site's ecological constraints. Effective utilization of scarce and seasonally intermittent water supplies required technical expertise to ensure optimum system functionality from the use of long-distance interconnected pipeline transport and distribution systems (Figure 8.1) that guided water throughout the city's densely populated urban core and surrounding agricultural districts.

Examination of three of Petra's water conveyance pipeline systems using Computational Fluid Dynamics (CFD) analysis permits discovery of the rationale behind design selections utilized by their hydraulic engineers. Solution of fluid dynamics equations through CFD finite-difference means (Flow Science 2018) describes water flow patterns associated with hydraulic structures and reveals damaging hydraulic phenomena within pipelines that would be observationally

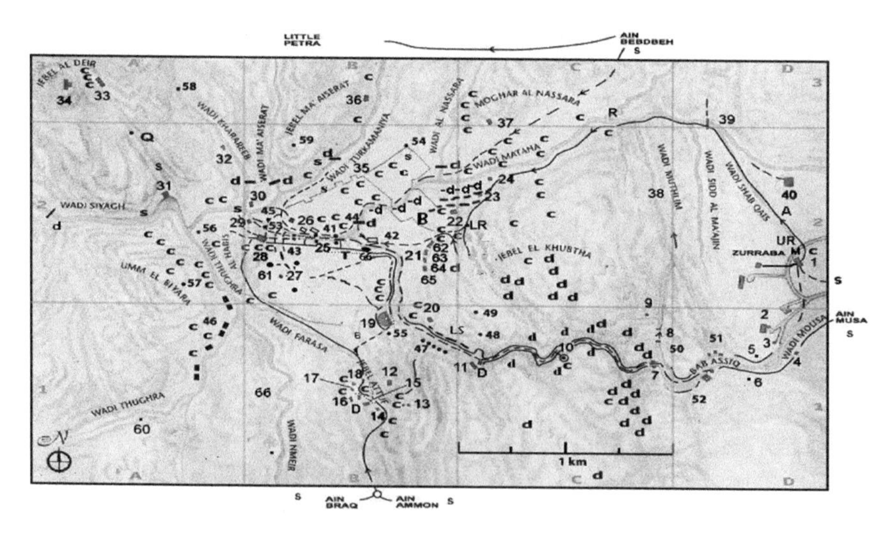

Figure 8.1 Site map of Petra (feature numbers in Appendix B).[1]

familiar to Nabataean engineers from past experience to guide their pipeline design/selection process to avoid system failures. CFD results demonstrating flow patterns within pipelines show what Nabataean hydraulic engineers intended (or avoided) in their hydraulic designs and permits insight into their civil engineering knowledge base. Given the low survival rate of documents and descriptions of hydraulic phenomena from ancient authors, and even from the few surviving documents, the descriptions of hydraulic phenomena is given in prescientific terms with little correlation to modern terminology to understand ancient author's descriptions and meanings of hydraulic phenomena. The analysis of flow patterns within piping systems using CFD methodology therefore present a viable recourse to calculate, visualize analyze problems Nabataean hydraulic engineers encountered and solved that is latent in the remains of their piping networks. Within the archaeological remains of these piping systems lay insights into the technical processes that lie behind Nabataean engineering decision making based on their observational, interpretive and empirical knowledge of hydraulic phenomena. CFD analysis which presents solutions to equations governing fluid motion therefore provides a graphic vision of problems encountered and solutions developed by Nabataean water engineers and helps to more fully understand what lies behind field observation of archaeological remains relevant to Petra's water supply and distribution systems.

Additionally, descriptive terms from ancient texts describing hydraulic phenomena given in prescientific terms unfamiliar to western notations now can be associated with actual phenomena through CFD modeling of the water flow patterns through water conveyance structures described in ancient texts.

The water infrastructure of Petra

To begin the discussion of Petra's water system development and their progress toward utilizing all possible water resources to meet population demands, the solid and dashed lines of Figure 8.1 detail the known supply and distribution pipeline systems leading water into the city's urban core. Numbered locations denote major site features listed in Appendix B. Shown are major and minor catchment dams and multilevel stepped dams (-d, d, D), cisterns (c), water distribution tanks (T) and springs (s). The (-d) dams located across streambeds are stone barrier structures built to hold large quantities of rainfall runoff water for redistribution to urban structures and/or localized agricultural areas. Dams denoted (d) are minor catchment structures that stored water in mountainous areas to prevent descent to the Petra basin and/or have channels leading water to cisterns.

Solid lines represent original pipeline paths originating from distant springs; in many cases only barren troughs now exist that once carried interlocked terracotta piping elements. The superimposed grid system (A, B, C; 1, 2, 3) define 1.0 km^2 grid boxes to enable the location of numbered features in Appendix B. Figure 8.1 represents the present knowledge state of pipeline connections/joining structures extrapolated from discovery of pipeline segments. Due to erosion and soil deposition landscape degradations over 2000 years, as well as reuse of piping elements by later inhabitants of the area, many terracotta pipeline sections are missing from their original holding troughs, obscured by subterranean placements or yet to be discovered, requiring extrapolation between available visible pipeline segments to represent the entirety of pipeline paths shown in Figure 8.1. This figure represents the current state of research from several sources (Bellwald 2007; Bellwald et al. 2002; Ortloff 2003; Schmid 2007) and personal exploration and is, at best, a first approximation to the entirety of the total pipeline system as much of the site remains unexcavated and subject to landscape change from soil erosion/deposition events from flood episodes and landslide events over millennia.

Further subterranean pipelines buried for defensive security purposes may exist to supplement the known pipelines shown in Figure 8.1 as recent on-site discoveries appear to add new information. In this regard, Site 66, Figure 8.1, represents a large newly discovered (in 2016) platform with a centrally located temple structure yet to be excavated to determine its role in Petra's administrative and ceremonial life (Parcak and Tuttle 2016). A water supply system serving this newly discovered complex remains to be discovered but surely exist given that all ceremonial structures presently known are accompanied by elaborate water systems for both display and practical supply of potable water. Additionally, as many of the pipelines are intentionally built subterranean and as yet unknown due to limited excavations performed in the urban center and distant reaches of Petra, Figure 8.1 can only be considered a first approximation of the total piping network. Some major monuments located at high altitudes exclude the possibility of canal water supplies; in this case cistern located nearby provide the water supply.

Petra's urban core lies in a valley surrounded by rugged mountainous terrain. Figure 8.2 demonstrates a section of the mountain terrain behind the ~10 m high Kasr el Bint Temple (Feature 29, Appendix B) located at the base of Jebel al Deir for perspective on the rugged mountainous terrain encompassing the urban core of Petra.

Seasonal rainfall runoff passes into the valley through many streambeds and is drained out primarily through Wadi Siyagh in antiquity (Figure 8.1, A-2).

Although present measurement of Wadi Siyah slopes preclude drainage presumably resulting from water transferred sediment deposits over millennia, in earlier times this was the main site drainage channel. The main water flow into the city (Bellwald 2007) originated from rainfall runoff from Wadi al Hay to the north, Wadi al Hudayb to the south and from the watershed area supplying runoff into the Wadi Mousa streambed; combined surface flows from these sources passed through the Siq (Figure 8.1: D-1 to C-1; Appendix B feature 10) which is the ~2 km long narrow entry passageway into the urban core of Petra (Figure 8.3). The Siq is a natural, narrow passageway through the Jebel el Khubtha mountain range varying from 5 to 10 m in width; the north and south walls of the Siq are near vertical with maximum heights up to 80 m.

By construction of a diversion dam and tunnel (8, Figure 8.1; feature 8 in Appendix B) in the 1st century AD, water was deflected from the Siq entrance and led to Wadi Mataha (Figure 8.1, C-2, 3) that linked up with the section of the Wadi Mousa streambed within the city center; water then drained into Wadi Siyagh (Figure 8.1, A-2).

Figure 8.2 Kasr al Bint Temple located at the base of the Umm el Biyara mountain.

This dam and tunnel construction served to divert silt laden floodwater away from entry into the Siq that would compromise the structural integrity of many of the monuments within Petra's urban core region. Runoff water from southern sector Wadis Thughra, Nmeir, Farasa and northern sector Wadis Kharareeb, Ma'Aiserat and Turkamanya (Figure 8.1) drained runoff into the Wadi Siyagh away from the city center (Bellwald 2007a).

Water supplies to the city in the 1st and 2nd centuries BC were through a slab covered, open channel dug into the floor of the Siq conducting water from the distant Ain Mousa spring into the urban center of the city perhaps as far

Figure 8.3 Interior region to the ~2 km long Siq bounded by high (~60 m) near vertical walls.

as the Temenos Gate (Figure 8.1, B-2; feature 43 in Appendix B). Later 1st- and 2nd-century AD builders replaced this water supply system with a long pipeline (~14 km) carrying water from the Ain Mousa spring in an elevated channel supporting a pipeline along the north wall of the Siq extending along Colonnade Street (23, Figure 8.2.q.1). A trace of this system in the vicinity of the south Nymphaeum along Colonnade Street may exist from GPS data (Urban et al. 2012).

Later 2nd-century AD construction (Ortloff 2014a) revealed an integrated approach to site water system management demonstrated by new and novel features. These include surface cisterns to capture rainfall runoff, deep underground cisterns (one located in the eastern part of the site (Appendix B feature 28, Figure 8.1), multiple pipeline systems sourced by different springs and storage reservoirs, floodwater control through diversion dams and tunnels, pipeline supply system redundancy to ensure water delivery from multiple spring and high level reservoirs; Figure 8.4, Appendix B feature 1 shows the high level Zurraba Reservoir supplied by a branch of the Ain Mousa spring that supplied water to high population concentration parts of the city, drought remediation purposes and large water supply delivery for large caravan arrivals into the city. Major pipelines, primarily on the north wall of the Siq, provided sediment particle filtration/removal basins and served a sophisticated hydraulic function described in detail in later parts of this chapter.

These features reflected the need to bring high quality potable water into the city center and serve city hillside occupation zones above the valley floor. First- and 2nd-century AD developments demonstrated continual evolution of the Petra

Figure 8.4 The high level Zurraba Reservoir.

water system and reflected application of acquired technologies from archaic and/or exterior sources integrated with indigenous developed hydraulic engineering innovations to respond to the increased water needs of the city as a trade center.

The means to capture and store a fraction of rainfall runoff through dams and cisterns, to build flood control systems and build pipelines and channels to deliver water from distant springs and to manage these assets to provide a constant water supply to the city is key to understanding Nabataean contributions to hydraulic engineering and water management practices. While water storage was key to the city's survival, springs internal and external to the city (Figure 8.1: Ain Mousa, Ain Umm Sar'ab, and Ain Braq, Ain Dibdiba, Ain Ammon, Ain Beidha and Ain Dibidbeh) supplied water channeled and/or piped into the city to provide the main water supply while several spring-supplied reservoirs exterior to the city (M-Zurraba, Figure 8.1) provided additional water storage reserves. The smaller M reservoir at a lower elevation connected to the Zurraba Reservoir or another branch of the Ain Musa spring no longer exists due to recent town expansion. While the Ain Mousa spring was the main water supply to Siq pipelines, only a fraction of the Ain Mousa spring source was used for this pipeline. Additional water was used to charge several reservoirs above the city through elevated pipelines (Figure 8.1:2-D; primarily the M-Zurraba reservoirs) to maintain a reserve water supply for drought alleviation as well increased water demands for arriving trade mission caravans. Additional on-demand use of reservoir water for industrial ceramic and metal working workshop areas as well as for ceremonial use to supply *triclinium* ritual use for tomb celebratory functions is indicated. Of the many springs used in antiquity, the Wadi Siyah spring near the quarry (Figure 8.1, Appendix B feature 31) remains functional for local inhabitants living in remote sections of the ancient city together with the Ain Mousa spring which now provides water for the modern tourist town of Wadi Mousa. Several minor reservoirs above the town have been recorded (Ortloff 2014b:254) but no longer exist due to town expansion.

Water management of the Siq pipeline system

The main city water supply originated from the Ain Mousa spring about 7 km east of the town of Wadi Mousa (Figure 8.1, D-1) combined with the waters from the minor Ain Umm Sar'ab spring. The flood bypass tunnel (Appendix B feature 8, Figure 8.6) at the Siq entrance together with an entry dam and elevated paving of the Siq floor reduced flooding into the urban center from rainy season runoff into the Wadi Mousa River. Part of the early slab-covered, open channel water supply to the urban core of Petra now lies under pavement construction attributed to Aretas IV and later Roman builders as this system was replaced by the pipeline water supply system from the Ain Mousa spring source. Recent excavations reveal ~5 m of flood silt/gravel debris covering the original hexagonal slab paved area in front of the el Kazneh Treasury (Figure 8.7, Appendix B feature 11) as well as four open surface water basins between Sig piping segments / typical of those shown in Figure 8.5. These basins were distributed along the Siq pipeline and give clues as to the hydraulic technology available to Nabatean water engineers.

With increasing water needs for increasing population in early centuries AD, an elevated ~1 km long, north-side Outer Siq pipeline extension (Figure 8.8; LS Little Siq, Figure 8.1 1-B, C) extended the utility of the Siq pipeline system (Ortloff 2003, 2008, 2014a, 2014b, 2014c, 2014d) to supply potable water to further reaches of Petra's urban core. Here this pipeline connections provided water supply to structures located in the Figure 8.1 B-2 district as well as providing pipeline water supply to the Nymphaeum (Appendix B feature 42, Figure 8.1); further pipelines emanating from the Nymphaeum to the Temple of the Winged Lions (Figure 8.9), major tombs fronting the Wadi Mataha housing structures and an elite palace structure in early stages of excavation likely existed although details await further excavations. Continuance to a bridge across the Wadi Mataha led water to the south side of the site.

Owing to destructive pipeline hydraulic instabilities, (transient pressure waves, flow intermittency, internal pipeline hydraulic jumps, transient turbulent drag amplification zones, partial vacuum regions affecting the flow rate), hydraulic problems affecting the design of pipeline systems required development of advanced technologies to produce stable inner pipeline flows that matched a fraction (or all) of the supply spring source output. The pipeline also needed to be designed to produce the maximum flow rate the pipeline could sustain. Further considerations involved limiting full flow (pipeline cross-section fully occupied by water) hydrostatic pressure within a pipeline to limit leakage at the thousands of terracotta pipe-joint connections along long lengths of pipelines. For a typical pipe element length of ~0.35 m and total pipeline length over ~14 km from the Ain Mousa spring origin location to the end of the Siq and Outer (Little) Siq

Figure 8.5 Section of the ~2 km long Siq north side channel supporting a pipeline basin to reset head for sequent pipeline section; four basins exist along the Siq pipeline.

pipeline with a total height drop well over ~50 m (data derived from the contour *Map of Petra*), high hydrostatic pressure in a full-flow pipeline would increase the likelihood of pipeline element connection joint leakage at ~42,000 pipeline connections. Solutions to mitigate many of these design/operational problems provide insight to Nabataean water management expertise as discussed later in the chapter.

CFD calculation (Flow Science 2018) of full-flow volumetric flow rates in 14 cm diameter (D) piping were made for internal wall roughness $\varepsilon/D > 0.01$ (where ε is the mean square internal pipeline roughness height); this ε/D value holds for roughness typical of terracotta internal pipeline walls demonstrating casting corrugations and erosion usage pitting (Morris and Wiggert 1972) for typical Reynolds number Re $\sim 10^5$ values. For pipelines at 2, 4 and 6 degree slopes for 1.0 m and 3.0 m supply head, Figure 8.10 reveals that past ~400 m, pipeline length, an increase of input head does not substantially increase the flow rate

Figure 8.6 Siq entrance flood water diversion tunnel.

(Ortloff and Crouch 2002). Past this length range, even a 3X head increase does not substantially increase flow rate.

The implication for the ~14 km long Ain Mousa supply pipeline to the Siq and Outer Little Siq extension is that only a fraction of the estimated (Oleson 1995, 2002, 2007) maximum seasonally varying Ain Mousa spring output of ~1000–2000 m^3/day; ~0.01–0.025 m^3/sec), could be transmitted through the pipeline under full-flow conditions due to the pipeline's internal flow resistance. The internal pipeline flow rates shown in Figure 8.10 are based upon a low pipeline wall roughness of $\varepsilon/D \sim 0.01$; higher $\varepsilon/D \sim 0.07$ values that account for pipe element connection socket roughness that reduce flow rates substantially leading to Figure 8.10 curves being shifted downward by a factor of ~3 for typical Re ~10^5 pipeline Reynolds number values.

On this basis, the observed Siq rough internal wall pipeline at 2 degree slope permitted 0.023 m^3/sec flow rate and a one degree pipeline slope permits a 0.017 m^3/sec flow rate corresponding to ~1.5 and ~1.1 m/sec mean velocity in the 14 cm diameter pipeline with corresponding flow rates of ~80 m^3/hr and ~60 m^3/hr respectively. Thus the frictional wall resistance to the water flow combined with the long length of the Ain Mousa-Siq-Outer Siq pipeline along its ~2+ km length permits only a fraction of the Ain Mousa spring flow. Additional flow from the Ain Mousa spring was likely used to recharge upper-level reservoir systems for on-demand water usage. Given the ~10,000 m^3 capacity of the main Zurraba reservoir (Figure 8.4) and that of several additional minor reservoirs, the additional Ain Mousa flow rate capacity not used for the Siq pipeline system served to provide additional water storage for on-demand and drought remediation use.

Figure 8.7 The el Kaznah Treasury.

Figure 8.8 Little Siq pipeline extension from the Siq to habitation sites and royal compound sites in the Wadi Mataha area.

Figure 8.9 Temple of the Winged Lions.

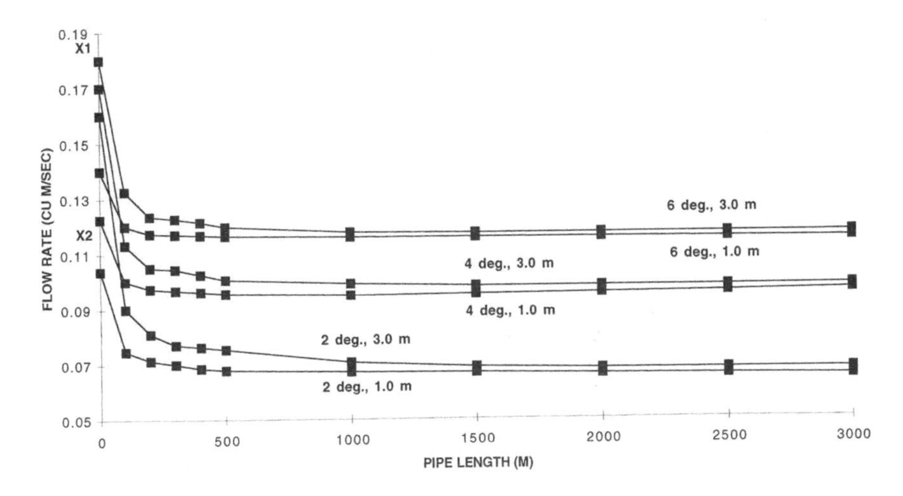

Figure 8.10 Allowable volumetric flow rate vs. pipeline length for 2, 4, 6 degree declinations and 1 and 3 m supply head values for ε/D ~ 0.01 wall roughness.

On the Siq north side pipeline, several open basin features (Figure 8.5) are found at intervals along the pipeline. Each basin permits an accumulation of water serving as a head tank for its downstream piping length segment. Figure 8.4 reveals that hydrostatic head substantially influences flow rate for pipeline lengths less than ~500 m.

The head is reset at the origin of each of the piping lengths between basins so that flow through each piping segment is only sensitive to its local head value. This modification led to partial flow (pipeline cross-section partially occupied by water) in Siq pipeline segments reducing overall flow resistance. As pipeline flow extending ~11 km from the Ain Mousa spring to the Siq entry point was mostly subcritical full-flow due to the high pipeline internal wall flow resistance and low ~2 degree slope, the lowered resistance of Siq pipeline segments at ~3 degree (hydraulically steep) slope between basins induced partial flow in distal portions of the ~11 km Ain Mousa supply pipeline thus slightly raising the overall system flow rate due to lowered back pressure. In simpler terms, lowering the frictional flow resistance in the end regions of a pipeline permits higher flow rate to be achieved through the entire pipeline system.

The Siq north side open basins (estimated to be four) (U. Bellwald, personal communication) additionally served to provide drinking water along the Siq and access for cleaning of soil/silt particles entrained in the flow from the Ain Mousa spring that enhanced potable water quality. Atmospheric pressure in the air space above the partial flow water surface in the Siq pipeline branches reduced pipe joint leakage as would occur under a pressurized, full-flow condition design. To achieve a piping design based on a 14-cm pipe diameter at the declination angle of the Siq, ideally partial, critical flow at unit Froude number (Fr = 1) at critical depth

(y_c/D ~ 0.8–0.9) would provide the maximum flow rate the piping can convey. As open basins were sequentially lower than their predecessor upstream basins due to the declination slope of the pipeline, each basin water height supplied head for its adjoining downstream piping section. Partial flow existed throughout the Siq piping sections where flow entering a ~3 degree slope pipe section from a basin transitions to a partial-flow, critical depth (equal to the normal depth) when Fr ≈ 1. This result indicates that the Froude number for the Siq pipeline flow is on the order of unity at the permissible ~96 m^3/hr flow rate. This subtraction from the larger Ain Mousa spring flow rate permitted surplus water for charging of high-level reservoirs above the city as well as for high water demands for agriculture in adjacent areas to the city. The thick sinter buildup in the now visible open bottom-most part of Siq piping supports the deposition pattern resulting from partial flow conditions existing in the Siq pipeline segments.

In the area immediately west of the Treasury, an Outer Siq pipeline (Figure 8.1, LS, B, C) continued the Siq pipeline and was set at a slightly lower angle than the Siq pipeline. Given the Siq pipeline flow rate, a further consideration arises as to the stability of the flow in piping sections within the Siq as well as for water arriving to the Nymphaeum (Figure 8.1, Appendix B Feature 42) through the further pipeline extension. If flow delivery is pulsating into open basins supplying the Nymphaeum, then spillage and sloshing occurs amplifying flow instabilities that compromise the aesthetic display fountains in the Nymphaeum and induce forces within the pipeline promoting joint separation and leakage. For an examination of flow stability, computer models were made for a ~1220 m section of the Outer Siq piping extension with a declination slope slightly less than that in the Siq pipeline; this length represents the distance from the exit of the Siq pipeline through the Siq to the Nymphaeum (42, Figure 8.1). Previous research (Bellwald 2007a, 2007b) has indicated a further south-side Nymphaeum across the Wadi Mousa streambed from the north-side Nymphaeum served by an extension of the south arm of the Siq pipeline; traces of elevated reservoirs providing pressurized water to the south Nymphaeum are given in the literature (Ortloff 2003, 2014b). Computer solutions (Figure 8.11) for 0.305(A), 0.610(B), 1.54(C), 3.05(E) m/sec water velocity were made with observed ε/D values typical of the interior surface of piping elements. Figure 8.11(D) results are for a smooth interior pipe element wall at 1.54 m/sec velocity to demonstrate the effect of wall roughness on flow patterns at the same water velocity as 1.54(C). For an illustrative example, water at ~96 m^3/hr is accepted into the Outer Siq piping – average pipeline water velocity is then ~1.73 m/sec. Given the flow is slightly supercritical (Fr ≈ 1.6) at this velocity on a hydraulically mild slope, the flow approximates critical depth for cases (B), (C) and (D). For flows close to critical, the normal depth is close to the critical depth (Morris and Wiggert 1972). As partial flow exists in the Outer Siq extension section (similar to that in the Siq piping) and has the same ~96 m^3/hr flow rate at about the same declination angle as the Siq piping, near critical flow conditions exist throughout the Siq and Outer Siq extension piping – but some transition from full-flow to partial flow occurs in the Ain Mousa sourced piping near the Siq entrance location due to lower Siq, Outer Siq downstream flow resistance. Figure 8.11 shows

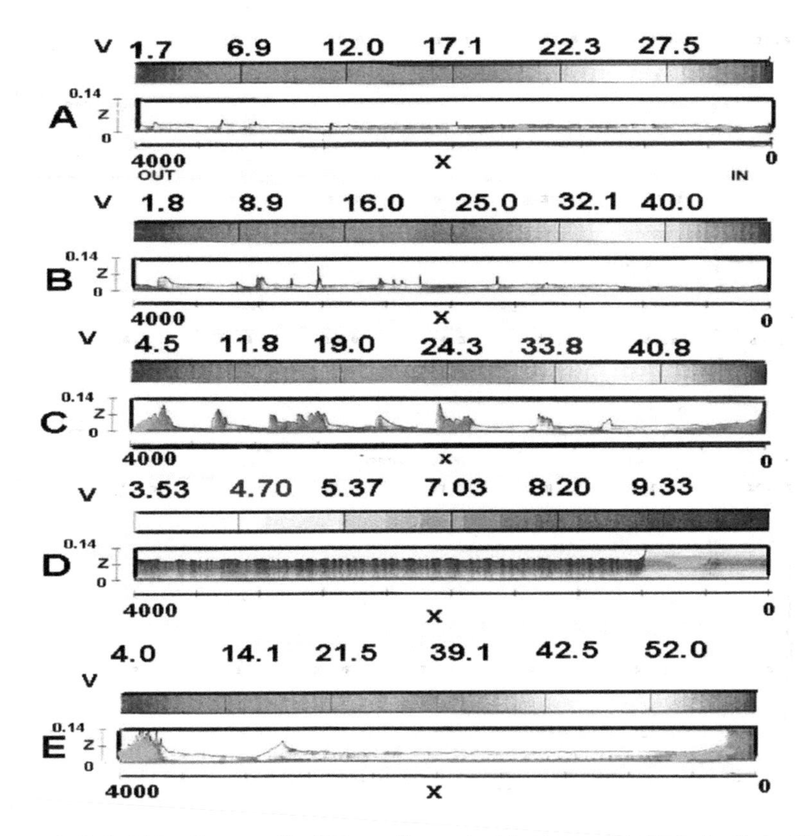

Figure 8.11 Stability diagrams for different flow velocities (cases A to E) in the Outer Siq extension pipeline. V denotes velocity (m/sec), X is the length of the pipeline $0 < X < 4000$ m and z is the height above the bottom of the 0.14-cm diameter pipeline – flow direction is from left to right.

transition to lower water depths from the entry boundary condition for all supercritical (B), (C), (E) and (D) cases; for the subcritical case (A), water is at a low level due to low flow velocity. Figure 8.11(B) at ~1.2 m/sec approximates the flow effect at a mean velocity of ~1.75 m/sec and indicates that only minor, random height excursions occur in water flowing within the pipeline – this indicates that the water supply system through the Siq extension to the Nymphaeum is stable – which is advantageous for proper operation of the Nymphaeum's water display system. At twice this velocity, Figure 8.11(C) indicates large height fluctuations in the water surface indicating unstable flow delivery occurring in the pipeline system. This indicates that the flow rate selected from the Ain Mousa spring by Nabataean engineers was a deliberate design choice to achieve stable flow in the total system. Comparison of Figure 8.11(C) to (D) indicates that internal pipeline wall roughness amplifies large scale turbulent eddies that contribute to flow unsteadiness; this effect is included in the (B) result. Tracking the distal end of the Outer Siq pipeline reveals it is intercepted by an open basin indicating a further feature to ensure flow stability for flow continuing on to further pipeline branches. In summary, the Siq and Outer Siq system was designed to optimally transfer ~96 m^3/hr with partial flow conditions to eliminate large hydrostatic pressures that would incur pipeline joint leakage at high hydrostatic pressure under full flow conditions. Additionally, open basins along the Siq pipeline sequence permit removal of sand and soil debris that would ultimately clog the system and destroy its function. This cleaning feature provided better quality potable water to Petra's population (assumed to be in the 20,000–30,000 range).

The Wadi Mataha pipeline system

The water supply systems of Petra employ springs (s) both within (Ain Siyagh) and outside city limits (Ain Dibidbeh, Ain Mousa, Ain Braq and Ain Ammon) to bring water to city monuments (Figure 8.1) through pipelines. Reservoirs fed by springs (M-Zurraba, Figure 8.1) as well as other springs and reservoirs no longer in existence due to modern urban expansion provided supplemental water on-demand for arriving caravans or other utilitarian purposes. Use of pipelines composed of interlocking, socketed terracotta pipe elements sealed at joint connections by hydraulic cement are e typical of pipeline construction. Nabataean pipe element designs are typically ~0.35 m long with an inner diameter range of ~14 to 23 cm. Later Roman modifications to existing Nabataean water systems in the Great Temple (Joukowsky 1998) area (Figure 8.1) include small-diameter lead piping and Roman standardized pipe diameter designs (Hodge 2011). For the Wadi Mataha pipeline, construction detail of the trough on the Jebel al Khubtha western mountain face (C-2, Figures 8.1 and 8.12) carrying the pipeline remains largely intact.

The initial presence of a pipeline, as opposed to an open trough channel conveying water (in the Wadi Mathha pipeline for example), is evident by noting that supercritical flow accelerating on a hydraulically steep channel encountering lateral bends would form surface height excursions and hydraulic jumps causing channel overflow. In an open channel design, only a trickle flow from the upper reservoir could be contained in the channel without spillage negating the original purpose to provide on demand high volumes of water flow. Some slab cover sections of the Wadi Mataha channel trough are evident from and typical of Nabataean methods to limit trough leakage. By surrounding embedded pipelines with stabilizing soil to limit pipeline motion at multiple connection joints, leakage can be reduced.

If slab covering over a water carrying trough was in place only then the slab covered channel would form an equivalent closed channel with a larger rectangular cross section subject to analysis by CFD methods. Independent of piping or a slab covered channel, Figures 8.10 and 8.11 govern admissible flow rates for full flow conditions originating from the M-Zurraba reservoir.

The pipeline along the eastern face of Jebel al Khubtha originates from the M-Zurraba reservoir complex. Typical reservoir maximum flow rates from Figure 8.10 are on

Figure 8.12 The elevated Wadi Mataha channel on the western face of the Jebel el Khubta Mountain (C-2 location in Figure 8.1); water supply is from the upper Zurraba reservoir.

the order of ~0.05 m^3/sec which would satisfy on-demand water supply requests. Lesser flow rates would be possible by using a variable area orifice from the reservoir. Piping from the high level reservoir proceeds to a lower reservoir elevated above ground level near the Sextus Florentinus Tomb area (Appendix B feature 22, Figures 8.12 and 8.13) then proceeds southward toward a proposed royal compound area (Schmid et al. 2014) (Figure 8.14, J) then on to monumental royal tomb architecture near the Palace tomb and a nearby fountain (Bellwald 2007a, 2007b) and on to the Nymphaeum for further water distribution to other parts of the city.

One major purpose of the Wadi Mataha pipeline was to bring water on demand from the upper to the lower reservoir and then on to nearby royal compound structures and subsequently to a pipeline transferring water to the Nymphaeum and multiple basin/reservoir areas created by multiple dams along the Wadi Mataha streambed (-d, Figure 8.1). This system provided water to cisterns associated with ceramic workshop areas and for celebratory *triclinium* functions associated with monumental tomb structures (Figure 8.1: 62, 63, 64 and 65), royal compounds (Figure 8.1, region **B**) and additionally provided upper reservoir water as a drought remediation measure. Resent research (Schmid et al. 2012) indicates that several royal palaces (*basileia*) exist (Figure 8.14) in the terminal area supplied by the Wadi Mataha pipeline (Figure 8.1, region **B**). In conjunction with water from the Siq Outer pipeline extension directed into this area, ample water supply to royal

Figure 8.13 Lower elevated reservoir terminus of the Wadi Mataha pipeline located near to the Sextus Florentinus Tomb (leftmost C-2 location in Figure 8.1) – pipeline continues from the reservoir to the Nymphaeum and a further reservoir and pipelines.

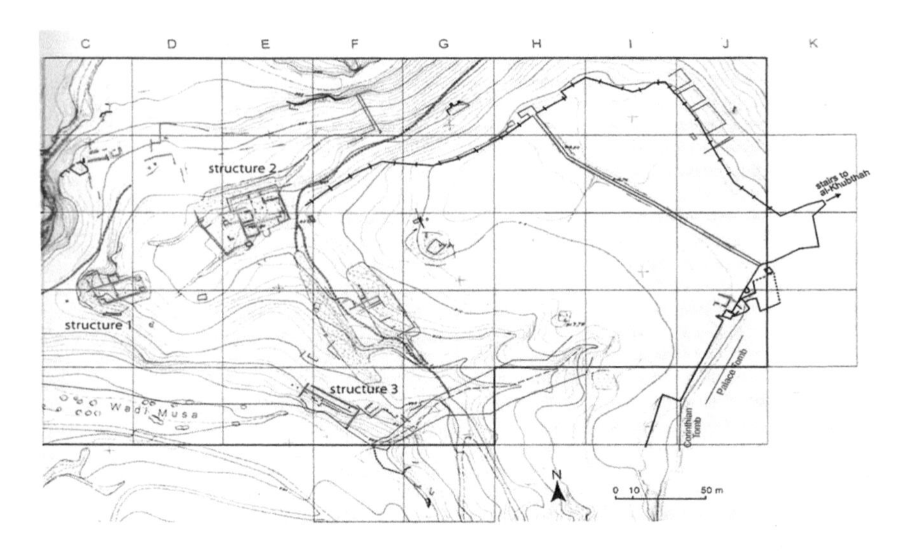

Figure 8.14 Royal compound **B** area west of the Wadi Mataha pipeline system (with permission from R. Schmid et al. 2012). Royal compound structures (red) 1, 2 and 3 and pipeline segments indicated.

compounds, even in drought periods, was assured as royal compounds necessities dictate. Pipeline connections to the multiple royal compounds await excavation but given nearby multiple water sources from the Wadi Mataha and Ain Mousa springs via the Siq north pipeline extension, as well as the upper reservoir water supply, ample redundant water supplies from multiple sources would be available to serve palace structures. Water from the reservoir/basin continued by pipeline past the Nymphaeum southward to city center areas and ultimately by a bridge across the Wadi Mataha to the commercial and fountain area (south Nymphaeum) district to discharge water into the Wadi Siyagh. The present-day bridge in the al Bint area contains visible ancient pipeline elements remaining after modern modifications to the earlier bridge. On this basis, another earlier bridge would serve to provide redundant water supplies to other parts of the city served by alternate pipelines. This in effect would permit cleaning outages and redundant water supplies to critical parts of the city.

Given the slope (~2.8 degrees) and construction detail of the entire Wadi Mataha pipeline, the question arises as to the engineering behind the selection of a pipeline slope that yielded the maximum flow rate given a variety of possible design options. Given the necessity to construct an elevated trough carrying the pipeline on the near vertical face of the Jebel al Khubtha mountainside and the surveying required for the pipeline slope over the pipeline length angle, a preliminary engineering analysis was necessary preceding construction. Analysis of alternate pipeline slope choices that maximize the flow rate while minimizing leakage from the many connecting joints between piping elements was key to a successful design. Details of flow patterns within the pipeline connecting the upper to the lower reservoir were analyzed by computer simulation to provide insight into the decision-making rationale underlying Nabataean engineer's choice of a particular pipeline slope.

Starting from an elevation of ~910 masl near the Mughur al Mataha and Jebel al Mudhlim areas (Figure 8.1, point **R**), the pipeline on the face of Jebel al Khubtha Extends ~1.5 km to the lower reservoir near, but offset and above, the Sextus Florentinus Tomb (Figures 8.12 and 8.13) area located at ~770 masl. The upper water source is from the M-Zurraba reservoir connection supplied by the Wadi Mousa spring branch. A ~15 m lower reservoir elevation above the tomb base at 770 masl yields an observed pipeline declination slope of ~2.8 degrees measured from the horizontal. While this slope was the ultimate choice of Nabataean hydraulic engineers, an alternate shallower slope choice, say of one degree declination, would require a yet higher elevation lower reservoir requiring a high support base built onto the near-vertical Jebel el Khubtha mountainside. While this construction had the benefit of allowing the pipeline to extend further to the south to provide water to urban habitation areas, this design option was redundant given water supplies through the Siq and Outer Siq providing water to the same area of the city. For the one degree declination slope option (slope measured from the horizontal origin point **R** of the pipeline as it switches to the western face of the mountain (Figure 8.1), the elevation of the lower reservoir would be much higher above the ground surface requiring an extensive base construction support.

Beyond structural difficulties involved in this construction, computer modeling of water flow patterns in a long pipeline at one degree negative slope reveals a problem. A CFD model (Figure 8.15) consists of a leftmost upper reservoir connected to a lower rightmost reservoir by a pipeline at one degree slope; calculations qualitatively duplicate hydraulic phenomena present in pipeline flow from the upper reservoir (**UR**, Figure 8.1 to the lower reservoir **LR**). Figure 8.15 results demonstrate the pipeline cross section is fully occupied by water given submerged pipeline exit condition into the lower reservoir. A design with the pipeline exit above the lower reservoir water level would still support mostly full-flow conditions over

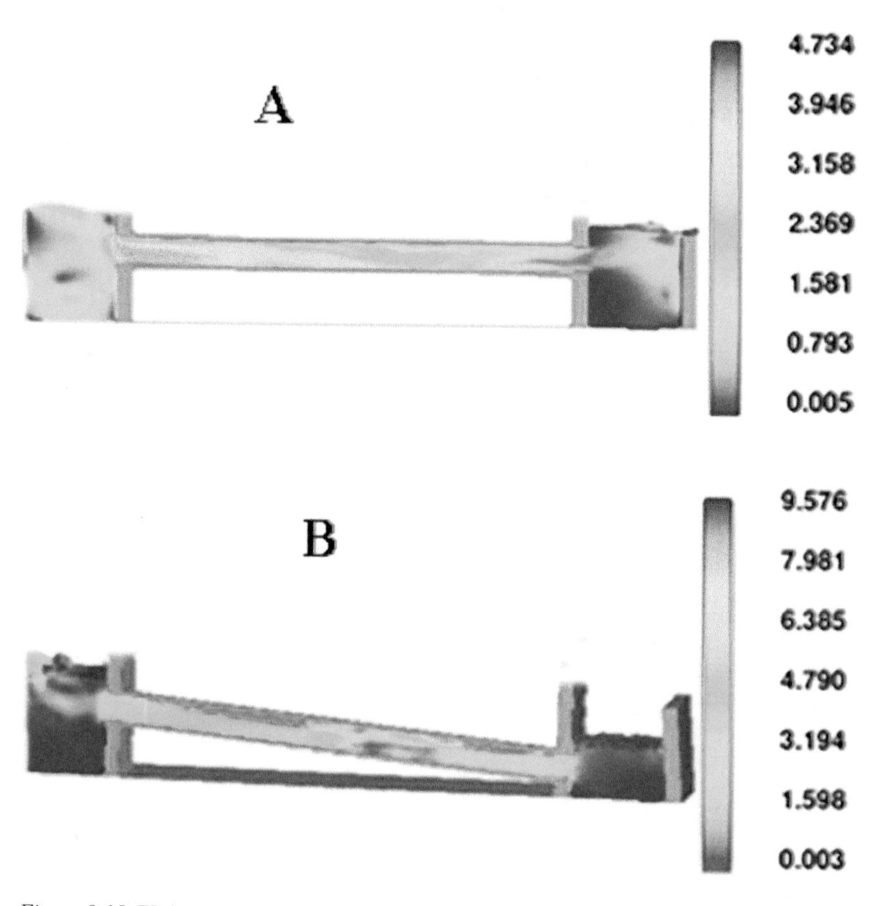

Figure 8.15 FLOW-3D internal pipeline flow representation (A) demonstrating subcritical, full flow using input volumetric flow rate conditions from Figure 8.10 results at one degree declination slope of the Wadi Mataha pipeline; (B) pipeline flow representation demonstrating mixed supercritical and subcritical flow regions with an intermediate hydraulic jump (HJ) at four degrees pipeline declination; (C) pipeline representation demonstrating near critical, partial flow from location C to D at an intermediate angular declination.

most of its length but be subject to ingested air slugs moving counter to the flow direction causing unstable flow conditions. A volumetric flow rate of ~0.05 m^3/sec enters the pipeline from the upper reservoir; this is the permissible full-flow rate for $\varepsilon/D \sim 0.1$ smooth pipe roughness over long pipeline distances (Figure 8.10) at one degree negative slope. Increasing the upper reservoir head does not result in a proportionate increase in volumetric flow rate for long pipelines over 400–500 m length as indicated by Figure 8.10. Results indicate full-flow for a submerged outlet pipeline under high hydrostatic pressure near the end reaches of the pipeline – this increases the probability of leakage due to full flow conditions under high hydrostatic pressure (Figure 8.6a). Based on the combination of leakage problems and the difficulty of building a high elevation reservoir with steep connecting pipeline to the Nymphaeum, Nabataean hydraulic engineers rejected this pipeline design as impractical. A further design option involves a steeper pipeline slope choice of four degrees with a volumetric flow rate is adjusted to 0.08 m^3/sec. The intuitive (but incorrect) observation that the steeper the pipeline slope, the faster will be the water velocity and thus serve the on-demand rapid water delivery requirement best is an observation that may guide the steeper slope choice. The steeper slope design option would result in a lower reservoir height from ground level but would require an additional length of piping to carry water on to the tomb structures (Figure 8.1: 62, 63, 64 and 65) as water would be delivered to ground level cisterns distant from these structures. Further piping to the north Nymphaeum and the nearby basin would require pipeline construction which, at low slope, may further impede the flow rate on to further water distribution centers. Based upon easier design option construction of a low level reservoir on the el Khubtha mountainside, this option appears to have positive benefits for workshop areas north of the monumental tomb structures; however hydraulic analysis of this design option reveals a problem. For a steep pipeline slope, the internal flow develops a partial vacuum air space above the rapidly flowing (partial flow) water surface (Figure 8.1 6B). The flow height approaches normal depth (Morris and Wiggert 1972) for supercritical flow in a hydraulically steep pipeline. This result is only approximated as flow resistance is amplified by the high partial flow velocity that converts flow from supercritical ($Fr > 1$) to subcritical ($Fr < 1$) through a hydraulic jump within the pipeline. Froude number is defined as $Fr = V/(g\ D)^{1/2}$ where V is the average flow velocity, g the gravitational constant (9.82 m/sec^2) and D the pipe diameter.

While partial-flow in the upper supercritical portion of the pipeline helps eliminate pressurized full flow conditions that promote leakage, the full-flow, post-hydraulic jump region within the downstream reaches of the pipeline promotes leakage due to hydrostatic pressure. The transition of supercritical to subcritical flow through a hydraulic jump is caused by a combination of back pressure resulting from submerged pipeline discharge into the lower reservoir and amplified flow resistance from the high velocity partial flow. Since water in the lower reservoir may undergo sloshing and transient height changes during this process, the hydraulic jump position can be oscillatory due to air ingestion countering a partial vacuum region upstream of the hydraulic jump contributing to sloshing in the lower reservoir and pipeline forces that compromise connection joint stability. The net effect of this design option is that a lower portion of the pipeline flow experiences full-flow conditions under hydrostatic pressure promoting connection joint leakage. While a design option exists to have the pipeline outlet above the lower reservoir water level under free fall conditions, supercritical, partial flow within the total length of the pipeline is unlikely to exist as internal pipeline resistance is amplified at high (supercritical) water velocity sufficient to cause full-flow conditions to develop through a hydraulic jump. The computer model verifies the formation of a hydraulic jump (**HJ**, Figure 8.7b) showing the interaction between incoming supercritical water from the upper reservoir to the lower reservoir and shows the jump establishing a pressurized, full-flow water region extending downstream from its location in the pipeline. Provided the lower reservoir maintained constant height by limiting flow into piping directed toward the Nymphaeum and basin regions, the hydraulic jump position can be stabilized; nevertheless, the full-flow, pressurized water region constitutes a potential pipeline leakage problem. Given potential hydraulic instability and leakage problems associated with a steep piping slope choice (regardless of the pipeline exit submerged or with a free-fall delivery condition), this design option was not implemented as an oscillatory hydraulic jump causes pressure oscillations inducing separation forces at pipeline connection joints amplifying the potential for leakage. Additionally, this pipeline design does not support the highest flow rate possible despite its steeper slope.

Given the ~2.8 degree pipeline slope declination design option selected by Nabataean engineers and given an allowable input flow from the upper reservoir of 0.062 m^3/sec for full-flow conditions into a long stretch of the 3.5 km pipeline upstream of point **R** in Figure 8.1, two criteria must be met for acceptance of this design option: (1) an easy-to-build and maintain low reservoir that provides observation of pipeline exit flows to check for flow problems; (2) a stable, high volume flow having an atmospheric pressure airspace above the water surface to eliminate hydrostatically pressurized leakage. Figure 8.7c results show that these criteria are satisfied by the selected design. Theoretically, a flow approximating critical flow at $Fr \approx 1$ yields the highest flow rate possible with flow occupying a fraction of the cross-section of the pipe with an airspace above the water surface. Calculations for a 0.062 m^3/s full-flow volumetric flow rate (Ortloff 2003) based on a ~2.8 degree slope and upper reservoir hydrostatic head values from one to three meters indicate that average velocity in the pipeline is $V \sim 1.5$ m/sec with $(g\ D)^{1/2} \approx 1.5$ m/sec leading to $Fr \approx 1.0$ indicating that a near critical flow condition\ is approximated at the ~2.8 degree slope. Given near critical flow observed at ~2.8 degree slope (corresponding to the observed slope) and observing computer results (Figure 8.15C) that indicate that this condition produces an air space over the water free surface approximating $y_c/D \approx 0.8$, an approximate critical flow condition exists at the ~2.8 degree slope. The significance of this is that the 0.062 m^3/s volumetric flow rate is the maximum value that this long pipeline design can sustain due to water wall frictional effects and that the critical flow water height in the pipeline, although close to full-flow conditions, leaves an atmospheric pressure airspace over the water surface over a long length of the pipeline to limit leakage.

Therefore among the choices of pipeline slope available to Nabataean engineers, the ~2.8 degree declination slope was selected. This slope option produced the maximum transport volumetric flow rate possible which is what an on-demand water system is set to accomplish.

The Ain Braq pipeline system

The upper reaches of Ain Braq water system (Figure 8.1) consists of a trough cut into bedrock and surface soils that earlier supported a pipeline; its continuance to the plateau north of Wadi Farasa (Figure 8.1) by a supply line (D) may have been supported by a siphon or, more likely, a pipeline originating from the theater frontage area (Figure 8.16) given land contours shown in Weis (2014) that would support a buried canal path. The plateau in the Weis's (2014) work shows many water installation structures that are consistent with a substantial water supply system; additionally, the water system extends to the Great Temple area and its water installations.

Figure 8.16 The theater.

A survey of the trough from the Ain Braq spring to the siphon area reveals troughs in hills (Figure 8.1) that would render an open channel water supply improbable; although the original pipeline is now missing, a pipeline within a carved stone passageway design was the only one possible to convey water under pressure over the many hills to the city center. Adjacent to this section, a channel carved in the adjacent mountain side shows that a contour canal carried water from the spring further to sites shown in Figure 8.1. Some traces of an attempt to bridge a large gully by a siphon are indicated in Figure 8.17; this design choice likely proved difficult to achieve resulting in an alternate design with a trough carved into the adjacent mountainside supporting a pipeline that sent water to the Lion Fountain (Figure 8.18) Appendix B feature 14) and sites further north. Figure 8.1 shows surface water distribution features in the Wadi Farasa area (B-1, Figure 8.1) together with the subterranean pipeline-cistern (A) connecting to the

Figure 8.17 Traces of the subterranean pipeline trough supporting pipelines (B, C), the flow distribution control center (E) and subterranean pipeline-cistern (A) located near the siphon end of the Ain Braq pipeline located on the Wadi Farasa hilltop plateau location All of these elements connect into the flow distribution center system (E) supplying water through dual pipeline branches to the Zanthur housing area and the Great Temple/Paradeisos in the city center.

Figure 8.18 The Lion Fountain.

distribution junction (E). Again, further exploration of possible contour pipeline paths threaded through mountainous areas to lower sites is necessary to establish the water supply source. A contour map of the hill area south of the Great Temple (Figure 8.22) area indicates the possibility of a water supply from the southern Siq channel passing by the theater (Figure 8.16) but changes in the landscape over millennia have made connections difficult to establish.

Starting from the hilltop junction (Figure 8.1), one of the pipelines (B) from (E) led water to the Zhanthur housing district (Weis 2014:136) while another pipeline branch (C) provided water to the Great Temple/Paradeisos area (Bedal 2002; Joukowsky 1998, 2004) through subterranean channel (B, C), Figure 8.19. Combining these elements, a FLOW-3D computer model (Figures 8.20, 8.21 and 8.22)

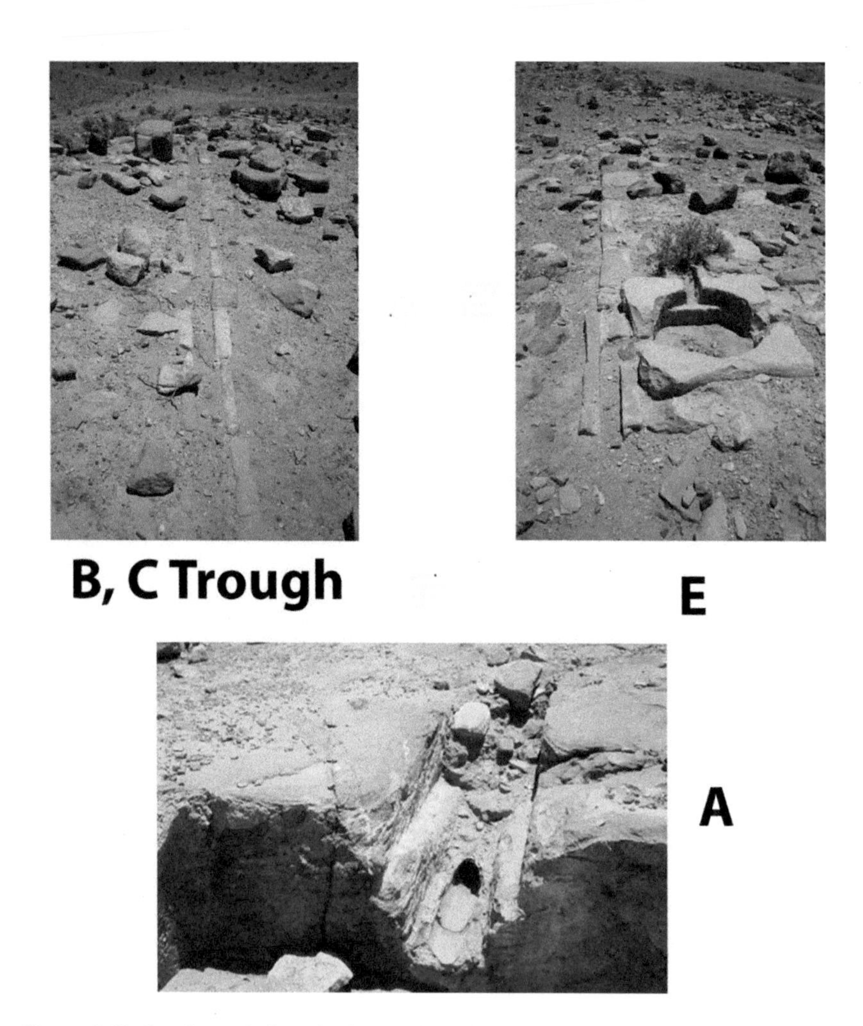

Figure 8.19 Canal trough, junction box and pit pipeline located on an elevated plateau supporting the Ain Braq pipeline to the Zanthur housing district.

utilizes an input water velocity consistent with Figure 8.4 for full flow conditions in a siphon (or pressurized pipeline). A level portion of the pipeline exists before the pipeline branches continue on to steeper terrain to supply downhill destinations. Assuming an Ain Braq supply branch, the pipeline-cistern system shown in Figures 8.19, 8.20 and 8.21 served to manage transient flow excursions as Ain Braq spring output can vary depending on recharge history from seasonal rainfall duration and intensity as well as water diversions into other pipeline branches upstream of the distribution center and dual pipeline branches.

The CFD model represents a logical compilation of the elements shown in Figures 8.18 and 8.19. In Figure 8.19, surface elements show a pipeline trough B, C

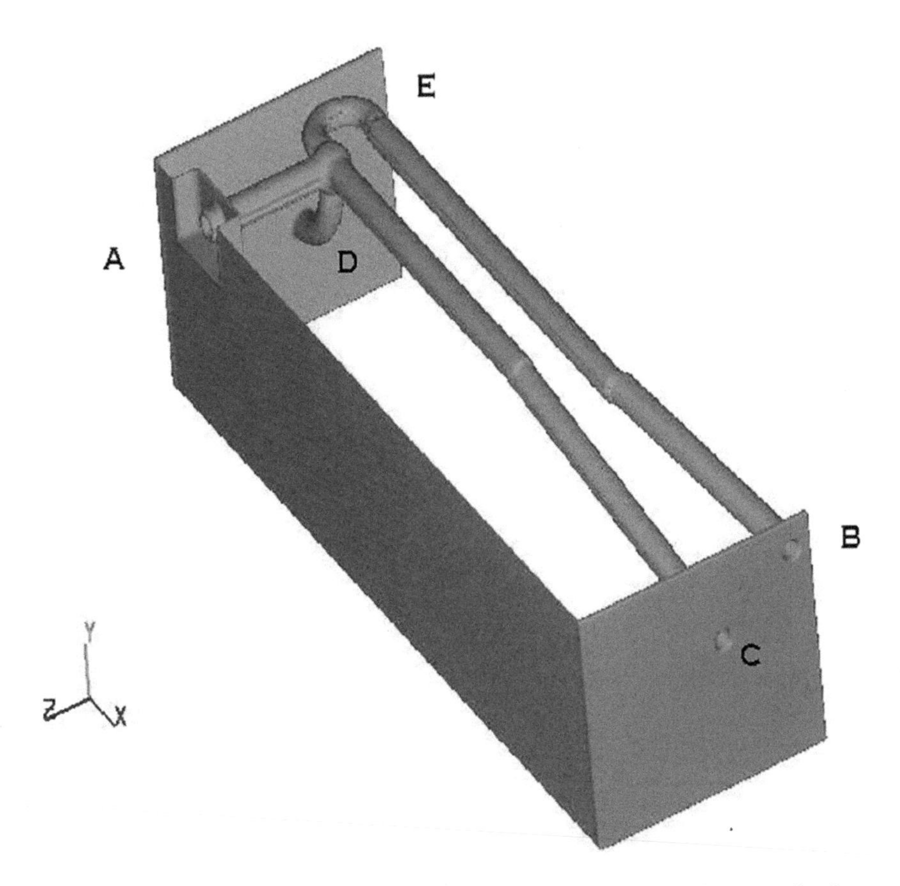

Figure 8.20 FLOW-3D model uniting Figure 8.19 features to compose the water distribution system – D is the end connection of the water supply from frontal theater piping to the distribution center E; A is the pipeline-cistern, B and C pipelines deliver water to the Zanthur housing district and Great Temple/Paradeisos areas of the city center.

to a lower elevation housing district, E is a junction box connecting different pipelines and A is a pit displaying a pipeline entrance connecting to adjacent pipelines. To construct the various elements into a logical CFD model, Figure 8.20 is next composed. In Figure 8.20, water enters at station D (from one of two possible water sources as previously mentioned) then transported down steep pipelines E to housing destinations B, C. The pit pipeline A connects to pipelines E. Questions arise as to the hydraulic engineering function of this arrangement.

For one function, the pit pipeline A served to divert flow away from the dual downhill branches if these were blocked during repair or if the distribution center was blocked to divert water to other pipeline branches of the A in Braq system. The blockage then diverts water to pit A in Figure 8.20. The flow into the pit

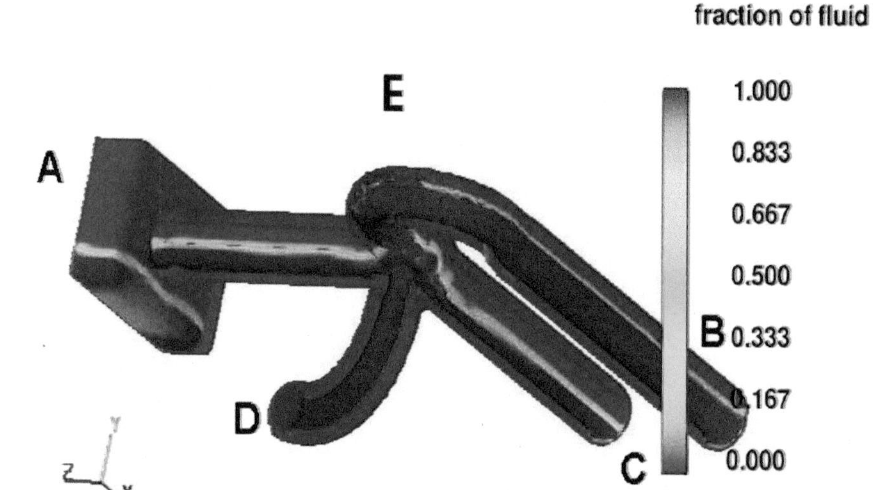

Figure 8.21 FLOW-3D Ain Braq internal pipeline fluid fraction (f) results indicating atmospheric air transfer from the pipeline-cistern A to relieve partial vacuum regions in pipeline branches C and D. f = 0 denotes air; f = 1 denotes water; intermediate f values denote water vapor.

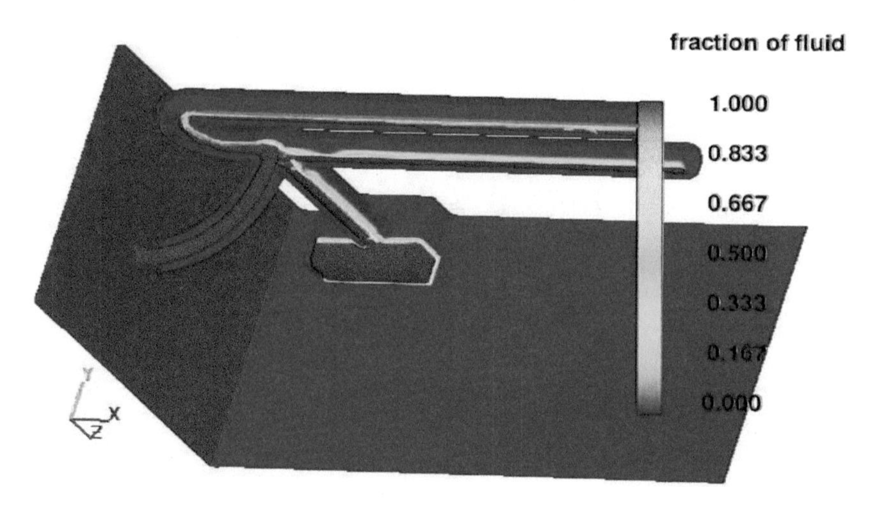

Figure 8.22 Alternate view of the Figure 8.20 model showing air above the water partial flow.

continues until a pressure balance of water accumulated in the Figure 8.19 pit D reservoir A in Figure 8.20 balances with the pressure in water supply pipeline A in Figure 8.20. An automatic system shut off follows once the lower exit ports B and C (Figure 8.20) are blocked and the hydrostatic pressure in the reservoir matches the pressure from the supply pipeline A.

Figure 8.23 The Great Temple.

A further sophisticated hydraulic engineering use of the pipeline Figure 8.20 configuration originates from the observation that subcritical full flow conditions exist in the supply pipeline D due to the water supply options previously discussed. Due to the hydraulically steep declination angle in dual pipelines E, input full flow past the junction area is converted to supercritical partial flow in branches B, C. This flow transition induces partial vacuum regions above the partial flow water surface. As the subterranean branch pipelines exits are submerged into reservoirs at their destination locations, then air is drawn into pipeline exit openings into the partial vacuum region. The intermittent air bubble stream proceeds upstream in D to counter partial the vacuum regions leading to flow intermittency and an oscillatory hydraulic jump in the dual pipeline branches. This hydraulic phenomena then affects the flow stability of the entire system leading to pulsating discharges into the B, C reservoirs as well as affecting the stability of the input flow at D. Additionally, pulsating flows cause internal pressure changes that can affect pipeline joint connections causing leakage. Now a cure to this negative condition was anticipated by Nabataean water engineers when air is supplied by the pit pipeline A shown in Figure 8.20. Air led into the partial vacuum region cancels the vacuum regions and subsequent flow instabilities in the dual subterranean pipeline branches (Figure 8.20). Figures 8.21 and 8.22. In these figures, fluid fraction ff = 1 denotes water; ff = 0 denoted air. These figures indicate the transfer of air into the E pipelines through the pit pipeline A.

Excavation of the Great Temple area over many years (and still continuing) by the Brown University team under the direction of Martha Joukowsky (1998, 2004) revealed that rainfall seepage from a hillside wall south of the temple was collected into a drainage channel (Ortloff 2009:266) that led to an underground reservoir on the east part of the site. A further subterranean channel starting from the surface entrance to the reservoir led water to the front northern platform part of the Great Temple site and likely served as an overflow drain channel activated when the reservoir was full.

Conclusions

Three water supply systems required different hydraulic engineering approaches to overcome problems and optimize water supply to Petra's city center. In each case, problems were overcome by engineering designs requiring empirical knowledge of fluid mechanics principles known in a prescientific format anticipating western science formalization some 2000 years later. While the Nabataean hydraulics knowledge base is not available from surviving literature from that period, its breadth may be gauged by use of CFD recreations of solutions that inform of thought processes available to Nabataean hydraulic engineers. As many of their solutions exhibit modern approaches, one can now attribute a role in the history of hydraulic science to Nabataean engineers.

Note

1 The "Appendix B" referenced in this chapter appears on page 365 of this book.

9 The Pont du Garde aqueduct and Nemausus (Nîmes) *castellum*

Insight into Roman hydraulic engineering practice

Historical background

The Pont du Garde aqueduct, built during AD 40–60 under the reign of Claudius, is composed of many individual hydraulic engineering components (Lewis 2001:181–188; Hodge 1992:184–190) that worked collectively to deliver water to the Roman city of Nemausus (Green 1997; Sage 2011) now the present-day city of Nîmes in southern France. Water from the Fontaine d'Eure spring at Uzès was conducted to a regulation basin at Lafoux with an over-capacity diversion channel to the Alzon River designed to deliver a maximum of 40,000 m^3/day to the ~50-km-long channeled aqueduct leading to the Pont du Garde aqueduct/bridge spanning the Gardeon River. A further extension of the channel to a tunnel delivered water to a basin distribution center (*castellum*) 17 m above the city of Nemausus; water delivered to the *castellum* then was conducted through 13 pipelines to city center reservoirs and site locations. The aqueduct flow rate of 40,000 m^3/day (Hauck and Novak 1988) was delivered under the 1.2 m wide by 1.10 m high basin's rectangular sluice gate into a 5.5 m diameter, 1.0 m high basin wall supporting ten *centenum-vicenum* ~30-cm inner diameter pipelines distributed around the basin's periphery together with three basin floor pipelines.

The Pont du Garde aqueduct/bridge crossed the Gardeon River near the town of Vers-Pont du Garde in southern France and was a key element of the ~50 km long aqueduct water transport system (Figures 9.1 and 9.2) providing water to Nemausus. While the straight-line distance between the spring source and the terminal distribution *castellum* was ~25 km, the channel path selected by Roman engineers was a winding route measuring ~50 km because of construction difficulties associated with the mountainous Garrigues de Nîmes direct route path (1). Roman surveyors selected the longer channel path to avoid difficulties associated with building numerous tunnels and bridges through mountainous terrain that would accompany the shorter length path that led directly from the spring source to the Nemausus *castellum*. In addition to construction difficulties associated with the mountainous and deep gorge terrain in the northern section of the proposed aqueduct path, further routing changes proved necessary to circumvent the southernmost foothills of the Massif Central known as the Garrigues de Nîmes. These foothills would prove difficult to cross with the shortest length water channel as

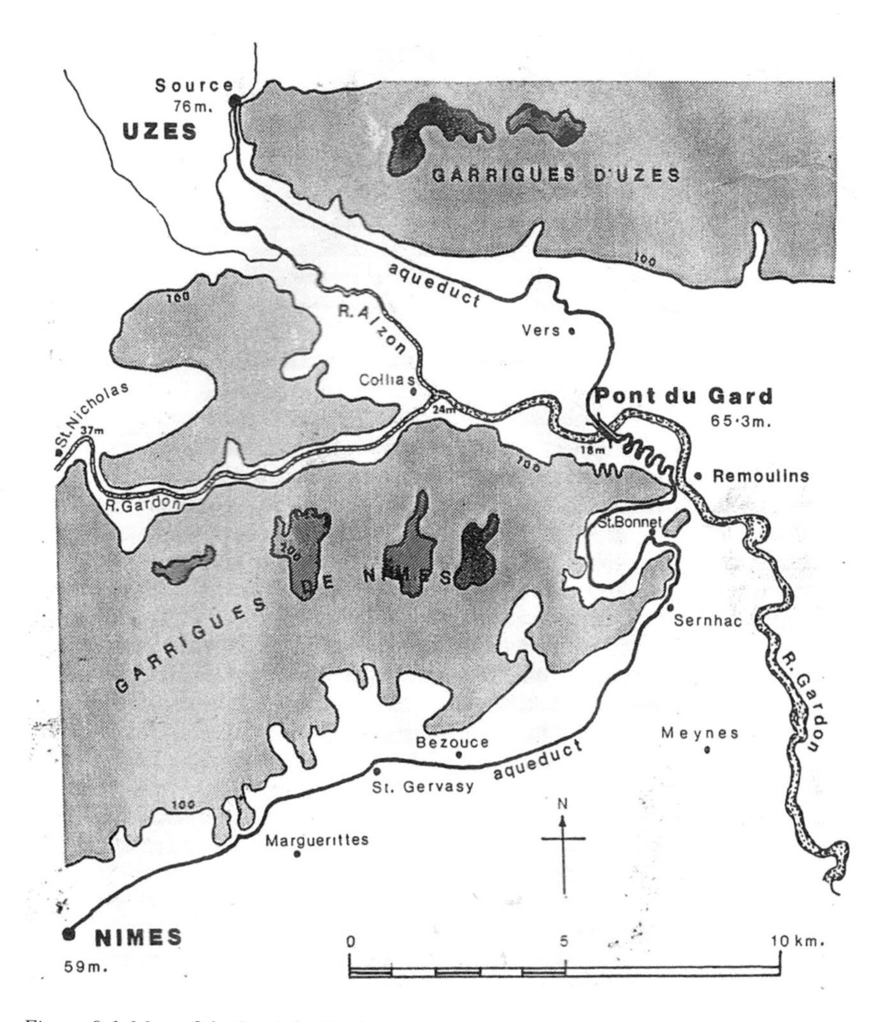

Figure 9.1 Map of the Pont du Garde aqueduct water system from the Fontaine d'Eure at spring to the Nîmes *castellum*.

they were covered in dense vegetation and indented by deep valleys requiring the construction of many small bridges and tunnels. In this area alone, passage through a long section of hills and ravines would require a tunnel between 8 and 10 km long depending on the starting point. A diversion course around the eastern end of the Garrigues de Nîmes mountain range (Figure 9.1) was the only practical and economic way of transporting water from the origin spring to the city to reduce construction time and minimize labor costs. The aqueduct design selected followed a long, low slope, winding covered channel path free of major construction obstacles along most of its route whose completion included the Pont du Garde aqueduct/bridge (Figure 9.2) spanning the gorge of the Gardeon River;

Figure 9.2 The Pont du Garde aqueduct/bridge.

past the aqueduct/bridge, a covered continuation channel and tunnel led water to the terminal *castellum* located above the city. Details of the aqueduct geometry determined the design of the receiving *castellum* and its pipeline water distribution system as detailed in discussions to follow. The aqueduct was designed and built to carry a given flow rate – the challenge to Roman engineers was to design the *castellum* to transport the large aqueduct input flow rate through a limited number of pipelines (13) in the most efficient manner. The innovative *castellum* design devised by Roman engineers to accomplish this end is described in later sections and gives a penetrating look into Roman hydraulic engineering practice.

In the 1st century AD, Nemausus was a prosperous Roman colony whose resource base consisted of Rhone Valley agricultural fields and vineyards to support trade and export to central Rome. The colony's prosperity was mirrored by population growth from 20,000 to 40,000 over a short time span leading to official

Roman city status. At the foot of Mount Cavalier, the Nemausus fountain neither sufficed to provide the city with its daily need of potable drinking water nor sufficient water for the baths, fountains, temples, government and commercial sector buildings and the many garden areas that Roman cities incorporated as part of corporate Roman city design practice. Based upon increased water needs for the city, foresight and planning to start the early building of an aqueduct water supply system from the Eure Uzés spring source to Nemausus anticipated the future water needs of the city. It is estimated that about 70% of the aqueduct channel pathway was constructed as excavated stone lined trenches with slab or arched roof covering with the remainder largely in the form of short length tunnels and small bridges. As a major construction challenge, the Gardeon River valley crossing required engineering innovations involving the design and construction of the multitiered Pont du Garde aqueduct/bridge to transport water across the river gorge area.

The planning and construction of the aqueduct has been credited to Augustus' son-in-law and aide, Marcus Vipsanius Agrippa, around the year 19 BC. At that time, Agrippa was serving as *aedile*, the senior magistrate responsible for managing the water supply of Rome and its colonies. Espérandieu, writing in 1926 (Hodge 2011), linked the construction of the aqueduct with Agrippa's visit to Narbonensis and newer excavations suggest the construction may have taken place between AD 40 and 60. Earlier built tunnels bringing water from local springs to the city dating from the time of Augustus were bypassed by the builders of the newly conceived aqueduct due to the projected increased water needs of the growing city population; coins discovered in the outflow pipeline catchments in Nemausus are no older than the reign of the emperor Claudius (AD 41–54) to more securely date construction times. On this basis, a team led by Guilhem Fabre argued that the aqueduct must have been completed around the middle of the 1st century AD. It is estimated (Hodge 2011) to have taken about fifteen years to build, employing between 800 and 1000 workers.

En route to the *castellum*: the Pont du Garde aqueduct/bridge

The Pont du Garde aqueduct/bridge (Figure 9.2) has three tiers of arches, stands 48.8 m high and descends only 2.5 cm over its length of 274 m (a gradient of 1 in 18,241). The entire water channel from the spring source to the Nemausus *castellum* descends in height by only 17 m over its entire ~50-km length indicative of the challenge in surveying precision that Roman engineers encountered and achieved (Lewis 2001). Figure 9.7 indicates average slopes in meters per kilometer (and equivalently in degrees) over long sections of the channel (Hodge 2011); the source is the Fontaine d'Eure spring at Uzès and the C notation denotes the *castellum* terminal location. Although Figure 9.7 indicates different mean constant slopes over long stretches of the aqueduct channel, local slope variations existed along channel path lengths due to landscape surface irregularities and accuracy limits of surveyor's instruments. The aqueduct was estimated to carry ~40,000 m^3/day of water to the fountains, baths, gardens, temples and homes of the citizens of

Nemausus (Hauck and Novak 1988). The water supply system may have been in use as late as the 6th century, with some parts used for significantly longer times, but lack of maintenance after the 4th century led to accumulation of mineral deposits (sinter) and debris that eventually limited the flow of water (Hodge 2011).

Figures 9.3, 9.4, 9.5 and 9.6 show construction details of the *castellum* located above the present-day city of Nîmes. From the individual 10 *castellum* basin wall ports shown, 10 terracotta pipelines directed water to individual fountains, *nymphea*, baths, temples, gardens, commercial sector and administrative buildings and private homes around the city (Hodge 2011:284). Figures 9.3, 9.4 and 9.5 show that pipeline pairs were originated from the *castellum* basin outer wall – the reason for this is at present obscure. Figure 9.6 shows three pipeline entrance ports located on the floor of the *castellum* basin. These ports may have served a drainage function for repairs and cleaning of the *castellum but*, more practically, served to provide water to sites that required a continuous water supply such as the numerous city water supply basins serving as the main source of potable water for the citizenry.

The Fontaine d'Eure spring, at 76 m above sea level, is 17 m higher than the *castellum* water distribution basin above the city of Nemausus; this provided a sufficient gradient to sustain a steady flow of aqueduct water to the population of the Roman city. The aqueduct's average gradient is only ~1/3000 but the local slope varies widely along its course (Lewis 2001:184–186) being as small as ~1/20,000 in some sections (Figure 9.7). The average gradient between the start and end of the aqueduct (0.34 m/km) is far shallower than usual for Roman aqueducts being only about a tenth of the average gradient of some of the aqueducts

Figure 9.3 Front view of the Nîmes *castellum*.

Figure 9.4 View of the front of the Nîmes *castellum*.

supplying Rome. The reason for the disparity in gradients along the water system's route is that a uniform gradient would have meant that the Pont du Garde aqueduct/bridge would have an extreme height and present a formidable construction challenge given the limitations of Roman construction technology in early centuries. By maintaining a steeper gradient along the channel path ahead of the aqueduct/bridge (Figure 9.7), Roman engineers were able to lower the height of the aqueduct/bridge by 6 m to a total height of 48.7 m above the Gardeon river bed – still exceptionally high by Roman standards but within acceptable limits. This height limit governed the profile and gradients of the entire aqueduct and created a slight depression in the middle of the aqueduct/bridge due to the weight of the multi-tiered stonework. The gradient profile before the aqueduct/bridge is relatively steep descending at 0.67 m per kilometer but descends by only 6 m over the remaining channel length to the *castellum*. In one channel section, the winding route between the Pont du Garde aqueduct/bridge and St. Bonnet (Figure 9.1) required an extraordinary degree of accuracy from Roman engineers who had to survey for a decline of only 7 mm per 100 m of the aqueduct.

It is estimated that the aqueduct supplied the city with a minimum of 20,000 m^3/24 hours to a maximum ~40,000 m^3 of water per 24 hours and that water took 28 to 32 hours to flow from the Uzés spring source to the city (Hauck and Novak 1988). The different limits in flow rate and delivery times reflect the seasonally variable spring discharge rate dependence upon the groundwater recharge rate from infiltrated rainwater and received groundwater from surrounding infiltration

Figure 9.5 View of structures in front of the Nîmes *castellum*.

areas. The *castellum* was designed for use at the maximum flow rate that the Uzés spring source could supply on a steady basis – this being necessary to provide sufficient water for the large city population. Average water velocity was on the order of ~0.5 m/sec according to one estimate (Hauck and Novak 1988) corresponding to the total maximum flow capacity of ~40,000 m^3/day. Aqueduct water arrived in the *castellum* at Nîmes – an open, shallow, circular basin ~5.5 m in diameter by ~1.0 m deep (Figures 9.3, 9.4 and 9.5) and would have been surrounded by a now

Figure 9.6 Interior view of the Nîmes *castellum* showing three floor drainage ports.

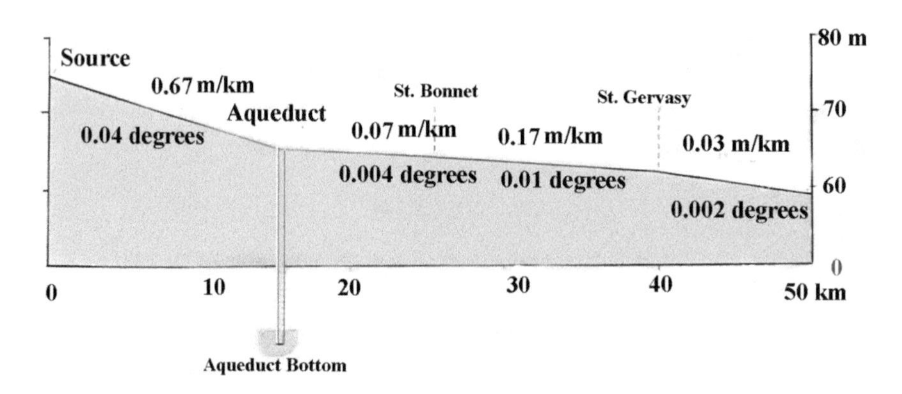

Figure 9.7 Average slopes of sections of the Pont du Garde aqueduct in m/km and equivalently in degrees.

lost balustrade within an enclosure under a small but elaborate pavilion. When the *castellum* was first excavated, traces of a tiled roof, Corinthian columns and a fresco decorated with fish and dolphins were discovered in fragmentary condition. As to details of the construction of the *castellum*, Figures 9.3 to 9.6 indicate that adjoining curved stone slabs lined the receiving basin inner rear wall. For the 10 basin wall pipeline ports, it appears that large adjoining stone blocks were carved in arcs then pierced by circular openings to accommodate pipeline insertion. The three floor ports appear to be carved through the leveled stone floor with carved passageways permitting pipeline attachments (Figure 9.5). The basin floor appears to have an outer rim upon which the arched basin wall blocks were placed. Details of sealant used to reduce leakage are not currently known but many types would be available ranging from asphalt to mixtures of volcanic grains mixed with cement frequently used in underwater structures. Due to variable seasonal water output from the Uzés spring exceeding the aqueduct design flow rate, a regulation basin at Lafoux with a diversion channel to the Alzon River (Figure 9.1) maintained the maximum flow rate to the aqueduct water system on the order of ~40,000 m^3/day. A (now lost) movable sluice gate located at the rectangular *castellum* water delivery opening (Figures 9.3 and 9.4) regulated basin water height, as well as water velocity, into the 10 pipeline ports located around the inner circumference of the containing basin (Figures 9.3, 9.4 and 9.5) and the floor ports shown in Figure 9.6. The flow area expansion from the entrance port to the basin significantly lowered the water entry velocity to the multiple pipelines. Conjecture as to the design of the entrance sluice gate structure (Hodge 2011:286) prevails with no current resolution as to the design intent of its function to regulate *castellum* water height and velocity – this issue is now addressed in the discussion to follow. A series of holes penetrating the top plate of the entrance structure exist (Hodge 2011:286, Figure 200) but as to the sluice plate lifting mechanism controlling the water entrance opening height, no consensus yet exists. The regulation of the water flow rate from the aqueduct to the distribution *castellum* was an important Roman design consideration to avoid basin over spillage not only from upstream reaches of the aqueduct carrying channel but also from the low wall height *castellum* (Figures 9.3, 9.4, 9.5). These considerations were inherent to the design of the *castellum* water system based upon the maximum estimated input flow rate. As the aqueduct carrying channel cross section was rectangular with a constant width, large variations in flow rate from intercepted local rain storm runoff that would cause water height changes in the low slope channel were anticipated by Roman hydraulic engineers by the local height increases of the channel walls where over spillage was likely. Similarly, changes in channel slope resulting in water height changes were anticipated by Roman engineers through variation of channel wall height changes.

The *castellum* floor ports (Figure 9.6): hydraulic engineering design considerations

Aqueduct water entered the *castellum* through a rectangular opening 1.2 m wide by 1.10 m high (Figures 9.3, 9.4, 9.5) and large circular holes in the basin containment wall, each ~40 cm in diameter, give indication of the pipeline dimensions that directed water into the 10 basin wall pipelines. The three floor pipeline ports (Figure 9.6) when open, presuming a continuous aqueduct water supply, would induce a vortex over each floor entrance port inducing rotation of water entering the three floor pipelines. Presuming pipeline elbow transitions below the basin floor as Figure 9.6 indicates, water rotation would amplify through each individual elbow. The rotating flow, given its passage through low angle pipeline slopes necessary to reach city destinations, ultimately would transition from full entry port flow to partial flow within the adjoining sloped pipelines due to flow gravitational acceleration that would increase pipeline flow velocity and lower the water height. To define the full and partial flow terms, a pipeline cross-section area fully occupied with water is denoted as full flow; a pipeline cross-section area partially occupied with water is denoted as partial flow. For high speed flow in highly sloped pipelines with significant internal pipeline wall roughness, a downstream hydraulic jump may occur in the region between the floor port entry full flow and the post hydraulic jump, full flow region. Between these full flow regions, a partial vacuum exists over the intervening partial flow region. Figure 9.9 B illustrates this condition. Unless the partial vacuum region is relieved by pipeline top holes to admit atmospheric pressure, flow delivery instabilities arise as air enters the floor pipeline inlet by means of an air-entraining vortex extending to the water basin surface (Lugt 1983:37,44; Figures 3.24, 3.35) together with air entering at the pipeline exit (for either submerged or free overfall conditions) to relieve the partial vacuum region. This effect causes transient, oscillatory motion of the hydraulic jump region resulting in *castellum* basin water level oscillations resulting in unstable flow entering into pipelines. This effect is largely governed by pipeline slope, diameter, internal pipeline wall roughness, rough pipeline segment connection joint roughness and input flow rate and hints of the complexity that Roman engineers contended with to produce stable pipeline flow to city destinations. As Roman engineers had concern about flow and pressure instabilities that induced pipeline vibrations (Nielsen 1952:111–170) that loosened connection joints between pipeline elements to cause leakage, pipeline flow stability concerns related to *castellum* design was a major problem to be addressed and eliminated by an advanced *castellum* design.

As a special case, if interior pipeline wall surface roughness is minimal and for low flow rate entry to floor ports maintained by partial opening of the floor port covers, then water flow may continue as partial flow into the basin floor pipeline entrances and the adjoining pipelines over large distances. This partial floor opening design option helps eliminate pressurized pipeline joint leakage with the further possibility of producing a maximum, stable flow rate if careful surveying of pipeline slope can be achieved to maintain continuous partial flow up to the delivery location. As this result depends upon *castellum* design, there is more to consider by Roman engineers to achieve this positive result to eliminate an interior pipeline hydraulic jump and its effect on flow instability that induces pipeline joint leakage. Therefore it is probable that for whatever use the three floor ports were intended, given problems associated with pulsating flow delivery, certain pipeline slopes would be selected by Roman engineers to limit unsteady hydraulic jump formation and maintain partial flow throughout with free overfall exit conditions to a reservoir. How this was accomplished by ingenious *castellum* and pipeline designs is described in a subsequent discussion related to the Main Aqueduct. Although it has been suggested that the three floor ports were mainly used for flooding the amphitheater for mock naval battles, the allowable flow rate through these three bottom ports alone is far below the input flow rate from the aqueduct as a later discussion verifies. The three floor ports may have served the purpose of continuous flows to important sites but are inadequate to carry the 40,000 m^3/day aqueduct flow rate without several of the 10 basin wall pipelines in use as subsequent discussion indicates. If aqueduct flow is diverted by blockage and flow diversion to the Alzon River for aqueduct cleaning and repair functions, then the bottom ports would well serve to completely drain the *castellum* basin as the 10 basin side wall ports are elevated from the basin bottom and not able to fully drain the basin.

The 10 *castellum* wall ports (Figures 9.3, 9.4, 9.5): hydraulic engineering design options

Allowing for pipeline wall thickness of at least ~2.54 cm, then several standard Roman pipeline sizes (Hodge 2011:297) are candidates for the pipelines emanating from the basin wall ports as well as the basin floor ports. These are the (120A) *centenum-vicenum* with a diameter of 22.83 cm and inner cross-section area of 409.4 cm^2 and the larger (120B) *centenum-vicenum* with an inner diameter of 29.5 cm and a cross-section area of 686.6 cm^2 (Bennett 1961). To include known Roman pipeline entrance flow devices in the discussion to evaluate their design inclusion merits, *calices* mounted in a horizontal pipeline section are considered as they are typical of Roman practice (Hodge 2011:295) for flow measurement and can be used to regulate and/or limit flow rates when used as chokes. A table of *calyx* sizes and an illustration of a *calyx*-pipeline connection (Hodge 2011:295, Figure 297) in the 50, 80 and 100 digit sizes with diameters of 27.8, 45.5 and 57.4 cm, respectively, may have been considered to regulate amounts of water flowing in different pipelines to different destinations with different prescribed water needs. (Chapter 10 of this book provides new information on the relation between designated flow rate markings on calices and actual *quinaria* flow rates.) While pipeline types have known standards, large bronze *calices* placed directly into the 10 basin wall entrance holes were a design option (see Chapter 10 for further *calyx* discussion). Given that *calices* work only under full flow conditions, their use by Roman engineers in the *castellum* pipeline entry ports may have appeared useful for precise flow rate delivery to destination sites given that full flow at basin wall pipeline entrances could be maintained by means of a horizontal pipeline element before pipeline slope continuance converted full to partial flow in the pipeline extension. This design option would require that the sluice gate would be fully open and that basin wall height exceeds the observed wall height shown in Figures 9.3 to 9.6. The higher basin wall height design would require a higher elevation of the tunnel supply line and an even lower aqueduct slope leading to the tunnel. Sustaining entry full flow into basin wall pipelines would rely on the increased hydrostatic pressure head and the ability to maintain constant water height in the elevated wall height basin just to support *calyx* usage. The advantages of full entry flow incorporating *calices* placed at the start of horizontal piping branches with markings to indicate the output flow rate had yet a further disadvantage beyond height reconfiguration of the inlet tunnel. *Calices* used at the entrances of all 10 basin wall pipelines theoretically provided the sum of their flow rates and therefore were vital to match the aqueduct flow rate. As *calyx* sizes appropriate for the ~30 cm inner diameter pipelines give erroneous flow rates based on the nozzle diameter rather than the square of the diameter appropriate to the cross-sectional area of the nozzle, the correct flow rate prediction (full flow rate = flow velocity times flow cross-section area) would ultimately be a problem if installed due to the inaccuracy of flow rate measurement. If this design option were pursued with *calyx* installations, then major redesign and reconstruction of the *castellum* would follow upon proof tests using the full aqueduct flow rate. As the existing *castellum* design is vastly different from this described design, this indicates that Roman engineers were aware of the inaccuracy of *calyx* water measurement devices and the changes in the aqueduct slope and tunnel elevation underlying that basin design. As the existing tunnel entry slope is already small (0.002 degrees, Figure 9.7), an even lower slope design would severely challenge surveying accuracy measurement capability as well as the transfer of the 40,000 m^3/day flow rate aqueduct supply.

Apparently the concept of water velocity and its measurement were recognized as important to determine flow rates but such considerations were not readily available to Roman engineers due to lack of precise time measurement devices (Hodge 2011). From these considerations, it was apparent that this *calyx* based design option was not practical. Based on the complexities of this design option for the *castellum*, Roman engineers would ultimately need to choose a simpler design that solved all the problems associated with this design option. What then was the practical Roman *castellum* design that solved all the problems mentioned to match input and output flow rates?

As no traces of actual pipelines or *calices* exist at the present *castellum* site, pipeline connection details, as well as the *exact* pipeline diameter used, remain conjectural. Nevertheless a reasonable estimate of pipeline diameter may be made for flow rate estimation purposes based upon the *castellum* retaining wall diameters shown in Figures 9.3, 9.4 and 9.5 and the floor ports shown in Figure 9.6. While the three floor ports likely were closed for normal daily usage, an early 19th-century investigator (Auguste Pelet) suggested that the three drains were used to supply water for *naumachia* (mock naval battles) at the nearby amphitheater. Based upon the preceding discussion, for the three large ports located in the *castellum* floor used to enable the nearby amphitheatre to be flooded rapidly, this function would require the addition of several of the 10 wall outlet ports to work in conjunction with the floor ports to accommodate the 40,000 m^3/day aqueduct flow rate. The *naumachia* function would necessitate that valves were available in other wall origin pipelines to redirect the entire flow to the amphitheater.

That such large valves were in the Roman engineer's purview is demonstrated by Herschell (1973). For the present analysis, it is assumed that all 10 side wall ports were in continuous use but not the three bottom ports at the same time – this conclusion underlies Table 6.1 results to follow.

Since the *castellum* was elevated above the lower Roman city and few traces of the multiple water destinations and connection pipelines now exist, it may be assumed that pipeline lengths were on the order of a fraction (or more) of a kilometer from the *castellum* to different city destinations. Figure 9.8 shows a typical Roman arrangement of pipelines from the *castellum* to a lower reservoir after Herschell (1973) to regulate flow to destination sites; several of the 10 pipelines leading from the *castellum* to a reservoir are indicated with multiple distribution pipelines controlling flow to city sites. Each city destination may have had time variable flow rate requirements (particularly baths and private houses) from cisterns and stilling basins so that an overflow cistern was necessary to captures excess water flows and direct water to collection basins serving gardens, pools and storage basins that did not require steady water input. Similar designs to Figure 9.8 were common in use at the Roman site of Pompeii (Hodge 2011). Based

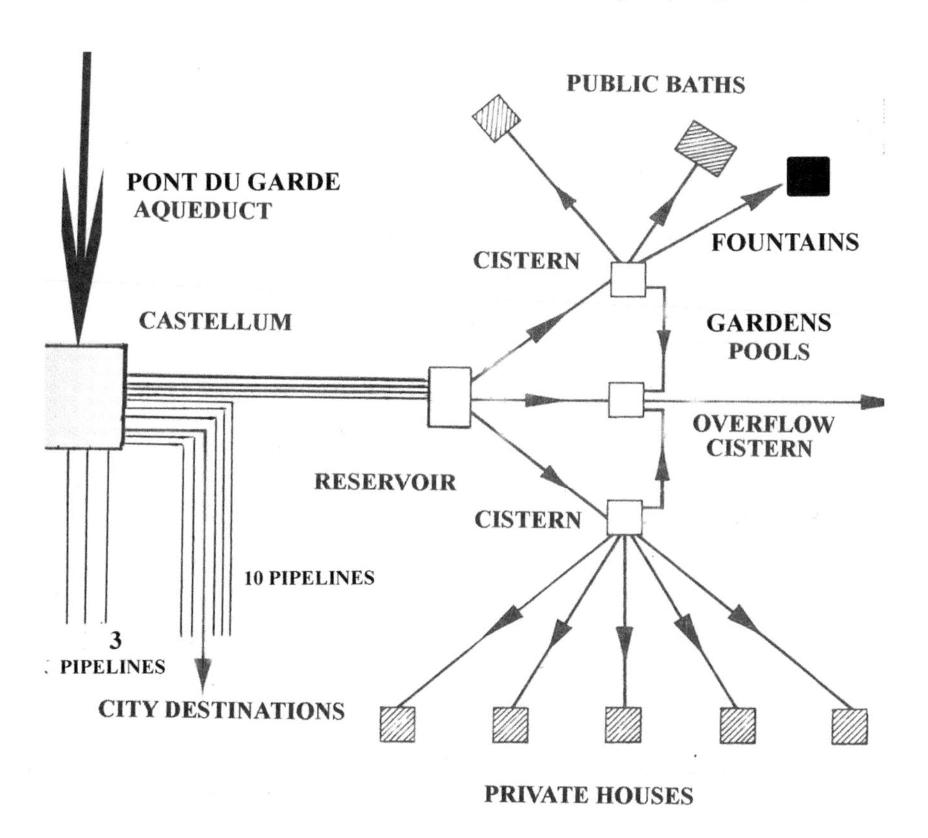

Figure 9.8 Aqueduct supplied water to the *castellum* showing the pipeline redistribution network structure typical of Roman cities.

on this design complexity for a city water distribution system, emphasis on stable pipeline flow delivery would be an important consideration that enhanced minimal maintenance and thus constituted a prime consideration inherent to the *castellum* design.

The early writings of Vitruvius and Frontinus on Roman hydraulic engineering practice (Bennett 1961) are replete with pre-scientific notations of hydraulic phenomena related to flow velocity, flow rates, time and hydrostatic pressure that were used for flow measurement. On this basis, there is much to recommend Roman hydraulic engineering practice as largely based on an observational recording basis based on pipeline slope effects on flow delivery as opposed to results that could be derived from calculations. The basic problem in determining flow rate was the accurate measurement of time and water velocity which eluded precise Roman definitions as indicated by Roman water administrator's book descriptions. Such problems therefore rendered *calyx* measurements inaccurate and not useful for *castellum* inclusion. In this regard, earlier Greek hydraulic engineers demonstrated progress in measuring average water velocity and flow rates (Ortloff and Crouch 1998) appropriate to fountain and water outlet designs at Priene.

To this point, the discussion has posited different hydraulic phenomena associated with pipeline flows for the basin wall ports and the floor ports. To understand pipeline flow phenomena dependent upon pipeline slope and basin attachment choices, discussion is next focused on the hydraulic positives of an alternate design choice. Here the *castellum* design contains elements of advanced Roman hydraulic technology thoughtful of the effects of pipeline slope choices and the effects derived from knowing how to control their flow rates. By the analysis to follow, design elements noted in the *castellum* are revealed that were known to Roman engineers as to how to design an optimum water delivery system that effectively and efficiently matched the aqueduct input flow rate to the *castellum* output flow rate and avoided flow instability problems. Key to an optimum design was knowledge of the effect of pipeline slopes emanating from the *castellum* to promote high flow rates and flow stability.

Toward the optimum *castellum* design

Many Roman pipeline designs transferring partial flow include top hole openings to eliminate the partial vacuum region (Ortloff and Crouch 2002; Ortloff 2009:311, 312) but use of this feature for the present case is not known due to absence of pipeline remains. In the absence of pipeline top holes, air ingestion at the pipeline exit port or from the basin water surface occurs in the pipeline to counter the partial vacuum region resulting in an unstable flow delivery rate (Ortloff 2009:317). From these destabilizing effects, pulsating forces acting on pipeline joints promote leakage and erratic flow delivery. This condition would exist for either free or submerged pipeline exit conditions into a reservoir. If pipeline flow delivery to a reservoir (Figure 9.8) was erratic, induced flow oscillations would cause erratic transmittal of water flow to other distribution basins (Figure 9.8). As these considerations were known to Roman engineers, their *castellum* and pipeline design choice must necessarily reflect a design that would eliminate erratic flow oscillations and that this problem involved knowledge of preferential pipeline slopes in some manner. A further consideration, likely known to Roman engineers, was that maximum pipeline flow rates are associated with partial flow – not full flow – and this is best initiated with partial flow into pipeline entrances and subsequent pipeline lengths.

This design option chosen is revealed by the closeness of pipeline entrances to the top rim of the *castellum* basin. The water height entering pipelines is controlled by lowering the sluice gate to precisely control the entry water height to the pipelines; this function lies behind the previously noted (Hodge 2011) elaborate control mechanism of the sluice gate and lies behind production of the highest pipeline flow rate when the slope equals the critical slope (Henderson 1966; Morris and Wiggert 1972). The physical significance of establishing critical flow conditions in pipelines emanating from the *castellum* lies in the fact that when pipelines are set at a critical slope, this minimizes the energy expenditure to transport the flow at the highest flow rate. The lower the energy expenditure to transport pipeline flow, the less reliance on supplemental ways to increase flow rate such as an elevated *castellum* basin wall and increased water height to provide additional hydrostatic pressure to increase flow rate. The resulting flow geometry under critical slope conditions is illustrated by Figure 9.9 Case C. As partial critical flow can be maintained over the entire pipeline length given smooth interior pipeline walls, atmospheric pressure exists over the partial flow. Pipelines set at the critical slope therefore are free of hydrostatic pressure that would cause water leakage under a full flow conditions. Additionally, critical flow conditions produce the highest pipeline flow rates (Morris and Wiggert 1972). Figure 9.9 Case A (shallow slope) and Figure 9.9 Case B (steep slope) designs at other than the critical slope would be subject to leakage and flow instability problems. Pipeline designs that sustain flows close to partial critical flow over long distances would largely eliminate flow stability concerns from hydraulic jump creation as only a small water contact area with the interior pipeline wall exists under partial flow conditions thus lowering flow resistance effects. Here use of large diameter piping in the *castellum* had the advantage of maintaining partial entry flow continuing into the sloped pipelines as less interior pipeline surface was wetted from the high velocity, low height water stream thus reducing water-wall frictional effects. Such considerations related to the effects of pipeline slope to produce stable exit flows were known to Roman engineers from observation of the many hydraulic engineering projects they implemented. In this regard, the pipeline system at Ephesus (Ortloff 2009:314–320) is a prime example.

With the advantage of flow stability derived from a critical flow design, use of downstream accumulators, water towers, settling basins and open basin reservoirs and settling tanks (Figure 9.8) used to stabilize flow conditions between pipeline exit flows to different destinations with specific flow rate needs (Hodge 2011:244) can be minimized. In the discussion to follow, the English unit system is used as this underlies many of the empirical hydraulic relations used in the analysis to follow.

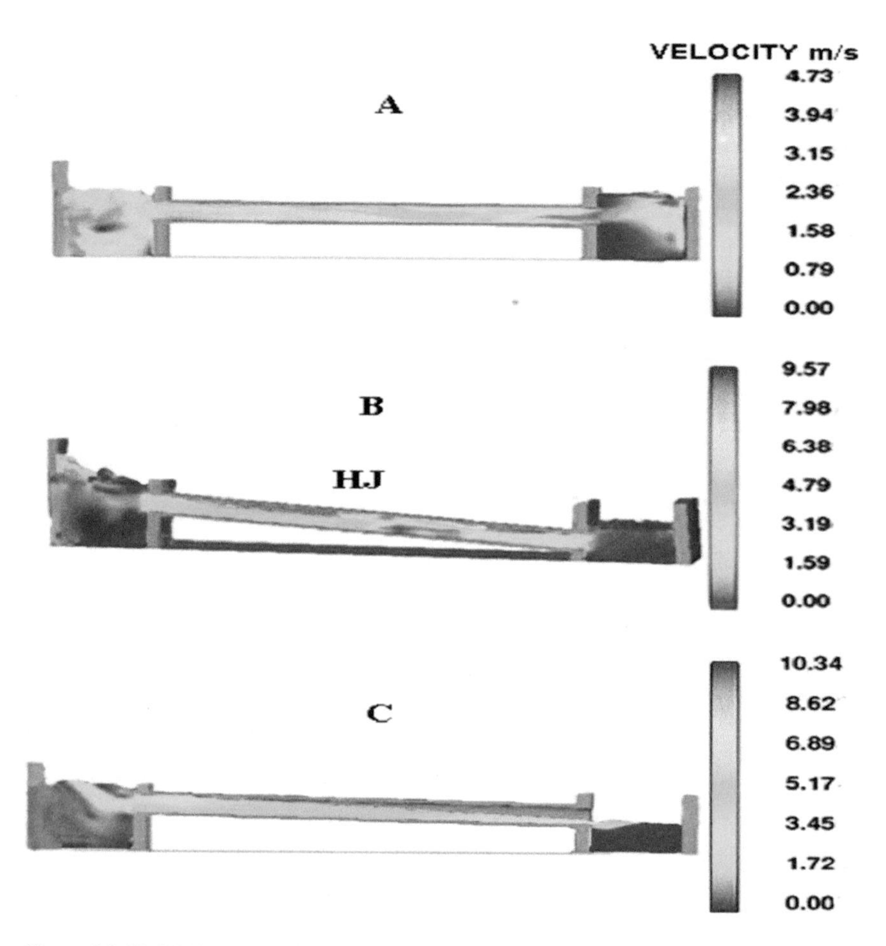

Figure 9.9 FLOW-3D CFD calculation results for pipeline Case A, B and C. Velocity in m/sec.

From the aqueduct input flow rate of 40,000 m^3/day to the *castellum*, the average flow rate for a single pipeline is 1.63 ft^3/sec; for 13 ports open, the average single pipeline flow rate is 1.26 ft^3/sec. From Henderson (2012:51 his Figure 2–15), for D the pipeline diameter (~ 30 cm) and g the gravitational constant (32.2 ft/sec^2), $Q/D^2 (g D)^{1/2} = 0.29$ for 10 open pipelines and 0.22 for 13 open pipelines. From Henderson's Figure 2–15, $y_c/D = 0.5$. This means that the y_c critical depth, which can be regulated by the open height position of the sluice gate, is half of the pipe diameter and that basin water rapidly enters the 10 wall pipeline entry ports at half the pipeline diameter height. This holds even when basin water height is slightly above the top of the pipelines. Now it is clear why the regulated height of the sluice gate in front of the basin entry port has such an elaborate height control mechanism: the gate opening height position can be set to automatically set the pipeline water basin entry level to the critical y_c depth. Here y_c is the water depth

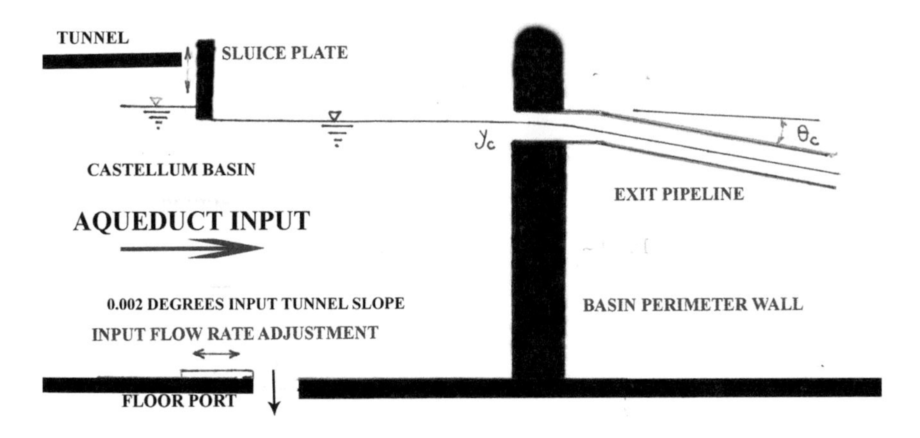

Figure 9.10 Basin exit wall flow dynamics (not to scale).

from the bottom edge of an entry pipeline to the water surface (Figure 9.10). This sets up the continuance of critical flow in pipelines set at the critical slope. Figure 9.9 Case C illustrates this condition as partial flow rapidly originates past the inlet under a critical flow pipeline slope.

The critical, partial entry velocity for 10 pipes open is $V_c = (g\, y_c)^{1/2} = 4.01$ ft/sec and 5.8 ft/sec for the 13 ports open case. Here hydrostatic pressure for the basin floor ports leads to higher entrance velocity. The Froude number, defined as $Fr = V/(g\, D_m)^{1/2}$ is $Fr_{10} \approx 1$ for the 10 port open case and $Fr_{13} \approx 1.3$ for the 13 port open case. The important conclusion is that wall port entry Froude numbers are either near critical ($Fr \approx 1$) or slightly supercritical ($Fr > 1$). The importance of this design feature, as regulated by the sluice gate height position, is that with critical ($Fr = 1$) and near supercritical ($Fr \approx 1$) entry port flows, this eliminates downstream resistance influence that may induce flow instabilities and alter the input flow rates to the pipeline ports. This is a most important design feature of the *castellum* and its sluice gate regulation mechanism. This critical design feature, not present in the alternate design previously examined, excludes upstream resistance effects from internal pipeline wall roughness, elbows, bends, slope changes and internal pipeline hydraulic jumps that would alter the pipeline input flow rate. In modern hydraulic terminology, upstream influence derived from downstream resistance obstacles does not occur for $Fr > 1$ critical or supercritical flows. With pipeline entrance critical flow established, to sustain critical or near critical flows to eliminate upstream influence that would alter the input flow rate to pipeline entry ports, pipeline slopes must be at a critical angle θ_c. Here slopes are defined as negative angles from the horizontal. The θ_c angle is derived from the Manning equation (Morris and Wiggert 1972; Henderson 1966; Bakhmeteff 1932) where n is an empirical resistance constant indicative of worst case internal pipeline wall roughness (given as $n = 0.034$) and connection joint roughness accumulated from the thousands of rough piping sections of ~0.3 m length that comprise long pipelines from the *castellum* to destination sites. Here R_h is the hydraulic radius given by the cross-section area of the critical flow (A_c) divided by its wetted perimeter. The critical pipeline angle θ_c is given by:

$$\theta_c = tan^{-1}\left(n\ V_c / 1.49\ R_h^{\ 2/3}\right)^2 = tan^{-1}\left(n\ Q / 1.49\ A_c R_h^{\ 2/3}\right)^2$$

Substituting, for the 10 port open case, $\theta_c \approx 3.3°$ and for the 13 port open case, $\theta_c \approx 7.1°$ with an average value of ~ 5.2°. This means that for the 3 basin floor pipelines closed, pipeline slopes emanating from the *castellum* should have declination slopes in the range of ~3.3° or higher; for all 13 ports open, the pipeline declination slopes should be a bit higher at ~7.1° to maintain partial critical flow. Note that for smooth internal pipeline walls and connections, the θ_c values would be lower. While final destination site locations may be distant from the *castellum*, at very least pipelines emanating from the basin should have slopes in the $\sim 3.3° < \theta_c < \sim 7.1°$ range to maintain all the critical flow advantages mentioned. Based on the surveying accuracies obtainable by Roman engineers (Figure 9.7), such slope accuracies are easily within their surveying capabilities. As critical or near critical flows are produced within this pipeline slope range, the design features are (1) the flow rates in pipelines are the maximum possible as given by critical flow (Morris and Wiggert 1972; Henderson 1966), (2) pipeline flow stability is achieved and (3) pipeline joint leakage is minimal due to the presence of atmospheric pressure above the partial flow. Note that Figure 9.9 Case C schematically shows the pipeline internal flow pattern under critical slope conditions. The use of this design with its many positive features is essentially given by the design of the *castellum* with pipeline entry ports close to the top rim of the receiving basin.

Possible pipeline types served by the *castellum* (Cases A, B and C)

As the critical pipeline slope configurations are observed as having flow stability and high flow rate benefits, the next task is to examine both critical and off-design, non-critical pipeline slopes that may occur if destination sites require higher or lower slopes to reach. As different pipeline slopes yield different flow rates, the task ahead is to determine what slope choices associated with different destination uses produce flow rates to match the 40,000 m^3/day input aqueduct flow rate.

Three possible pipeline configurations (A, B, C) determined by their slopes originating from the *castellum* basin wall are examined using FLOW-3D 2017 Computational Fluid Dynamics (CFD) models (Figure 9.9). The use criteria involving different pipeline slopes is determined by computing the output flow rate from all 10 pipelines configured at different A or B or C slopes to determine if the total output flow rate is lower, matches, or exceeds the input aqueduct flow rate.

Figure 9.9 CFD model results show plane views of centerline interior pipeline flows for different pipeline slope conditions. The *centenum-vicenum* pipeline with a diameter of ~30 cm is used for the CFD model. The LHS model region represents the *castellum* entry port with a full open sluice gate; the model RHS shows a submerged reservoir catchment at the end of a pipeline with a bottom drainage hole leading water to a destination site or intermediate reservoir. Although only a short pipeline length is illustrated in the models, the results are typical of flow patterns within longer pipelines as once a uniform flow profile is established, it continues over a long distance. Figures are characterized by an average full flow 1.63 ft/sec input velocity to a single basin wall pipeline; individual inlet velocities to pipelines are slightly different due to their relative locations with respect to the supply inlet.

The first figure (Case A, Figure 9.9) shows flow velocity conditions for a near level pipeline leading from the *castellum*. This configuration would provide water to housing located at approximately the same height as the *castellum*) and/or water supply to upper level reservoirs designed to store water at night to later supply, through additional pipelines, baths and discharges to sites with large, immediate water demands exceeding the aqueduct supply rate over a given time period. Flow velocity is low in the near level pipeline due to full flow wall friction effects and, for the low velocity subcritical ($Fr < 1$) flow, upstream influence exists so that distant exit reservoir conditions play a role in determining the delivery volumetric flow rate. This usage would require upper level reservoirs to store water so that when fully charged, valves on pipelines to destination baths would open and have drainage rates higher than the aqueduct supply rate. Once water was delivered, then valves were closed and reservoirs refilled. This cyclical use could be made consistent with bath water change timing provided near horizontal piping has the capability to transfer water at the aqueduct supply rate – a question considered in the next part of the castellum analysis.

The second figure (Case B, Figure 9.9) represents pipeline slopes exceeding the optimum pipeline critical slope. Pipelines at this slope are consistent with the height difference and the distance between the *castellum* and some of the nearby flat areas of present-day Nîmes that once held the streets of Nemausus. Initial entry port flow is full flow and, at a downstream location, partial flow develops in the pipeline until a hydraulic jump (HJ) occurs due to large wall-water friction effects converting supercritical partial to subcritical full flow. A smooth internal pipeline wall would delay the appearance of a hydraulic jump but for the present example case, very rough pipeline internal walls and connection joints are assumed. For significant internal wall roughness, an internal hydraulic jump is created isolating a partial vacuum region between full entry and post hydraulic jump flow regions. If hole openings were placed along piping top regions over the partial vacuum region, then achievement of a stable flow rate would be enhanced. Again, as no extant pipelines exist, the presence of top holes is conjectural but well within Roman technology as seen on Ephesus pipeline designs, those at the Laeodocian site and other Roman sites (Ortloff 2009:311–312; Lewis 2001). Pipeline designs with Case B flow characteristics without a terminal stilling basin would be devoted to lower priority sites that do not require a stable water delivery rate such as gardens, reservoirs, latrine flushing channels and intermittent household use. Other uses may include pipeline water transfer to fountain houses (Ortloff and Crouch 2002; Ortloff 2009:318) that have multiple chambers supporting different hydrostatic head values to transfer water at different flow rates to different destinations; such terminal fountain houses may have been be part of a city flow network. For short pipelines at steep slopes higher than $\theta_c < \sim 7.1°$ the altitude drop from the *castellum* would be consistent with close-in sites.

For display fountains, nymphaea, high status administrative buildings, elite residential areas, Figure 9.9 Case C critical flow designs are preferred as all these destinations would benefit from a high velocity, stable, high delivery flow rate without the use of intermediate distribution reservoirs and stilling basins (Figure 9.8) and thus have an immediate economic benefit to reduce labor costs and system complexity. Case C represents the optimum critical slope condition for which the volumetric flow rate is the maximum possible and, as an atmospheric air space exists over a long stretch of the pipeline length, pressurized pipeline leakage is reduced to a minimum thus producing lower maintenance requirements.

This pipeline choice can be used for city sites reachable by a steeper slope range $\sim 3.3° < \theta_c < \sim 7.1°$ depending on the number of basin entry ports open. This consideration helps city planners place structures requiring large flow rates and helps determine the placement of main reservoirs from which additional pipeline branches emanate to destination sites. Given that the deliberate use of the Case C pipeline design was likely within Roman hydraulic engineer's hydraulic knowledge base, it may be surmised that the lower priority pipeline slopes of the Case B type required a stilling basin attachment before distribution to other destinations (Figure 9.8) and would be of secondary use while the higher priority pipelines requiring a high, steady flow rate directly to a destination site of the Case C type are preferable. As the slope difference between Cases B and C is small and direct use of a Case C design with a precise slope has the constraint of direct access to a destination site by a pipeline of that slope, most probably Roman engineers constructed pipelines as close as possible to a critical slope design to obtain the many benefits listed.

Based upon CFD results, an estimate of the volumetric flow rate is next made for each of the Case A, B and C pipeline designs for 10 basin ports open and results compared to the aqueduct input flow rate of 40,000 m^3/day. If the input aqueduct flow rate exceeds any of the Case A, B and C 10 pipeline outlet flow rates, then such pipeline configurations are not feasible. If the input aqueduct flow rate is equal to the total out flow from 10 pipelines for pipeline configurations given in Case A, B or C, this gives indication of the probable pipeline usage. Here the pipeline slopes range from ~1.0 degree (near horizontal) for Case A, ~5.2° mean slope for Case B, and a critical angle slope for Case C of ~4.1°. The main purpose is to compare the flow delivery rate from 10 pipelines to the aqueduct flow delivery rate for the A, B and C cases. When a match is achieved, then the usages described are viable options for the pipelines emanating from the *castellum*. Table 9.1 summarizes the CFD computed output flow rates for A, B and C pipeline configurations:

For Case A, full flow exists (Figure 9.9 Case A) in near horizontal pipelines with a submerged exit into a reservoir. For Case B, full flow into the pipeline

Table 9.1 Flow rate determination for 13 and 10 *castellum* ports open cases

Type	*Flow velocity (m/sec)*	*Flow volume (m^3/sec)*	*13 ports open 10*	*ports open (m^3/day)*
A	0.20	0.02	~22,400	~17,200
B	0.51	0.04	51,500	~39,600
C	1.22	0.12	~52,000	42,000

entrance transitions to supercritical partial flow (at the asymptotic normal depth) on a steep pipeline slope; a hydraulic jump (HJ) is formed within the pipeline induced by the deceleration of flow by wall frictional effects together with submerged exit flow into a reservoir. For Case C, critical flow exists in a pipeline at a ~4.1° slope yielding the maximum flow rate. The high water velocity is consistent with a small flow height thus lessening the water contact area with the rough interior surface of the pipeline – this largely minimizes the creation of a hydraulic jump from water-wall frictional effects. The pipeline exit flow is assumed to be free fall into a receiving reservoir.

Evaluation of pipeline use for Cases A, B and C

From Table 9.1, Case A low slope pipelines appear to be of minor (or no) use as the total of 10 pipelines open (three bottom ports closed) permit a much lower output flow rate (17,200 m^3/day) through all pipelines than the input aqueduct input flow rate of ~40,000 m^3/day. Even with all 13 ports open, the output flow rate of 22,400 m^3/day is well below the input aqueduct flow rate. The interpretation is that the input aqueduct flow rate far exceeds the capability of Case A near horizontal pipelines to transport such high flow rates. The use of many near horizontal pipelines filling high level reservoirs then appears not to be the main design intent of the *castellum*.

For three bottom ports closed, the calculated Case B flow rate is ~39,600 m^3/day which is close to the estimated aqueduct flow rate of ~40,000 m^3/day. This close flow rate matching produces a steady water height in the *castellum* that guarantees steady flow throughout the water distribution system. This close match signals the Roman engineer's design intent of the *castellum* to provide water to city distribution locations by pipelines of slopes in the $\sim 4.1° < \theta_c < \sim 7.1°$ range. Although a hydraulic jump may occur due to wall roughness effects that transition supercritical to subcritical flow it may largely eliminated if Roman engineers utilized smooth inner wall pipelines cognizant of the benefits to promote flow stability. The near flow rate match indicates the hydraulic technology to make a *castellum* design to closely match the input aqueduct flow rate in advance of the building and testing of installed pipelines. For a 0.25 km long pipeline sloped at ~4°, the altitude drop from the *castellum* to the city area is ~17.5 m which is a reasonable value given a downhill walking tour survey from the *castellum* to the city center.

Case C critical flow pipelines appear to have the delivery capacity close to the ~40,000 m^3/day aqueduct flow rate and preferably would be in use as there is only a minor slope difference between Cases B and C. It is likely that Roman engineering practice included recording of test observations leading to knowledge of the benefits of near critical flow designs. Since Case C slopes would reduce the occurrence of an internal pipeline hydraulic jump, this design would be preferable, but not always achievable, due to surveying accuracy constraints or destination site requirements that dictate pipeline slopes. The critical slope on the order of ~4.2° may have played a role in locating city structures that demanded rapid, stable transfer of water – such as *nymphaea* and elite housing with internal water display structures. Given the ~4.1° pipeline slope and considering a height difference from the *castellum* to a potential city level reservoir, the pipeline lengths would be on the order on ~0.25 km; this may influence the placement of intermediate reservoirs.

From Table 9.1, the likely pipeline candidates in use were Cases B and C examples used in conjunction with three closed floor ports. The Case B flow delivery capacity approximates the aqueduct water supply water of ~40,000 m^3/day. Case B and C designs are practical for 10 basin wall entrances operating continuously as pipeline slopes on the order of ~3.3° to ~7.1° guarantee stable water transfer from the high elevation *castellum* to lower city level sites. The remaining three floor ports, if open, would rapidly drain the *castellum* as flow from all 13 ports exceeds the input aqueduct flow rate as Table 9.1 indicates. It is important to note that for Case B where the pipeline transfer flow rate approximates the input aqueduct flow rate, that a steady water height is maintained in the *castellum* basin close to the basin top rim – which was exactly the design intent of Roman engineers. For a case for which several of the pipelines are situated at the critical slope, then a mixed pipeline array of slopes more than, less than, and equal to the critical slope would suffice to match the input flow rate while maintaining constant water height in the *castellum*.

Conclusions

The challenge to Roman engineers was to eliminate sources of flow instability by a *castellum* and pipeline design that transferred input aqueduct water at the highest flow rate possible through the 10 basin wall pipelines. The *castellum* design demonstrates Roman hydraulic knowledge at work in many ways – particularly in use of a shallow, wide diameter basin with the retaining basin wall slightly higher than the top of the pipeline entrance ports (2.2.3). This design initiates basin input flow from the aqueduct water entrance port into pipeline entrances at critical, partial flow conditions; its continuance as critical, partial flow is guaranteed by critically and near critically sloped pipelines. Given Roman experience with flow instabilities associated with pipeline internal wall roughness, it is likely that selection of smooth interior walls was the design preference. As the 10 pipeline cumulative flow rate approximates the maximum aqueduct input flow rate for Cases B and C (Table 9.1), it is clear that the design intent of the *castellum* recognized advantages in selecting pipeline slopes to largely limit or eliminate hydraulic jump occurrence – this occurs when Case B near critical flow conditions apply. As the basin wall height is only ~5 to ~10 cm above the top of the pipelines; clearly partial entry flow into the 10 wall pipelines is the design intent of Roman engineers; this is made possible by the positioned height opening of the sluice gate. Given the pipeline critical or near critical slope range of ~4.1° to ~7.1°, CFD results indicate the 40,000 m^3/day aqueduct flow rate is adequately transferred at the near maximum pipeline flow rate by Case B and C designs. The critical and near critical pipeline slope designs under smooth inner wall surface conditions allow atmospheric pressure over the pipeline partial flow eliminating pressurized pipeline joint leakage that would otherwise occur under an alternate full flow *castellum* design. For all wall pipeline slopes at the higher critical flow angle of ~7.1°, the theoretical total pipeline flow rate can exceed the input aqueduct flow rate (Table 9.1) – but to maintain a constant water height in the *castellum* basin, a design of different pipeline slopes at near critical, lower and higher slopes outside of the ~4.1° to ~7.1° range may be used to match the aqueduct flow rate and maintain the *castellum* water height constant. As pipelines led to reservoirs within the city with further distribution pipelines to local cisterns coupled with overflow pipelines recycling water back to reservoirs (Figure 9.8), mixed slopes allow pipelines to extend to sites at different spatial and height locations within the city with different water demands.

The conclusion of the analysis is that Roman hydraulic engineers designed the *castellum* to match the input aqueduct flow rate by employing a critical and/or near critical flow conditions in all basin wall pipelines. This option presumes a series of lower level reservoirs at pipeline termination locations that support pipeline branches to different sites (Figure 9.8). As different pipeline flow rates occur within different pipelines with slopes at critical, near critical and higher and lower slopes to supply spatially dispersed sites with different water demands, this necessitates use of the three floor ports to accommodate the aqueduct input flow rate. Provided the 10 basin wall pipeline slopes can be maintained in the ~4.1° to ~7.1° slope range, this design option provides the most efficient and stable way to match the aqueduct input flow rate and eliminate maintenance problems.

The totality of the Pont du Garde aqueduct and *castellum* design demonstrates a coordinated engineering design of all subsidiary hydraulic components that supply the *castellum*. The total aqueduct and *castellum* design includes:

1. Flow rate regulation through a far upstream intersecting side channel to drain away flows exceeding the designed aqueduct input of ~40,000 m^3/day flow rate. The partially open bottom three *castellum* floor drains can also be used to remove excess water to achieve the design flow rate. The altitude of the *castellum* tunnel is chosen so that pipelines emanating from the basin wall are sloped to match the land contour to the city below – this slope is on the order of ~5° to 7°.
2. Aqueduct channel width, depth and slope dimensions designed to contain the tailored ~40,000 m^3/ day flow rate without spillage (Lewis 2001:185).

3 The low channel slope entry (0.002°) to the *castellum* is designed to slow water velocity and raise its height to the near top of the supply tunnel shown in Figures 9.3 and 9.4. A sluice gate (Figure 9.10) is used to produce the desired basin entry water height.
4 The ~40,000 m^3/day aqueduct input flow rate closely matches the output total flow rate of the 10 basin wall pipelines under Case B and C near and exact critical flow conditions. This is the intended design intent of Roman engineers.
5 Critical flow conditions at 10 pipeline entrances are produced by making the sluice gate opening height equal to half the pipeline diameter, this ensures critical entry flow to pipelines (Henderson 1966); critical flow in pipelines is continued by critical and near critical pipeline slopes (~4.1 ° to ~7.1°). Note that this pipeline slope range is about equal to the hill slope angle showing the design intent to place the tunnel at a specific height to achieve this result.
6 Production of critical or near critical flow conditions in pipelines eliminates the influence of downstream flow resistance elements (bends, chokes, pipeline angle change) that can propagate upstream to produce unstable, transient oscillations in the *castellum* basin.
7 Use of the critical and near critical flow pipeline slope design produces the maximum, stable flow rate possible in pipelines and reduces pipeline joint leakage as an atmospheric airspace exists over the partial critical flow eliminating pressurized full flow conditions.
8 An observer viewing the castellum basin surface during regular operation would see a smooth water surface free of any disturbances – this result proceeds from downstream critical flow in pipelines that prevent propagation of flow downstream originated disturbances back to the castellum.

Figure 9.10 summarizes the flow pattern obtained by regulating the sluice plate opening height to produce basin water height equal to the y_c critical flow height entering the pipeline; this height is equal to half the pipeline diameter. Note that the basin wall height up to the bottom of the pipeline ports is about ~0.5 m; the distance between the top of the pipeline entrance ports to the top of the basis is ~ 0.5 m. The tunnel sluice gate height is regulated to have entry flow into the 10 pipelines at half the pipeline diameter.

Critical flow and near critical is maintained in the pipeline by critical flow angle θ_c between ~4.1° to ~7.1° to produce the (1) to (8) benefits indicated previously. Note that if the pipeline angle is less than θ_c then downstream disturbances can propagate upstream to cause unstable basin oscillations that destabilize steady state behavior and cause spillage from the top of the *castellum* rim. For pipeline declination angles greater than θ_c, upstream disturbances cannot propagate upstream but the output flow from the 10 basin wall pipelines is less than 40,000 m^3/day. The integrated, coordinated design of all components of the Pont du Garde aqueduct reflect Roman engineer's hydraulic engineering knowledge and serve to add to the compendium of Roman practices and inventions thus far described in the open literature.

10 Roman *castella*, Calyx Quinaria and flow rate measurements

A wide variety of *castellum* types exists throughout the Roman world. From Vitruvius (*De Architectura*, Book 8, Chapter 6) is his commentary:

> When the water has reached the city, build a reservoir with a distribution tank in three compartments connected to the reservoir to receive the water and let the reservoir have three pipes, one for each of the connecting tanks so that when the water runs over from the tanks at the ends, it may run into the one between them. From this central tank, pipes will be laid to all the basins and fountains; from the second tank, to baths, so that it may yield an annual income to the state; and from the third, to public houses so that water for public use will not run short; for people will be unable to divert it if they have only their own supplies from headquarters. This is the reason why I have made these divisions, and also in order that individuals who take water into their houses may, by their taxes help to maintain the conducting of the water by the contractors.

A candidate for this type of *castellum* is to be found at the terminus of the Claudia and Anio Novus aqueducts where remains suggest a five chambers that divide aqueduct flow to separate destinations (Van Deman 1934) with different water flow rate requirements. The water distribution system at Pompeii shows multiple channel divisions supplied by a main aqueduct that led off the different parts of the city. A further candidate *castellum* of this type is provided by the water distribution system at Ephesus (Ortloff 2009:318) where an unexcavated site exhibiting pipelines to different destinations with different water needs strongly suggests a *castellum* of this type. Further *castellum* types different from the Vitruvian description include the Pont du Garde system previously discussed as well as *castellum* types found at Pompeii (Hodge 2011:283) described in this reference. Several of these include movable inlet sluice plates and stepped height weirs to distribute water to different destinations – here many different variations exist on the Vitruvian type described as excavations to date have revealed.

Of interest in *castellum* design is the use of calices to measure flow rate. Typical designs are shown in Figures 10.1 and 10.2 and indicate a pipeline constriction insertion device designed to measure flow rate; markings on the *calyx* indicate the

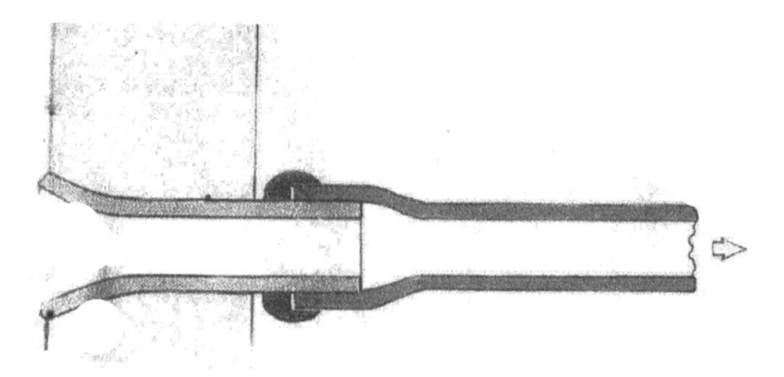

Figure 10.1 Metal-to-metal (lead) *calyx* connection joint with soldered seam binding.

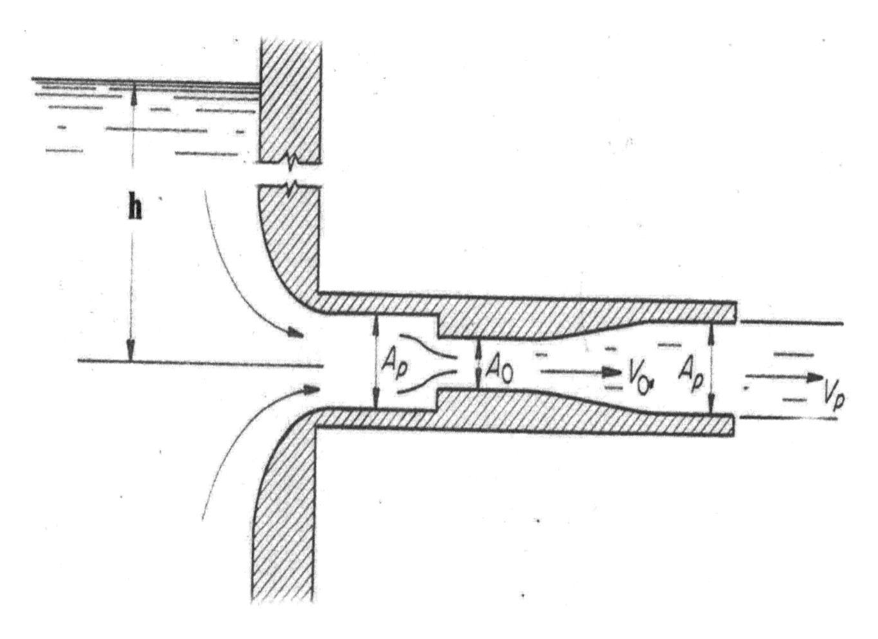

Figure 10.2 Notation for Equation (1) usage.

value of the flow rate in *quinaria* that the *calyx* allows through its orifice. As part of Roman water distribution networks, water from aqueducts was delivered to reservoir basins then to *castellum* systems which, for public use, contained *calyx* devices to measure flow rate with a tax appropriate to that flow rate. Here a typical design incorporates a bronze *calyx* fixture with an outer contraction diameter smaller than the inner lead pipeline diameter to measure the flow rate – for user taxation purposes, the flow rate is, according to Roman descriptions, dependent upon *calyx* throat inner diameter. Additional *calyx* use for terracotta or wooden pipelines necessarily incorporated transitional connection elements (Hodge 2011:316) or insertion baffles holding the centrally placed *calyx*. A user desiring twice the flow rate than that delivered to his site would order a replacement *calyx* of twice the throat diameter of a previously installed *calyx*. The accuracy of this claim is subject to investigation in the discussion to follow. Several of the *calyx* types and characteristics available to Roman water engineers are given in Hodge (2011:297) and serve as the basis for investigation of their flow rate properties. Several *calyx* sizes and types are listed in Table 9.1 using metric and Roman units. From the notation in Figure 10.2 and use of Sabersky et al. (1971:94) notation, the transfer velocity V_p and volumetric flow rate for Table 9.1 flow rate specifications and for different head values is given by:

$$\mathbf{V_p} = \left\{\mathbf{2gh} / \left[\mathbf{1} + \left(\mathbf{A_p}^2 / \mathbf{A_0}^2\right)\left(\mathbf{c_v}^{-2} - \mathbf{1}\right)\right]\right\}^{1/2} \quad (1)$$

where $g = 32.2$ ft /sec^2, and $c_v = 0.48$ ($c_v^{-2} - 1 = 3.34$) for smooth constriction *calyx* shapes (Sabersky et al.1971:92) and A_p, A_0 are areas defined in Figure 10.2. Table 10.1A is shown as a calibration data set to determine the effective qualitative value for one *quinaria*. From Hodge (2011:299), previous estimates of the flow rate value of a *quinaria* (Di Fenizio 1916) are ~40 m^3/24 hours or equivalently, ~0.04 ft^3/sec for low head values on the order of 14 cm. Comparison with Table 10.1A values for unit *quinaria* for a A_p = 1.0 ft inner diameter lead pipe indicates a value on the same order of magnitude (0.08 ft^3/sec) calculated from Equation (1) – but as Hodge (2011:462) indicates, Di Fenizio's 40 m^3/24 hour day flow rate estimate is intended as only a minimum value for the *quinaria*. Table 10.1B provides a further estimate of unit *quinaria* for a *calyx* inserted into a 0.5 ft inner diameter (A_p) pipeline to determine the effect of containing pipeline

Table 10.1A Quinaria estimates – 30 cm/1.0 ft pipeline diameter

Type Name	*Opening D A_0 (cm/ft)*	*Pipeline D A_p (cm/ft)*	*Head m/ft*	*Flow Rate ft³/sec*	*Exit Velocity ft/sec*	*Roman Capacity quinarias*
5 quinaria	2.31/0.08	30.5/1.0	0.10/0.33	0.08	0.12	1
12B duodenaria	5.56/0.18	30.5/1.0	0.10/0.33	0.35	0.45	6
40 quadagenaria	13.2/0.43	30.5/1.0	0.10/0.33	0.84	1.07	32
80 octogenaria	18.6/0.61	30.5/1.0	0.10/0.33	1.12	1.42	65.2
100B centenaria	22.2/0.73	30.5/1.0	0.10/0.33	1.32	1.67	92

Table 10.1B Quinaria estimates – 15 cm/0.5 ft pipeline diameter

Type Name	*Opening D A_0 (cm/ft)*	*Pipeline D A_p (cm/ft)*	*Head m/ft*	*Flow Rate ft³/sec*	*Exit Velocity ft/sec*	*Roman Capacity quinarias*
5 quinaria	2.31/0.08	15.3/0.5	0.10/0.33	0.08	0.40	1
12B duodenaria	5.56/0.18	15.3/0.5	0.10/0.33	0.17	0.89	6

size on *quinaria* flow rate estimates; again a 10 cm head is used consistent with Di Fenzio's head value. Here c_v is an empirical shape coefficient indicative of the smoothness of the inner surface of the *calyx* over its length, A_o is the pipeline inner diameter and A_p is the minimum inner diameter of the *calyx* (Figure 10.2). Note that a lesser head value than that stated by Di Fenzio would lead to near equivalency with his stated flow rate estimate. Values shown in Table 10.1B for larger diameter *calyx* openings reflect increased flow rates. For the same hydrostatic head (h), as the *calyx* opening diameter approaches the pipeline inner diameter, the intake and exit flows become equal. Table 10.2 values are shown for head values of 1.0 and 5.0 ft. The latter values reflect a head value from water sources from a reservoir while the lower head value reflects values from taps into shallow surface flows. As expected, as the head (h) increases, flow rates increase. The higher head values reflect that flow rates increase with increasing hydrostatic pressure so that an implanted *calyx* of a given diameter transfers different flow rates dependent upon the hydrostatic head. If Roman water bureau taxation is based solely on *calyx* diameter and not the head value, then flow sources from deep reservoirs clearly benefit the user.

Table 10.2 investigates the Roman proposition that doubling the *calyx* diameter doubles the flow rate.

The question as to the Roman practice of doubling the A_0 diameter of a *calyx* to double the flow rate is next examined. From Table 10.2 for the 5 *quinaria calyx*, the 0.08 ft diameter *calyx* has a flow rate of ~0.32 ft³/sec for 1.0 ft hydrostatic head in a supply reservoir. When the A_0 diameter is doubled to 0.16 ft, the extrapolated flow rate becomes ~0.65 ft³/sec which is close to the Table 10.2 value for the 12B entry. The Roman approximation of doubling the diameter to double the flow rate appears to be reasonable for the smallest size *calyx* for both the 1.0 and 5.0 head cases. For the 12B *duodenaria calyx*, when the 0.18 ft diameter is doubled to 0.36 ft, the flow rate becomes ~6.9 ft³/sec which again indicates that the approximation of doubling the *calyx* diameter to double the flow rate is still a reasonable assumption even for larger *calyx* sizes. Given the approximations involved in using Equation (1) which neglect turbulence effects at high Reynolds numbers as well as flow profile distortions prior to the *calyx* entry, the first order approximations given from Tables 10.1A and 10.1B indicate that the Roman approximate method doubling the *calyx* diameter to double the flow rate to estimate flow rate change

Table 10.2 Head change effects on *calyx* flow rates

Type Name	*Opening D A_0 (cm/ft)*	*Pipeline D A_p (cm/ft)*	*Head h ft*	*Flow Rate ft³/sec*	*Exit Velocity V_p ft/sec*	*Roman Capacity quinarias*
5 quinaria	2.31/0.08	30.5/1.0	0.305/1.0	0.32	0.40	1
12B duodenaria	5.56/0.18	30.5/1.0	0.305/1.0	0.71	0.90	6
40 quadagenaria	13.2/0.43	30.5/1.0	0.305/1.0	1.64	2.09	32
80 octogenaria	18.6/0.61	30.5/1.0	0.305/1.0	2.25	2.87	65.2
100 centenaria	22.2/0.37	30.5/1.0	0.305/1.0	2.62	3.34	92

is reasonably valid based upon Equation (1) approximations and holds even for significant head change. Based upon a calculated flow rate value of 0.08 ft^3/sec for unit *quineria* for an A_p = 1.0 ft diameter lead pipe and a value of 0.08 ft^3/sec for the A_p = 0.5 ft inner diameter pipeline, it appears that the same estimate for unit *quineria* exists independent of pipe diameter. Values indicated in Tables 2.3.1A and B are for a laminar flow condition – this condition is valid where the *calyx* is located close to the pipeline junction with the supply reservoir. For *calyx* locations located after a long length of pipeline (at the entrance to a public use building, for example), Reynolds number effects indicating turbulent flow conditions apply altering the head-flow rate conditions. Table 10.3 describes this condition and its effect on *calyx* flow rate measurements.

In all tables the Roman capacity (*quineria*) values listed are Roman approximations for the flow rates through different size *calyx* diameters. Based upon new estimated values of unit *quineria*, new flow rate values for the five types of *calyx* sizes are given in Table 10.3 using Table 10.1A and Table 10.1B data.

The Reynolds number is defined as Re = V D/ υ where υ = water kinematic viscosity at 70° is equal to F = 1.076 x 10^{-5} ft^2/sec, V is the mean pipeline water velocity and D is the pipeline inner diameter. Values for Re indicate turbulent flow conditions based upon the classic transitional Re = 2000 value for all cases. For a pipeline *calyx* located at the entry to a public use building being supplied by a 100 ft feeder pipeline from a main supply *castellum* reservoir, the supply velocity to the *calyx* is indicated in Table 10.3. For a friction factor (h_f) between 0.02 to 0.03 for smooth interior piping, the head loss h_f values, for calyx locations close to the supply reservoir and 100 ft distant from the supply reservoir, indicate a maximum reduction of ~15% for the extended location of the *calyx* 100 ft distant from the supply reservoir. Thus for typical urban water supply systems to public use systems characterized by smaller diameter pipelines on the order of ~0.5 ft, initial *quineria* estimates of flow rates for *calyx* locations distant from the supply reservoir are on the order of ~0.06 ft^3/sec close to Di Fenjzio's initial estimate of 0.04 ft^3/sec. The present discussion implies that there is variability in the water delivery rate depending on *calyx* location (close to the supply reservoir or 100 ft distant, or more, from the supply reservoir) so that the markings on the *calyx* as to its delivery flow rate are not precise under different *calyx* location positions but appear to be good approximations nevertheless for typical urban water supply

Table 10.3 Flow rate dependence on induced turbulent resistance

Type Name	*New Flow Rate ft^3/sec*	*Roman Capacity quinarias*	*Reynolds Number*	h_f *ft*
5 quinaria	0.08	1	1.1×10^4 (0.5×10^4)	0.6×10^{-4}
12B duodenaria	0.35	6	4.2×10^4 (2.1×10^4)	0.01
40 quadagenaria	0.84	32	1.0×10^5	0.01
80 octogenaria	1.12	65.2	1.3×10^5	0.09
100B centenaria	1.31	92	1.5×10^5	0.13
1.70 (maximum value with no *calyx* at 0.33 ft head)				

systems involving smooth interior surfaces (lead pipes) to households and public use buildings that have lower water supply demands than larger facilities that require much higher flow rates such as baths, fountains, *nymphaea* and public water supply basins that require larger diameter piping. Given that under laminar and low intensity turbulent flow conditions for typical water supply conditions to lower water consummation structures, that the marked *quineria* flow rate on a *calyx* provides a near consistent taxable flow rate value – and from prior discussion, even doubling the pipeline diameter to double the flow rate appears to be a reasonable approximation for taxation purposes. Based on what little is known about Roman water measurement technology and its experimental basis, it nevertheless appears that a suitable technology level was achieved that fairly accorded to the taxation rates for water usage.

From Figure 10.2, for high flow rates it would appear that water emanating from the contracted *calyx* inner diameter to a wider inner diameter pipeline would form a jet upon entering the wider pipeline section and thus continue as partial flow to a site destination. To avoid this unstable condition, downstream delivery of pipeline water, by design, would have submerged entry into a basin with further basin pipelines to multiple destination locations to assure smooth water delivery. This type of flow conditioning would be necessary for fountains and other water systems that require a steady input flow but not necessary for water devices with intermittent delivery conditions such as sink taps, toilets and taps to fill water containers. As the *quinaria* markings on a *calyx* (1, 6, 32, 65.2, 92) are basically best estimates as to their real flow rate capacity, Table 10.3 values are more likely the true flow rates associated with the markings. In summary, it now appears that Roman water technology has more experimental backing than previously reported or described in the works of Frontinus and Vitruvius – here Chapter 9 on the Roman Pont du Garde aqueduct and *castellum* gives indication as to the level of hydraulic technology available to Roman water engineers.

11 Observations on Minoan water systems in Crete

The water supply system of the Minoan site of Knossos in Crete, as well as for several other island sites, has been discussed (Angelakis et al. 2012:227–258; Mays 2010; Graham 1987; Willets 1977; Evans 1903). The palace structures at the Knossos site originate from several construction periods starting from Middle Minoan IIIB 1700/1650–1600 BC, Late Minoan IA 1600–1500 BC, Late Minoan IB 1500–1450, with final construction in the Final Palatial Period 1450–1390 BC, Late Minoan II and Late Minoan IIIA, 1390–1370/1360 BC periods. Figure 11.1 shows a theoretical reconstruction of the final phase of the Knossos site given latest excavation data and partial reconstruction details from earlier eras (Evans 1903; Papapostolou 1981:44–72). Due to abundant rainfall on Crete, runoff water is generally available for collection into cisterns and due to the high water table, wells and springs are a further source of water at many of the island sites. Of particular interest to hydraulic engineering study is the elaborate palace structure at Knossos given its last phase reconstruction configuration and its elaborate internal palace water distribution system.

Water distribution and control precedents observed within the palace structure represent many unique firsts not observed previously in the world archaeological record. Initially a spring at Mayrokolimbos (Gypsadhes Hill) located approximately 0.5 km south of the palace provided water to a terracotta pipeline system (Figure 11.2) that led to an internal palace piping system.

Given that this spring originates at an altitude of approximately 100 m and the palace at 90 m, the supply pipeline is at an average slope of 10/500 or 0.02. This low slope value together with deliberate internal roughness elements incorporated into the pipeline, likely guaranteed low flow velocity within the pipeline. The internal pipeline design induces a full flow at a given location within the pipeline given initial partial flow from the spring source due to deliberate internal pipeline roughness that provided pressurized flow delivery to palace locations at an altitude value corresponding to the full flow transition location. In other words, the delivery hydrostatic pressure to downstream location pipeline networks is given by the altitude difference between the final delivery station and the full flow transition altitude within the pipeline. Available literature sources indicate that localized bridge structures leading from the spring convey the pipeline over ground depressions to maintain a near constant slope throughout its entire length. This pipeline placement design works well when the upper spring source is active

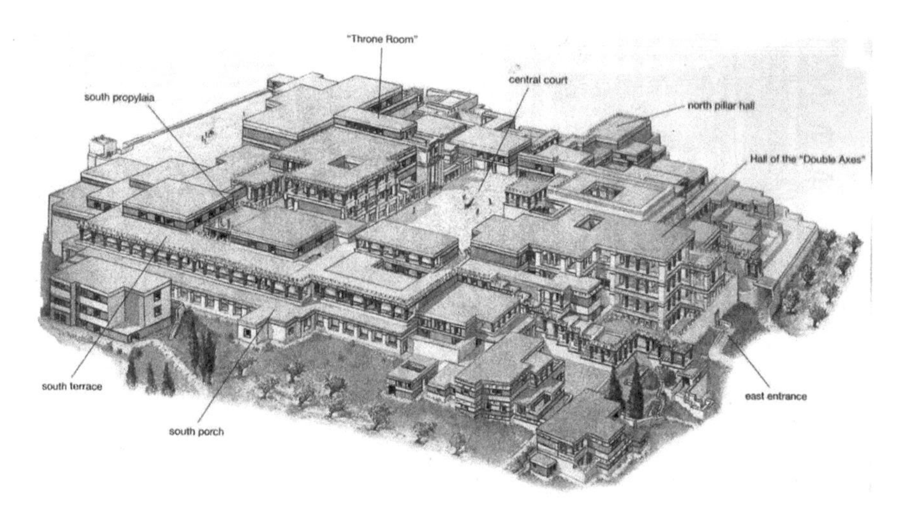

Figure 11.1 Theoretical reconstruction of the totality of the Knossos palace structure given early excavation data from Evans (1903) and later excavators.

but during drought conditions that lower the water table and spring flow rate, a lower altitude spring may have been in use with a corresponding new pipeline placement position at yet lower slope. After Roman administrative occupation of the Knossos palace site during 87 BC–AD 330, a more reliable water source was found at a higher flow rate output from the Karidaki and Paradisi springs accompanied by a tunnel/aqueduct complex 1150 m in length to provide water to the palace administrative structure. Figure 11.4 shows an aqueduct built by the Romans as part of their new, more reliable water conveyance system.

Of special interest is the internal structure of the Minoan palace period terracotta pipeline shown in Figure 11.2. Typical pipeline element lengths are on the order of 0.75 m and taper from a maximum diameter of 0.17 m to a minimum diameter of 0.086 m indicating the conical shaped pipeline design. Connection joints were cemented to limit leakage. From a hydraulic engineering standpoint, if subcritical, Froude number $Fr < 1$ full flow exists within the low slope pipeline at some station induced by internal pipeline roughness, then an internal toroidal ring vortex (Figure 11.3) exists in the flow at the small-to-large diameter connection locations. If this feature was purposefully intended by Minoan water engineers in their piping element design, then the presence of internal flow structures that induce flow instabilities and agitation was designed for the purpose to keep sand and soil debris particles in suspension until delivery to a destination settling chamber where particles could be removed to promote clean water transferred to intra-palace water distribution networks.

Figure 11.3 (Lugt 1983:92) shows the formation of a toroidal ring vortex at a low Reynolds number ($Re = 1350$) generated by an axisymmetric input stream at velocity V similar to that at the low diameter pipeline junction into the wider

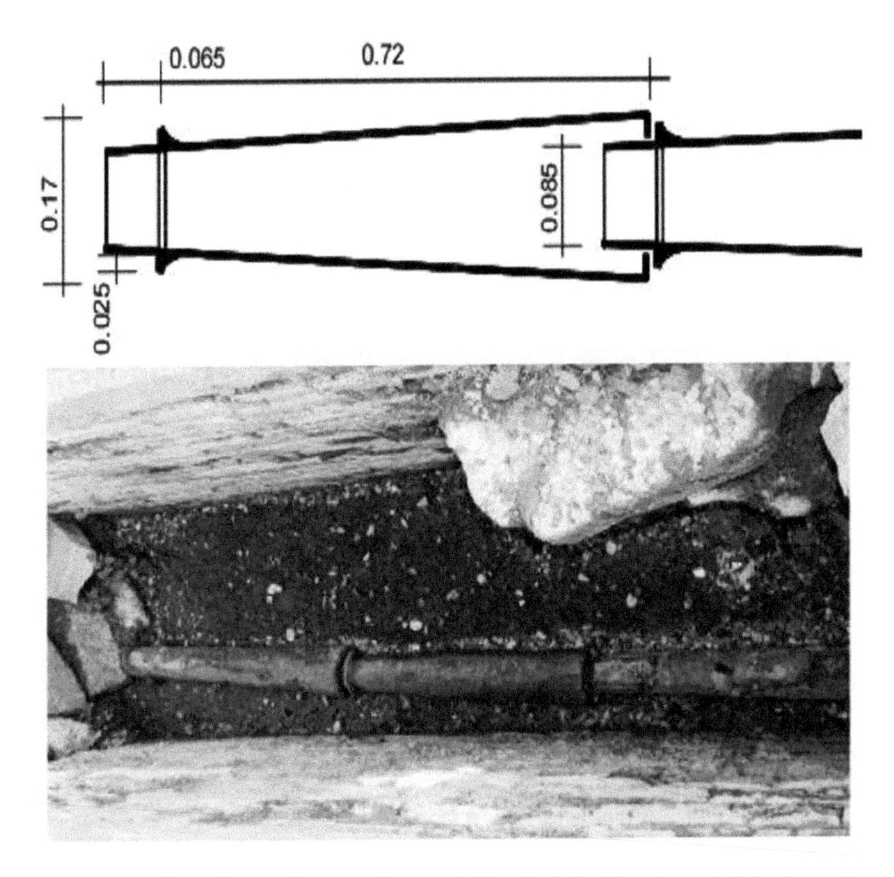

Figure 11.2 Dimensions and *in situ* placement of conical pipeline elements bringing water to the Knossos palace.

diameter part of the pipeline. For higher Reynolds numbers corresponding to higher spring source flow rates, the toroidal ring vortices may be continuously developed and shed (Lugt 1983:99) further increasing water agitation within the pipeline to keep debris particles in suspension until delivery to a downstream settling chamber. Thus for both low and high seasonal flow rates originating from spring water input that correspond to low and higher Reynolds number values, water agitation was purposefully present to keep debris particles in suspension until final water delivery to a downstream settling tank. This feature prevented debris accumulation that would otherwise clog parts of the piping system within the palace structure and yield, after particle removal, a potable water source. If full supercritical flow exists corresponding to high flow rates obtained from the spring source within the pipeline, then higher energy toroidal ring vortex strengths and flow instabilities are amplified yet further keeping particles in suspension until delivery to a destination settling tank. For both cases, the energy loss sustaining

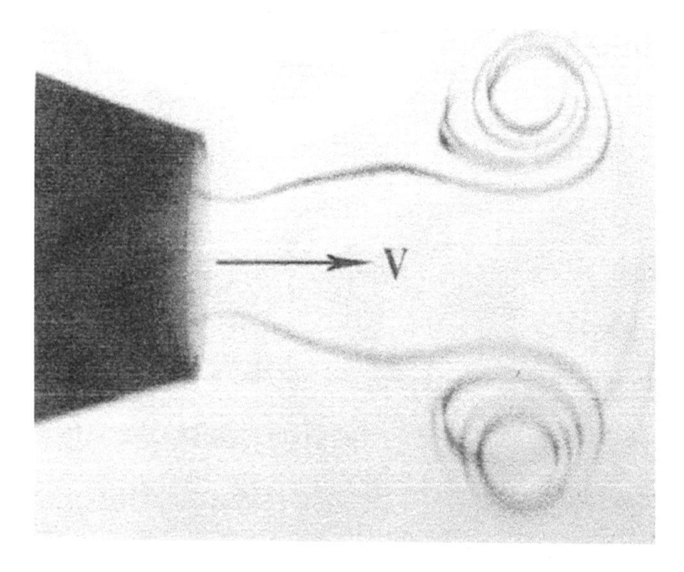

Figure 11.3 Streamlines within a steady toroidal ring vortex driven by flow through the axisymmetric contracted pipeline section shown in Figure 11.2.

a toroidal ring vortex is a function of its rotational speed (ω, radians/sec) and the spatial extent of the vortex ring structure; the resulting head loss resulting from stream energy transfer to drive the vortex ring structure per piping element multiplied by the number of piping elements then provides an estimate of the total head loss for the piping system. The flow from the smaller diameter entry end opening drives the rotation angular velocity (ω) of the toroidal ring vortex. This narrow end water stream is likely turbulent and constitutes a further energy loss increasing the total head loss of the flow. Whether the design features thus far described were intentional or not is a debatable question but then again, given the elaborate and historically unprecedented palace architecture and its elaborate water distribution network, hydraulic engineering on an advanced level would not be unanticipated. A further observation is that it is more difficult to manufacture a long length conical pipeline than a constant diameter pipeline – thus the extra manufacturing labor must have been for a reason like that described in the preceding text.

As previously mentioned, successive pipeline elements pose flow resistance derived from toroidal ring vortex creation at the junction region between pipeline elements. The energy (E) of water passing through a single pipeline element at velocity V may be expressed as:

$$E_{element} = m_1 V^2 / 2 - (m_1 - m_t)\omega_{av}^{\ 2} r_{av}^{\ 2} \tag{1}$$

where V is the water velocity passing through the circular narrow part of a pipeline element shown in Figure 11.2, m_1 is the mass of this water, m_t is the water mass

Figure 11.4 Roman aqueduct used to convey water to the occupied Knossos palace administrative center of Knossos in the 330 BC–AD 87 time period.

contained in the toroidal ring vortex, ω_{av} is the average rotational angular velocity of a toroidal ring vortex element and r_{av} is the radius from the center of a toroidal ring vortex cross section to its outer edge. The first term in Equation (1) represents the passage kinetic energy of the entry water stream through the narrow part of the pipeline element at a given time period after subtraction of the energy contained in the toroidal vortex which is represented by the second term in Equation (1). Here the toroidal ring vortex angular velocity is driven by the frictional contact of the V_1 input stream along their contact surface. Evaluation of this relation depends upon knowledge of the intercepted spring flow rate which unfortunately is not available given the fragmentary remains of the pipeline system as well as the spring flow rate available in ancient times. Within the palace structure itself, only fragmentary remains exist of the internal piping system making assessment of the totality of the hydraulic engineering knowledge base underlying the palace water distribution network difficult to reconstruct accurately.

One problem associated with design of the pipeline system is the variable seasonal spring output flow rate that decides the pipeline delivery rate. Here a viable water system design would incorporate a terminal collection reservoir at the terminal end of the piping length with a stable water height maintained by pipeline delivery flow matching water subtractions from intermittent use such as for latrine flushing, housing intermittent demand use, public water basins for household water collection delivery to public/private rooms as well as to storage facilities containing *pithos* water containers. Figure 11.5 taken from a Knossos burial sepulture painting source illustrates daily activities involving transfer of water (or other liquids such as olive oil or wine) to storage containers; such activities were vital given seasonal changes in availability of key fluids necessary for sustaining commercial activity.

Given that different palace locations required different flow rate deliveries to different rooms with different water requirements, sufficient hydrostatic pressure to support different flow rates to different pipelines emanating from a reservoir at different heights from the reservoir bottom must have existed. The requirement for sufficient hydrostatic pressure to produce different water flow rates to different locations within the internal palace piping network would derive from the intentional creation of large flow resistance within the supply pipeline to rapidly

Figure 11.5 Filling of a *pithos* storage container photographed from a museum sepulture painting.

transition any partial flow spring input to full flow conditions where the hydrostatic head available to the palace flow network would be determined by the partial to full flow transition height compared to the reservoir bottom height in first approximation. While surviving remnants of the palace water distribution system prevent knowledge of the totality of the water system network design, an internal reservoir with pipelines emanating at different heights from the reservoir to supply different flow rates to different parts of the palace structure with different steady and intermittent water needs would be both a logical and necessary part of the palace flow network. What does exist, however, of the palace water system at Knossos, are several drainage channels at lower elevations emanating from palace lower level structures (Figures 11.6, 11.7 and 11.8) that indicate the completion of the water distribution network.

Figure 11.6 Flow channel observed at Minoan Knossos as part of the water drainage network.

Figure 11.7 Rainfall runoff drainage channel observed at Knossos.

Presumably the presence of many rainfall runoff water collection channels observed at the palace site sourced some of the site water collection cisterns while drainage channels located at lower site levels promoted rapid drying of the site after rainfall periods as well as the transfer of waste water from palace rooms. In summary, given what little remains of the actual water supply and distribution system of the Knossos palace, intriguing elements exist from surviving parts of the original piping system to indicate a high level of water engineering science present in Minoan society.

The New Period Palace site of Mallia dating from Early Minoan III, Middle Minoan II times preserves a southwest located workshop area for ceramics, textiles and metal working that required an adequate supply of water for production of specific items. Open channel water canals (Figure 11.9) are found interior to

Figure 11.8 Slab covered drainage channel observed at Knossos.

workshop areas but no spring supplied aqueduct water supply origin is apparent from the excavated site.

At the site of Phaistos, well openings are found (Figure 11.10) shows an example) and likely provide access to an underground cistern supplied by a rainfall runoff collection channel (which is not apparent from the excavated site).

From later period and Middle Minoan I-II water channels observed at Gortyn, Phaistos, Gornia, Mallia, Ayia Triadha and Knossos, it is apparent that rectangular cross-section channels are the design norm (Figures 11.6, 11.7 and 11.8) at these sites appropriate for low flow rate subcritical flows. The Figure 11.9 channel located at Ayia Triadha (Late Minoan I to IB) hints of application of an advanced hydraulic technology as a contraction of the channel cross section occurs with a corresponding change in downstream wall height

Figure 11.9 Drainage water channel segment observed at Knossos.

of the channel. Given the low slope and slow velocity and height of the entry flow, this implies that for a given contraction width ratio of approximately 0.7, the higher wall height past the bend is required to contain the flow height within the channel. Of course, this result can be obtained by simple observation of the flow height change in the bend contraction section during operation that required an elevation of the wall height to contain the flow; never the less, this change in channel geometry indicates an awareness of changes in flow height dependent upon both entry flow velocity and flow height as well as the contraction width ratio effect of the channel. Only later in 19th- and 20th-century hydraulic engineering theoretical development was a flow height change related to initial upstream channel flow velocity, channel width and

Figure 11.10 Water channel cross-section shape change observed at the site of Ayia Triadha.

height and thechannel width contraction ratio and formalized in equation form by scholars as the use of Figure 11.11 indicates.

While this later formalism was certainly not within Minoan hydraulic engineering practice, some form of empirical hydraulic science derived from observation of water flow patterns was certainly available in early Minoan times as apparent from the pipeline and channel designs albeit in a formalism yet to be discovered. Given spring source flow rates, channel widths, heights and slopes were constructed in advance indicating knowledge of a counterpart relation similar to the semi-empirical Manning equation used in modern hydraulic engineering practice (Morris and Wiggert 1972).

As Minoan Crete was in involved in trade with major surrounding advanced Mediterranean societies, technology transfer from these sources may have played

Figure 11.11 Well located at the site of Phaistos.

a role in their water system designs – but given the uniqueness of the Knossos palace architecture with no architectural or water system precedents of the type observed at Knossos existing in other contemporary societies, it is likely that water system design innovation coexisted with the demands of the advanced architectural developments.

III

Hydraulic engineering and social structure in Asian hydraulic societies

12 Hydraulic societies of Southeast Asia

Ancient to modern

Hydraulic engineering and social structure in Asian hydraulic societies

As Wittfogel brought to the attention of scholars his observations of the connection between the source of domination of early (and later) Chinese dynastic societies over their subjects based on their sponsorship, creation and management of water resources for irrigation systems, flood control and transportation ends, it is instructive to review what he observed from the scant published literature of his time as the source of his inspiration and treatises on hydraulic societies. Typical of these systems is the Dujiangyan city irrigation system built in 256 BC by the State of Qin in the Sichuan province of China to be used as a flood control and agricultural irrigation project (Mithin 2012:150–175; Needham 1971; Gillet and Mowbray 2008; Payne 1959; Harrington 1974; Du and Koenig 2012:169–226). The project developed from the course of the Min River tributary of the Yangtze River by subdividing the Min River into multiple channels through the construction of weirs, levees and a major diversion gate (termed the Yuzul levee) in the shape of a channel contraction levee described as the Baopingkou Gate which is alternatively named Fish Head in other literature sources. This construction was part of a water control system controlling the silt deposition and flooding problem that compromised irrigated field systems and settlements in the Chengdu area prior to the watercourse construction. Notable in the hydraulic engineering detail of the construction is the Yuzul levee – an artificial island built of bamboo cages filled with boulders that divided Min River water into two channels. The inner channel was built deep and narrow and carried 60% of water flow to the irrigated areas in the Chengdu plain area in the winter dry season and 40% of the water flow in the summer months wet season. The wet spring season results from Himalayan glacier melt water that floods rivers and leaves destructive silt and sediment layers to compromise agricultural field settlements and supporting population settlements. The division structure termed the Fish Head is a fill structure largely composed of large stones obtained from excavation of a mountain pass necessary to support a subsidiary channel. The wide and shallow outer channel carries the remaining 60% of Min River flow in the spring wet season to divert silt and river sediments away from agricultural field system deposition. Along the inner canal is the Flying

Sand Fence which has a 200 m wide elevated opening channel designed to rejoin the inner and outer canals to convey the discharge of excess flood water into the outer canal from the inner canal. This feature is active during the Min River wet season when excess water in the inner canal is diverted to the outer canal; the combined flood flow is then led to the Yangtze River. Under flooding conditions, about 80% of flood waters are diverted from the inner canal to the outer canal while the additional 20% goes over the Flying Sands Fence with its 200 m wide opening into the outer canal. These values are estimates as actual values depend upon the degree of flooding.

In summary, under normal, non-flood stage conditions, by design 60% of the dry season flow goes through the Fish Head division separating the inner and outer canals; a Bottleneck structure on the outer canal directed any excess flow toward the Chengdu Plain with the remainder rejoining the Min River flowing toward the Yangtze River. In the summer dry season when Min River flow rates were high, 60% of the flow was diverted toward the Yangtze River at the Fish Head through the outer canal with lower amounts of flow diverted at the Bottleneck toward the Chengdu Plain. Thus an approximate, near constant flow toward the irrigated Chengdu Plain was maintained throughout the year to support multi-cropping agriculture. The ancient system is still in use to irrigate over 5300 km^2 and together with the Zhenggua Canal in Shaanxi and the Lingqu Canal in Guangxi, these systems constitute a major irrigation system with river transport constructs that served to inspire Wittfogel's interest in defining hydraulic societies. A comprehensive summary of further elaborate canal systems and unique water supply systems to cities and agricultural systems (Du and Koenig 2012:169–226) through Chinese history indicates advanced hydraulic technologies starting in early dynastic periods and continuing to the present day with mega-scale projects such as the Three Gorges Dam providing both flood control and electrical power to China.

Of major hydrological interest is the Grand Canal development (Payne 1959:59–62; Needham 1971; Coats 1984) over millennia. Early canal versions linking the Yellow and Yangtze Rivers via the Si and Shanyang Rivers were made by early dynasties. Later expansions linked the Haihe, Yellow, Yangtze and Qiantang rivers by later Sui, Yuan, Ming and Qing dynasties (Figure 12.1) with extensions and renovations of earlier phases of the canal up to modern times as indicated in Figure 12.2. The earliest linking constructions were carried out during the Shang and Zhou Dynasties (11th to 9th centuries BC with later large-scale constructions by the Han Dynasty (206 BC–AD 220) and later dynasties.

The earliest navigable waterway was the Hangou Canal linking the Yangtze with the Huaihe River authorized by the state of Wu in the 5th century BC. The second Honggou Canal extension linked the Huaihe and Yellow Rivers authorized by the state of Wei in the Warring States Period of the 4th century BC; later extensions were made in the Sui dynasty period in the 7th century AD where a Y-shaped canal ran from Hanghou in the southeast to Zhuojun (Beijing) in the northeast. Early motivation by dynastic rulers of the Shang and Zhou Dynasties recognized that east-west communications and trade were facilitated by east-west flowing Yellow and Yangtze Rivers but there was no north-south communication

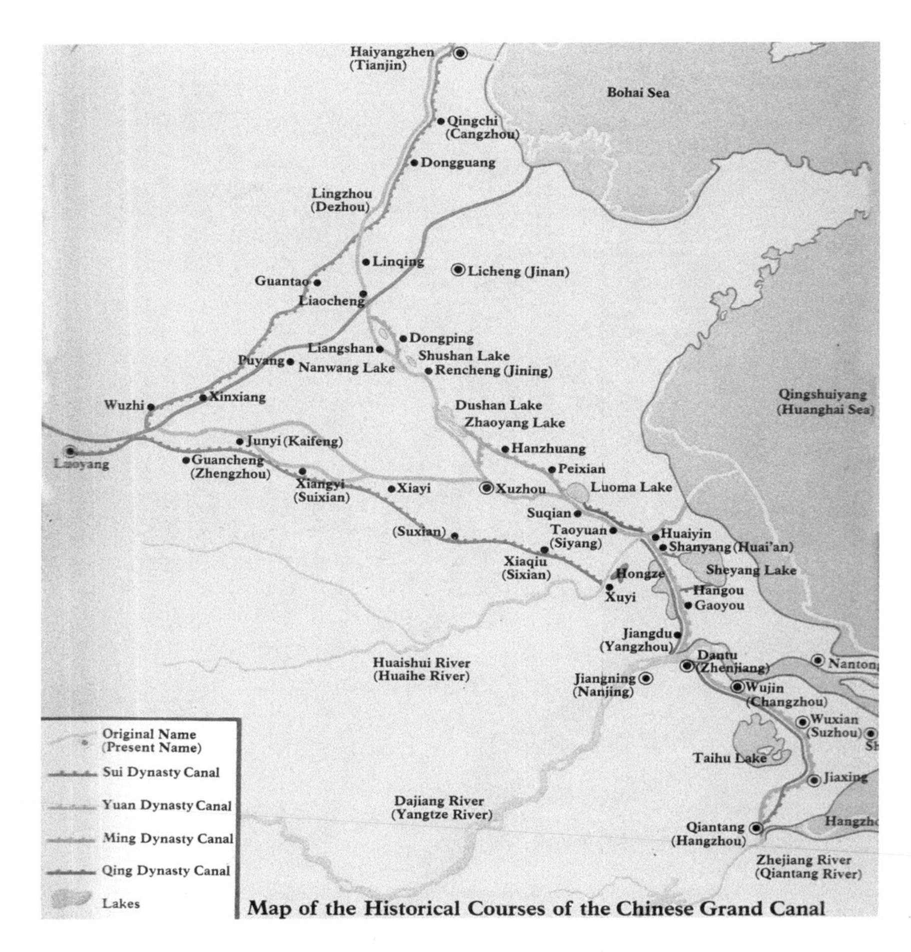

Figure 12.1 Canal systems constructed by early Chinese dynasties.

link canal to expand trade between north and south regions. A unifying canal to provide trade and communication between north and south provinces was necessary to unify regions under single dynastic control as envisioned by expansionist dynastic rulers vying for supremacy over neighboring states. The middle and lower reaches of the Yellow River had a well integrated economy, dense population and advanced production techniques but to the north and south of this developed area, society was undeveloped. To implement construction, the early rulers of the Shang and Zhou dynasties (11th to 9th centuries BC) started construction of a communication and trade canal network to integrate the north and south regions into the economy and bring prosperity to these regions (Figure 12.1). In the 13th century AD, dynastic control under the conquering Mongol ruler Kublai Khan establishing the Yuan Dynasty enlarged and constructed the Beijing-Hangzhou Canal to reach Peking (present-day Beijing) and the imperial palaces of

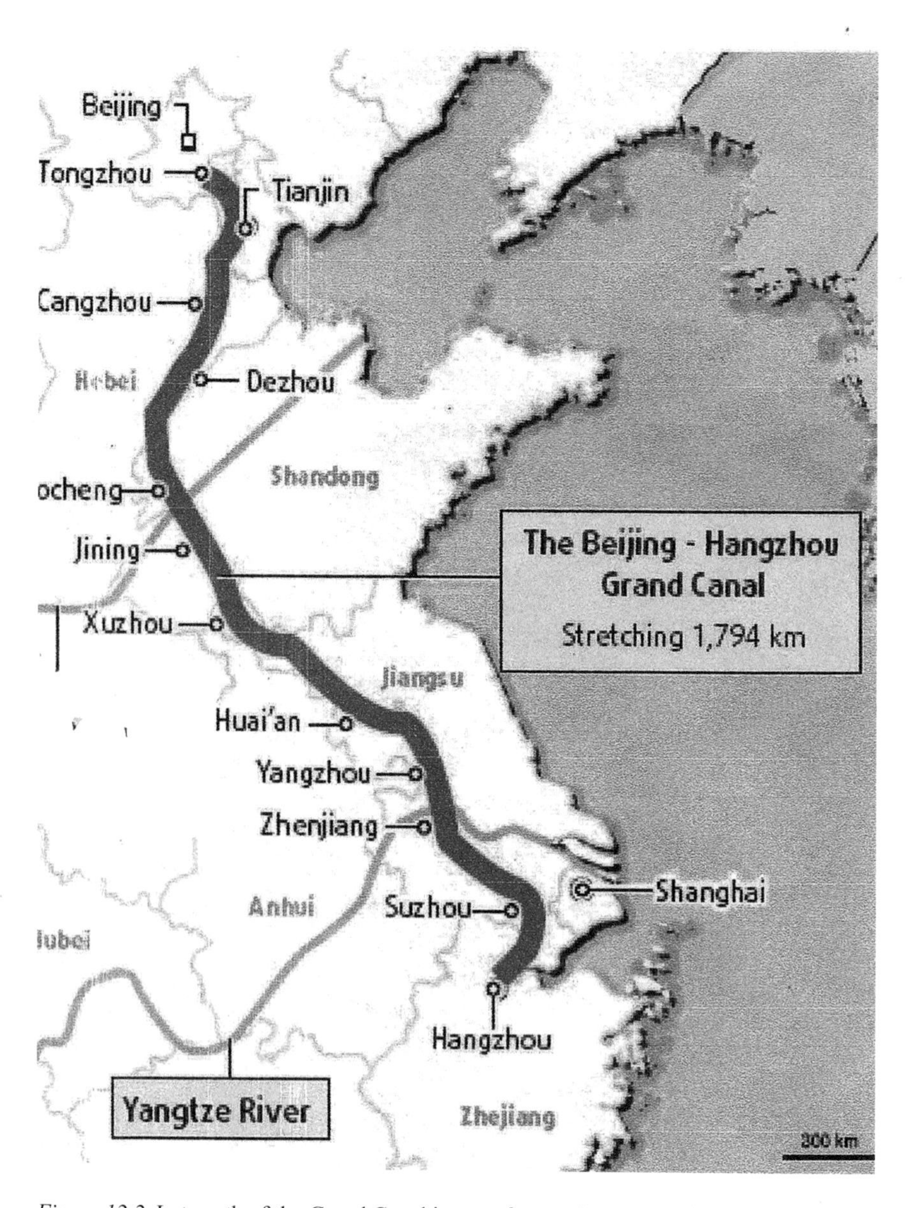

Figure 12.2 Late path of the Grand Canal in use after modern renovation.

the Forbidden City. This canal crossed five rivers to reach its Hangzhou terminal point and shortened its earlier course from 2500 km to 1728 km as Figure 12.2 indicates. Figure 12.3 indicates an internal Forbidden City waterway built under his control used to further embellish royal city precincts.

Figure 12.3 Waterways in the Forbidden City.

The final version of the Grand Canal now included 12,000 bridges over river tributaries and the main span of the Grand Canal – now ~2000 years in the making. The initial design philosophy was to take water from northern provinces with partially developed agricultural land resources devoted to grain production and provide a water transport system based on canal transport by barges to southern provinces with large tracts of agricultural land devoted to rice production. Trade of these and other commodities between different parts of the country would then be facilitated by barge traffic along the wide canal as well as unifying and integrating different regions under a single dynastic control. Agricultural production in the Jiangsu and Zhejiang provinces south of the Yangtze River then could be developed through use of the canal. The Qin and western Han Dynasties before the 1st century AD had imported grain from provinces along the middle and lower reaches of the Yellow River by land transport but after 7th century, later Tang and Song Dynasties transported grain from regions south of the Yangtze via the then existing version of the Grand Canal. Later Ming, Yuan and Qing Dynasties in the 13th to 19th centuries AD then consolidated their rule and power from this transformation with grain, financial income and manpower to support their power base (Coats 1984).

In summary, earlier canal segments date back to the 5th century BC with additional sections built during the Sui Dynasty (AD 581–618) and Tang Dynasty (AD 618–907) with further canal segment parts constructed in the AD 1271–1663 time period by Ming and Qing dynasties (Figure 12.1). The earliest canal, the Hangou Canal, linked the Yangtze with the Huaihe River in the 5th century BC; the addition of the Honggou Canal connected the Huaihe and Yellow Rivers was built in the 4th century BC. In a later time period, improvements focused on the last remaining 80-mile link between Cambaluc and the Yellow River – work started in AD 1289 and was completed in AD 1293. Installation of lock basins integral to the canal to transport boats from low to higher elevations (and vice versa) was invented during the Song Dynasty (AD 960–1269) – apparently a crane device was used to transfer smaller boats between lock water levels. This in conjunction with sloping dams that permitted winching boats from lower to higher river levels were also in use (Harrington 1974:44).

One 100 km section of the Grand Canal has over 60 lock basin gates to address complex topography issues for river barge transport. For barge transport from either lower to higher or higher to lower river levels, intermediate lock basin water was provided from subsidiary river channels to manage the staged transport of barges under difficult topographic conditions. In total, five river systems from north to south (Hai, Yellow, Huai, Yangtze and Chientang rivers) were crossed and ultimately integrated into the Grand Canal network that led to the imperial palaces of dynastic Peking. From the existing length of the canal system another 100 miles of subsidiary canals that led from Cambaluc in the north to Hangchow in the south crossed the Yangtze River at Chinkiang. The river crossing was facilitated not by an aqueduct by rather by use of subsidiary canal structures across intermediate marshlands. Further Grand Canal improvements were added during the Ming Dynasty (AD 1368–1644); later dynasties and modern 18th- to 20th-century Great Revolution Period improvements were continued to widen the canal from 100 m in city transits to over 300 m in countryside areas to promote commercial river travel between cities along its length. The channel widening effect serves to lower river current velocity making upriver boat travel for cargo and personnel transport more economically efficient in rural areas. Evidence of channel widening lowering river velocity and river turbulence levels is acknowledgment by early Chinese hydraulic engineers of the hydrodynamic effect of channel shaping on flow characteristics; as such, it is a first indication of this technology noted in Chinese water systems. Canal construction in different time periods was frequently interrupted by military conflicts between different kingdoms (Warring States Period 475–221 BC, followed by the Qin Dynasty), destructive flooding episodes (Yellow River 1920 BC major floods, 1930–1935 Yangtze floods among others), resource diversion from canal construction to court pleasures (Ching Dynasty construction of the Marble Boat and the Summer Palace in Peking), failed military adventures (Sui Dynasty, 7th century AD as well as river course changes and domination by foreign invaders (Mongol invasion, 13th century, involved defeat of the Jin, western Xia, Dali Kingdom and the southern Song Dynasty). With respect to river course change, the Yellow River changed its course in AD 1194 with the new exit into the Yellow Sea some 200 miles from its former exit into the Gulf of Pohai (Figure 12.1) and again in 1851–1855 finally exiting at a different location at the Gulf of Pohai. A new canal segment across the North China Plain and the Yellow River was then constructed to rejoin the ancient waterway at Huaian. From earliest to later modern time stages, canal construction has spanned millennia with many different canal paths devised by many different dynasties.

The modern canal version (Figure 12.2) has a total length of ~1115 miles (1782 km) from Beijing to the southern city of Hangzhou and has a shortened path from previous 2500 km canal versions as shown in Figures 12.1 and 12.2.

The north-south final version of a Beijing-Hangzhou canal was completed ~2000 years after the original of the earliest Hangou Canal and testifies to completion of early and later dynastic thoughts on integrating all parts of the county through a comprehensive system of waterways. Among the hydraulic engineering innovations were a dam built to divert water from the Wen River to direct 60% of its flow into the Grand Canal; additionally, in Shandong Province, four large reservoirs were built to regulate canal water levels to avoid changing the groundwater level that would alter the productivity of nearby agricultural field systems. Thus both hydraulic and hydrological engineering played a role both in the early canal versions as well as the later Grand Canal design as tributary river flows into the canal alter water available for irrigation farming and groundwater based farming. The hydraulic engineering of the Grand Canal lies in maintaining a small north to south current despite altitude changes en route – here additions and subtractions of water from intersecting rivers and the Grand Canal itself over its 1115-mile length by dams and diversion structures was a necessary part of its design and construction. Collectively, the Grand Canal has evolved over millennia to its present configuration; it now in its final version extends from Beijing to Hangzhou and connects the Yellow to the Yangtze River and serves as a major transport link between cities along its 1115 mile length. A further consideration of both ancient and modern Grand Canal planners was creation of an inland water route that connected many interior country cities and eliminated long distance sea-borne and land transport routes between cities; thus slow and difficult land based transport of unwieldy goods is avoided making material transport of agricultural and commercial goods more convenient and economical. As the longest man-made canal in existence together with its underlying and vital role in supporting historic dynastic as well as modern government rule together with major irrigation projects, the concept of the hydraulic state was readily apparent to Wittfogel. Beyond the Grand Canal and its long history, excavations of the Neolithic Liangzhu agricultural society's domain (Jiang Province, southern China) has discovered a large scale water management project involving eleven dams and multiple levees, canals, ditches and moats dated 5300–4300 BC in the Yangtze delta area; these hidden constructions were underwater up until 7000 years ago and are still partially submerged in occupation zone outer reaches. The water system's purpose was flood prevention, transportation and irrigation agriculture and is now regarded as the earliest elaborate and comprehensive water system yet discovered in China and indeed, in the world at that time. Its complex water system and adjoining city is seen as the model for later dynastic projects and indicates the vast

Figure 12.4 Chinese cloud storm demons and river demons.

Figure 12.5 The Sinusi water channel diversion of Yellow River water to the Grand Canal.

time period over which sophisticated water management was performed in China. This discovery reinforces the long history of hydraulic engineering accomplishments of early Chinese societies and the hydraulic society definition applied by Wittfogel.

Of course, accompanying technical matters related to ancient Chinese water systems are beliefs in water gods that control the destinies of societies. Appropriate to symbolism of the river deities are representations of the spirits that reside in water. In Figure 12.4, the rain dragon resides in clouds and the river dragon resides in river waves.

These two water deities tell the story of what nature offered to the Chinese and what human ingenuity could make of these offerings to better the human condition. Apparent in early Chinese society and continuing today in population elements outside of mainstream state dictated beliefs is the status of respect to

Figure 12.6 The Jiangnan Canal tributary to the Grand Canal.

ancestor's beliefs in the form of water deity representations. These representations continue not only as a treasured art style from past eras but also as real deities with temperaments that produce either floods and storms or good conditions for people depending upon reverence and sacrifices given.

Although history provides a long list of major hydraulic works conceived and built by different civilizations from early to late centuries (Payne 1959), the continued development and importance throughout history of canal systems to integrate and unify all parts of the Chinese mainland indicates that the hydraulic state was integral to the development of China. To quote Coats (1984), "the remains of many water control projects and navigational facilities along the canal, some of which are still well preserved, are today proving valuable aids in the study of hydraulic engineering in ancient China." Here the millennia of canal construction and the role of then Grand Canal in Chinese history was apparent not only to Wittfogel in his definition of the hydraulic state but now to many scholars interested in the history of water science.

Several pictures of the Grand Canal and tributary rivers along its length are offered as Figures 12.5, 12.6 and 12.7 to show the scale of the Grand Canal and the tributary river water supplement input and water diversion connections to adjacent rivers.

Considering that the many early canal versions shown in Figure 12.1 as well as later versions shown in Figure 12.2 resulted from human labor involving millions of enlisted workers directed by royal dynastic command, the scale and engineering

Figure 12.7 The Grand Canal through Hangzhou City with the Qiantangjiang River junction.

skill behind the construction of the many different Grand Canal versions at different time periods, its engineered connections to many adjoining tributary river systems, its five major river crossings, its many sluice gate to regulate commercial boat traffic and its many offshoot branches utilized for surface and groundwater agricultural purposes, the system is truly a monument to Chinese creativity on an almost unimaginable scale.

13 Conclusions

Overview of water engineering of ancient South American, Middle Eastern and Asian hydraulic societies

This book introduces Complexity Theory mathematical methods incorporating known economic laws to enable analysis of the hydraulic engineering technical base that underlies hydraulic societal development in ancient South American, Middle Eastern and Asian societies. Additional use of Similitude Theory new to archaeological analysis and Computational Fluid Dynamics (CFD) models to bring forward new discoveries and revelations on the hydraulic engineering capabilities of ancient societies comprise the content of the book's chapters. My earlier 2009 Oxford University Press book *Water Engineering in the Ancient World: Climate and Archaeological Perspectives on Ancient Societies of South America, the Middle East and South East Asia* introduced use of CFD techniques to understand the hydraulic engineering advances made by ancient New and Old World societies and considered effects of climate change on the sustainability of the ancient Andean societies of Peru and Bolivia. New discoveries of hydraulic engineering technologies at Andean pre-Columbian sites are described in Chapters 1–7 and add new perspectives on the hydraulic engineering used to cope with ENSO climate change effects (primarily extended duration drought and flood events) that affected their societal structure and survival – a topic of current relevance to world order given present concern with global warming effects and its effect on societal structure. The pre-Columbian South American experience with climate change has many lessons of immediate relevance to modern societies as climate change can seriously affect agricultural productivity and influence societal continuity. The archaeological record shows coping strategies (Chapter 3) of several major ancient Andean societies to sustain agriculture and societal continuity under extensive climate change events – in some cases several societies survived while in other cases societies disappeared from the archaeological record dependent upon success or failure of their hydraulic engineering capability and coping strategies to understand the limitations and strengths of their agricultural system designs. Results pertinent to climate change effects on societal continuance and sustainability have an extensive history as major Andean societies experienced dramatic ENSO (El Niño Southern Oscillation) climate change effects in the form of extended drought and destructive El Niño flood events that challenged their survival. In many cases, lessons learned from memory of past destructive climate/weather change events on agricultural and urban water supply systems were used to construct advanced defensive water control measures to ensure greater levels of security of replacement agricultural systems and their supportive irrigation canal systems as Chapters 3 and 4 detail. Chapter 4 defines a governing (Q) equation for societal sustainability and provides a quantitative measure from data abstracted from the archaeological record to gauge the survival or collapse of major Andean societies subject to major ENSO extended drought events. Given results from Chapters 1–7 that detail both the development and extent of hydraulic and hydrological science accomplishments of several major Andean societies, these developments can now be considered as the origins and motivation of technical source development vital to maintain societal continuity.

Societal complexity research underlying the ascent to elite hierarchical management creation from initial founding generational group origins has been documented (Rosensweig and Marston 2018) through resilience, legacy, subsistence threat, agricultural surplus, ecological advantage and various other socioeconomic theory and sociopolitical analysis paths; further paths (Prentiss 2019) toward this end are examined using mainly cultural continuity directives. To further examine paths toward development of hierarchical management structures, use of modern Complexity Theory described in Chapter 1 (Nicolis and Prigogine 1989) presents a new exploration path toward understanding societal complexity evolution by a mathematical model based upon societal consensus cooperation that initially leads to cooperative, collective societal transformations providing mutually beneficial economic improvements for all societal classes. Such paths originate in early Formative Period Peruvian societies absent of an elite hierarchical ruling class with later development of hierarchical management structures in the Early Intermediate Period, Middle Horizon, Late Intermediate Period and Late Horizon societies. Chapters 1, 2 and 3 seek to explain the evolutionary transition to more complex governmental structure by Complexity Theory and Similitude Theory methods and present new analysis paths to understand this evolutionary transition. Basic to this transition, as Similitude Theory demonstrates, are benefits in labor saving to increase agricultural productivity once cooperation among constituent groups is established and leadership roles defined to guide and continue new practices. Similitude Theory thus provides the analytic means to formally demonstrate the economic benefits of cooperative effort to unite different groups toward a common goal in terms of labor saving and agricultural productivity benefits. This theory provides insight into optimum field system designs to achieve labor saving together with higher agricultural productivity gains and shows from the archaeological record that such optimum field system designs were developed by many Andean societies both for coastal and highland agricultural systems. Similitude Theory methodology provides equation systems that measure the economic gains provided by cooperative use of optimized agricultural field system designs; these economic gains were apparent to populations of early societies and provided the incentive to organize to higher levels of organization that required large labor input and high level management of technical and labor resources to manage to expand agriculture to higher productivity levels. This in turn led to creation of a managerial class in charge of organizing labor pools, providing guidance on technical matters and organizing the logistical support to expand agriculture to support population growth. With this follows community binding rituals, unity feasting rituals, religious ceremonies and rituals to invoke divine guidance and support as well as other complementary display functions in the form of temples and ceremonial structures – a recipe well known in world society elite structures as lessons from history provide.

Agricultural failure due to extended drought or flood damage as examined from Complexity Theory demonstrate the path toward societal disintegration, fragmentation of the leadership class and ultimate political conflict as competition for diminishing resources persist. Results from Complexity and Similitude Theory demonstrate the climb to, and descent from, societal complexity progress. While climate change can serve as a catalyst to disrupt prior societal norms, oft times in history the initial cooperative consensus arrived at in early formative societies to develop leadership roles and an elite ruling class, subtly change the choice of a ruling class to either serve public interests – or their own interests – to determine the fate of a society. For the ancient Andean world, it appears from the many large scale agricultural and mega-canal system projects that human ingenuity and scientific discovery have produced that technical advances in the hydraulic and hydrological sciences provided the path toward progress. It appears that ruling classes, independent of their path to ascendancy, were vital to organize labor and logistics on a grand scale, provide technical direction and guide developmental advances in hydraulic and hydrological science all to increase societal sustainability while at the same time viewing expansionist policies in the form of satellite centers and territorial expansion as key to continuance of autocratic rule.

In summary, the early phases of hydrological science creation are discussed in Chapter 1. Complexity and Sustainability Theory models (Chapters 1 and 2) examine societal development under the influence of drought and ENSO climate change episodes that threaten agricultural collapse leading, in some cases, to extinction of several ancient Andean societies. The memories of prior events in the histories of several major Andean societies lead to creation of defensive measures to sustain agriculture (Chapters 3 and 4); Chapter 3 discusses the vital components that comprise societal sustainability under climate duress and present a sustainability Q equation to evaluate climate change events that threaten societal continuance. As seen in Chapter 5, major Andean coastal and highland drought events in the 6th and 10th to 11th centuries AD as well as major El Niño flood events determined from the archaeological record are examined to show the societal response of several major Andean societies. At this point, the intention of book chapters is to provide substance to major climate related events outcomes by new mathematical models that incorporate modern economic theory to replace speculation with mathematical proofs as to causes and outcomes of climate related events noted in the archaeological record. Chapter 2 details evolution of field system designs approaching theoretical optimum configurations indicative of continuing technical development activity encouraged by elite management sources to verify their importance role to the populace in the form of management oversight that ensures high levels of sustainability under external threats from climate change. The historical changes in societal complexity noted from the archaeological record are related to technical advances in agricultural science – Chapters 1–7 detail advancements in agricultural science particular to different major Andean societies over different time periods. Chapters 1–7 provide the basis for the classification of Andean societies as hydraulic societies based upon Wittfogel's studies of ancient Chinese society examples – some of which are described in Chapter 3. To further demonstrate the hydraulic science knowledge base developed by ancient Andean societies, examples of what the ancient (AD 600–1100) Tiwanaku society of Bolivia accomplished in groundwater based raised field agriculture and urban water supply systems is presented in Chapter 5; a further example of Inka hydraulic knowledge as observed at the site of Tipon is given in Chapter 6. In both example cases, use of modern hydraulic engineering theory and nomenclature provides explanation of their accomplishments that show correspondence to modern hydraulic engineering practice.

The many environmental challenges to agricultural continuity, given the wide ecological differences in coastal and highland settings, are met by many different types of agricultural systems (Figure 1.1) constructed to meet to the ecological challenges that vastly different environments present. Chapter 7 describes the Preceramic Period, Late Formative (2500–1600 BC) site of Caral located in the coastal Supe Valley which had one of the earliest agricultural systems based on an *amuna* groundwater system that maintained valley groundwater levels sufficiently high to promote extensive agriculture. This early example of a unique farming method exploiting high groundwater levels is still in use 4000 years later by present-day Supe Valley farmers. The Peruvian north coastal Chicama Valley site of Huaca Prieta-Paredones (10,000–4000 BC) indicates an even earlier agricultural technology development that incorporated a variant of raised field agriculture in wet coastal lowlands – later Late Intermediate Period examples of this technology are exploited further after ground saturation from El Niño flood events in the Santa Valley and later at the Middle Horizon site of Tiwanaku in the most elaborated form. Later societies continued progress in agricultural technology pertinent to their local ecologies and, as several sections of Chapter 1 indicate, hydraulic and hydrological engineering advances occasionally reached levels on par with modern technology levels. Together, all text sections of Chapter 1 show the wide variety of agricultural systems used in the Andean world and the application of sophisticated water control technologies supporting these systems which, in many cases, parallel modern hydraulic science levels. As secrets of *quipu* notation beyond accounting tasks become more available, some indication of notations used in hydraulic technology applications will likely be discovered consistent with water system design results given in Chapter 6 on the Tipon site.

Irrigation agriculture projects on a massive scale as performed by the Chimu, Inka, Wari and Tiwanaku Andean societies required much in the way of planning involving calculations to specify field system designs together with their water transport and distribution systems before committing vast labor resources to complete projects – this done in the same manner as present-day engineering practice where extensive planning is key to a successful outcome. Based upon extensive personal experience (50 years) in modern engineering project development for the defense, nuclear reactor, petroleum, chemical, ocean engineering and industrial agricultural industries, years of research, development and planning precede commitment of vast monetary and labor resources prior to the final construction phase of a project. Given the universal engineering mindset of project success that transcends time and cultural differences and welcomes new challenges as the source of new learning, ancient engineers shared an equivalent, counterpart path when challenged with societal sustainability and societal benefit advancement projects related to agricultural development. The history of science appears to be a continuous advancement path as new discoveries generate new higher-level questions that require new higher-level thinking. This was perhaps recognized early in time as expressed by the Roman writer Virgil as "*ut uarias usus meditando extunderet artes*" (that need and thought should useful arts devise) acknowledging what underlies creativity (Griffin 1990:213). It would appear that high levels of hydraulic science attainment as observed and extracted from the Andean archaeological record follow a similar path as that in western science albeit with vast differences in codifying technical matters. One aspect of Andean technology was focused on sustaining and expanding agricultural production by technical innovation; the path toward optimum agricultural field designs to yield the maximum output per unit field area is a logical consequence of a path toward understanding how nature provides examples of least energy water transfer processes – details of field system optimization for coastal and highland sites is along these lines as presented in Chapter 2. While it is known that other world societies developed intricate irrigation systems without accompaniment of an elite hierarchical management sector, most Andean societies appear to have elite architectural compounds discrete from secular parts of a city indicating existence and ultimately reliance on an elite management structure to develop and manage major projects to a successful conclusion. As all Chapter 1 sections indicate, all major Andean societies developed defensive measures to protect agriculture from nature's threats together with development programs to improve agricultural efficiency toward optimum productivity levels. This goal was implemented by an elite ruling class component of a society as demonstrated by the presence of royal compounds present at major Chimu, Wari, Tiwanaku and Inka sites. Together all the features thus far described define a hydraulic society in the Wittfogel sense and permit classification of major Andean world sites in this regard.

Wittfogel's classification of China's water system attainments, both through ancient and modern times, described by him as characteristic of a hydraulic society, is well substantiated as Chapter 1 sections indicate. Through multiple dynastic changes in ancient times to the present-day regime, intense concentration on new construction and modification of previously built canal systems for freight and passenger transport, flood control, field system irrigation and communication links to integrate formerly independent principalities continues to the present day. Dynastic regimes from ancient to modern times appear to dominate Chinese history with the populace serving the will of an emperor's vision of a perfectly structured society and the large scale projects (waterways, great walls, palatial structures, royal compounds, religious structures) necessary for elite class dominance while addressing practical defensive and offensive measures given the ongoing existence of neighboring hostile warring powers vying for territorial expansion.

As an example of a hydraulic science project consistent with past dynasty-scale hydraulic engineering endeavors that continues to the present day as a main source of Chinese prosperity, the newly constructed Three Gorges Dam on the Yangtze River is a further addition to the water infrastructure of China designed to alleviate Yangtze River flooding and produce electrical power (18,000 Mw). This project is not without warnings and criticisms as to imminent dangers posed by this massive construction project – a sign that a dominant state-controlled oversight committee chooses to overlook technical disadvantages when political advantages dominate state-controlled thinking. These disadvantages (Hvistndahl 2008) include landslides from increased dam water height (now at 156 m), hydrostatic pressure increase on upstream unconsolidated canyon side walls, waterborne disease increases, localized earthquakes (822 tremors noted seven months after dam water height increase), biodiversity and wildlife habitat reduction, drought occurrence in central and eastern China from lowering downstream river water height that strands shipping, downstream watertable changes altering farming productivity, decreased water delivery to the terminal port city of Shanghai decreasing water table height that allows inland salt water incursion, and other changes in environmental variables with as yet to be determined biological and populace health related problems. Water projects originated for political power demonstration ends need not always fulfill intended positive ends when unanticipated reality effects emerge. In this regard, the history of water engineering projects originated in all corners of the world in all time periods is not without successes and failures as the historical record indicates (Rosenzweig and Marston 2016).

Comparison of two hydraulic states (Andean and Chinese) thus far discussed in Chapters 1 and 4 reveal two different versions of the hydraulic state. For the Andean world, the final Inka version of societal structure is directed under the control of an elite management class. The population appears to dutifully accept their assigned roles in a structured society as long as sustainability and continuity was achieved under Inka elite class rule. The Inka ruling class consisted of some 40,000 royal sanctioned administrators directed by the Inka emperor in Cuzco over some 10,000,000 subjects living in coastal and mountainous parts of the empire stretching from present-day Ecuador to midway through Chile. Personal freedom of expression expressing contrary views of government set rules of societal conduct was voluntarily sublimated and conceded to the prescribed set of rules given by elite Inka management as apparently these rules and laws worked well to yield an acceptable and secure lifestyle for all elements of the Inka population. In Inka societal structure, there was an element of personal responsibility to achieve, but not transcend, what the elite royal management class proposed as the best societal structure and personal conduct roles possible. Many religious and rural lifestyle traditions persist from ancient to modern times in village life consistent with freedom of religious expression; personal initiative in city life to achieve higher personal status through education is encouraged with the underlying thought to modernize Andean society to world standards.

Chinese history shows trends of dominant control of society originating from early dynastic times and continuing to the present day with a near-deified ruling elite prescribing acceptable state-controlled thought processes denoted as loyalty to state control. The emergence of civilization is attributed to the intelligence of ancient sages and the evolution of a well-ordered state is essentially the work of human endeavor. In contrast to Andean societies and their belief in deities that play a major role to influence life events, Chinese history shows belief in a superior ruling class whose intelligence guides society to new levels of achievement under a strict moral behavior code. In this societal structure, there continues to be from dynastic times to the present a social hierarchy in which different classes of society know their place under supervision of an elite ruling class. The advent of Confucianism in early centuries BC made conformity a virtue and gave successive emperors the willing allegiance of all subjects in state directed goals. Natural disasters, such as floods and famine, were portents of the failure of a ruler who could only enjoy the mandate of heaven if his actions remained in harmony with nature. From precepts well ingrained into successive generations exposed to dynastic Chinese conditioning to direct the population to the will of beneficent leaders to prescribe correct personal behavior, continuance in form exists to the present day with rules appropriate to societal conformance. In the modern world characterized by personal goal achievement encouraged by western democratic institutions rather than state conformance prescriptions, tightening of Chinese state control prevails with reeducation camps to correct behavior back to state prescription rules. It may be speculated that a version of state control mechanisms to control dissenters to state dictates, perhaps in harsher terms, also existed in early dynastic times. State control now dominates with elimination of religious freedom, religion itself, expression of contrary views of government policies and the forced acceptance of the Marxian view of a perfect society. Here state dictates and power comes first, families come second and far down on the list of importance is the individual. While to some degree this is acceptable to many due to increased prosperity and elevation of many to middle class status, there is no evolutionary path toward expanding personal freedoms but the opposite as state control appears to strengthen with time. Whether dreams of world dominance follow, the world holds its breath.

The well studied Roman society is not classified as a hydraulic society although many water control and supply structures exist in the form of aqueducts and city water supply and drainage systems throughout Rome and its controlled territories. These water systems were viewed by Roman administrators as necessary elements of a civilized society rather than key to power of the ruling elite. The ruling elite were concerned more with civilizing the rest of the known world according to Roman laws and customs enforced by military dominance. This is best expressed in Virgil's words (*Aeneid*, VI, pp. 240, tr. John Dryden, Charles Eliot, ed., Harvard Classics, New York, 1909):

> *But, Rome, 'tis thine alone, with awful sway*
> *to rule mankind, and make the world obey,*
> *Disposing peace and war by thine own majestic way,*
> *These are imperial arts, and worthy thee*

Water control structures were a necessary component to cities built under Roman corporate plans that included baths, ceremonial display fountains, market agoras, drainage sewers, libraries, temples, separate ruling administrative class and secular private housing structures, coliseums, arenas, dedicatory structures to an emperor's military victories, multiple city water basins and public forum areas – among other elements of a civilized society according to Roman perceptions. Here the transient popularity of emperors and the ruling elite influenced their tenure with civil instability episodes both at Rome and abroad characterizing several periods of Roman history. The ability of emperors to provide a source of food staples (mainly imported from conquered territories) and coliseum entertainment episodes (Weiss 2014) provided by new sources of extracted wealth and tribute from conquered territories to support these enterprises decided the tenure of some of the emperors. Criteria for continuance of imperial leadership involved selection of loyal supportive civilian and military personnel, reassurance to the public of Rome's superior role to bring civilized standards to conquered territories, diversionary coliseum entertainment episodes, provision of facilities devoted to the needs and pleasures of the common man (baths, for example) and reassurance to the public of prosperity, military dominance and stability of governance to provide these means – all of which help deepen and guarantee the loyalty of the populace to the state leadership class that provided these amenities. The continued military superiority over barbarian societies expanded the empire to achieve world dominance; this provided by their gods' support and directives manifested through leadership skills bestowed upon their emperors. In many ways, what Rome had established in this regard persisted to a large degree in later European, Middle Eastern and Asian ruling classes as a commonly held standard of maintaining leadership through dominance over weaker societies and opposition elements within their own society.

As seen in Chapters 9 and 10, Roman ingenuity in technical areas related to design, construction and operation of water systems for city use was supported by new hydraulic technologies being continually developed, improved and utilized by successive administrations. Chapter 9 presents elements of the hydraulic technology base available to engineers in Rome as well as cities in foreign areas under their control. As described in Chapter 9, for the Pont du Garde aqueduct in southern France (at Nîmes) built under the reign of Claudius, hydraulic technology levels were in many ways comparable to modern practice albeit in a format yet to be discovered and described in Roman technical nomenclature. For Chapter 10 an examination of the value of one *quineria*, a measure of Roman pipeline flow rate as determined by *calyx* values, is determined. Chapter 11 examines the role of interior pipeline shape designs to reduce the transport of sediment debris to the Minoan Knossos palace structure. In summary, Roman engineers as well as those from Minoan Crete apparently mastered major water control technologies necessary for growth and expansion of their empires; key to this was a knowledge base that in many ways came close to modern hydraulic engineering practice and permitted a high level of civilization achievement that laid the foundation for modern western societies.

Of interest are differences between Rome and the Han dynasty to illustrate different types of governmental and social control structures. Rome had social mobility due to wealth and career progression; contemporary in time Chinese common class subjects remained under elite class rule with roles in society equal to those of their forebears. Of course, court intrigues provided access to higher state positions but this was only for privileged classes by birth or intrigue. Roman exchange was the focus in the public forum which was the locus of economic, social, political and religious activities. In China, the administrative, political, social and religious activities were centralized in walled dynastic palaces with no recourse to change from lower class population members. In both societies, the marketplace served for urban populations to mingle, socialize and trade opinions and thoughts on their lives and the structure of the society they lived in. In China markets emphasized the symbolic control of the state in trade, prices, measurement standards and state monopoly of manufactured goods while Roman markets established product development under individual enterprise and private entrepreneurship. In contrast to China, Rome recognized private property (although many conflicts arose in confiscating private land for veteran award lands and defining private property rights for newly enfranchised citizenry) and the complex interaction between patrons and marketing personnel that required individual bartering and interaction skills to proceed. In total, although both civilizations had vast water control systems vital for successful continuance of city life, both achieved high civilization standards through different administrative means. Thus, although both societies had vast water systems vital to city life, different societal structures evolved based upon different cultural goals and standards. The Wittfogel hypothesis thus is only selectivly applicable.

For the Nabataean society at their capital city of Petra, Chapter 8 describes the water system of that hydraulic society. As rainfall occurs in a short winter month time period, the classic Nabataeans (100 BC–AD 300) were adept at constructing numerous reservoir and cistern water storage facilities to maintain a supplemental water supply to several spring sourced pipeline systems for their urban population of 20,000–30,000. Contributing to the water supply to the city center temples, living quarters and marketplace areas were spring supplied pipelines most of which are no longer in operation except for the major

Ain Mousa spring now diverted from its original pipeline path through the Siq to supply the nearby town of Wadi Musa. Of note is the Nabataean urban pipeline network within city limits to ensure that key commerce and water display locations, such as the centrally located agora marketplace and the temple-pool complex (Bedal 2002) and *nymphaeum*, continue to maintain an adequate water supply from redundant interconnected pipelines sourced by different springs should the main supply pipeline be out of use for repairs or maintenance. Chapter 8 details the different hydraulic engineering technologies used for three main Petra pipeline systems – here different hydraulic technologies were applicable for different pipeline or open channel systems due to topographic and water supply transfer constraints. Of special note is the use of a subsidiary air transport pipeline intersecting a main pipeline that developed a partial vacuum region resulting from mild to steep slope transition within pipelines. This technology as seen at Petra is vital to cancel transient flow pressure instabilities that cause leakage at pipeline connection joints and to provide smooth water delivery flows to urban structures such as fountains, Nymphaea, display water structures and water basins for public use. The adjacent agricultural field systems to the Petra urban center were largely supplied by rainfall and had systems of *wadis* with successive stone dam barriers placed into their stream beds to trap rainfall runoff water to augment groundwater height necessary for different crop types. Using this agricultural strategy and given the slow decline in groundwater levels due to evaporation when infiltrated water arrives to deeper groundwater levels past the evaporation zone, multiple crop cycles permitted sustained agricultural output. In total, there appears to be a wide variety of water technologies available to Nabataean engineers both from imported foreign technology transfer sources and from trade route contacts together with indigenous hydraulic technology inventions particular to the environmental conditions of the site. One particular technology is the use of a critical flow design for the Wadi Mataha pipeline system – this advanced technology based on careful selection of a pipeline slope guarantees the maximum pipeline flow rate from a storage reservoir to counter drought remediation purposes, sustain flow in key city pipelines and supply water for arrivals of large caravans that place immediate large water availability demands on the city water supply system. Excavation of the Great Temple area over many years by the Brown University team under the direction of Martha Joukowsky (1998, 2004) revealed that rainfall seepage from a hillside wall south of the temple was collected into a drainage channel (Ortloff 2009:266) that led to an underground reservoir on the east part of the site. A further subterranean channel starting from the surface entrance to the reservoir led water to the front northern platform part of the Great Temple site and likely served as an overflow drain channel activated when the reservoir was full. The full extent of water supply systems internal to the Great Temple, including subterranean channels, has been mapped from excavated channel segments; thus superior excavation/conservation strategy practiced over decades by the Brown University team well preserves elaborately paved areas composed of hexagon shaped stone tiles in its original setting. From on site discussions with Brown team members performing excavations of this system during one of my site visits, mention was made of a pipeline branch that brought water to the upper southern reaches of the Great Temple. This branch originates from an elaborate water terminal location located above the Great Temple (Ortloff 2014a) that is designed to deliver water also to a housing district southwest of the terminal. The sophistication of this water delivery system is notable as it includes a pipeline branch allowing atmospheric air to counter a partial vacuum region created internally within in the steep angle descent branch pipelines as described in Ortloff (2014a). The marketplace area shows traces of an elevated pipeline supplying reservoirs above the site as Chapter 8 indicates – this done to provide a pressurized water supply to fountains that fill water supply basins in the marketplace area situated at a lower elevation. Roman occupation of the site past 106 AD indicate some pipeline additions using standard Roman pipeline sizes that were added to Nabataean pipelines to divert water to the marketplace and the Temple-Pool complex area. Some lead piping was found in the Great Temple platform area – most likely a Roman addition. There may have been a bridge across the Wadi Mataha streambed that carried additional water to the marketplace from the nearby water distribution *nymphaeum* to supplement water from south side spring sourced pipelines. As Petra is only fractionally excavated, and as much of the urban area pipeline system is either subterranean, no longer in existence due to flood erosion/deposition damage or long since removed for other purposes, the full extent of the pipeline system is yet to be discovered. Chapter 8, Figure 8.1 results represent the best estimate of the pipeline system from extant fragments – much remains to be investigated in the mountainous southernmost part of the site as destination pipeline segments imply channels carved into rugged mountain sides that supported pipelines. The physical difficulty of exploration in this area means that the full extent of all channels and pipelines in Petra remains for future investigators.

In final summary, many new aspects of water engineering practice and technology for societies of the ancient New and Old Worlds are brought forward in detail for the first time in this book to add to discoveries described in my earlier Oxford Press book 2009 *Water Engineering in the Ancient World*. The process of discovery of water systems underlying hydraulic technology from what remains after millennia of change from man's and nature's destruction and the supporting technology base of these systems is still in an early phase of discovery and much fieldwork remains to be done to understand the accomplishments of ancient water engineers. It remains enigmatic as to the discovery and use of alternate versions of modern hydraulic and hydrological science and its governing laws in as yet unknown formats by ancient civilizations as only results of application of these laws are apparent from the water structures in the archaeological record but not the laws themselves from written records. My work to discover ancient water technologies will continue from reservoirs of information gathered in past years at sites in South America, the Middle East and Southeast Asia together with recent travels to sites in South America, the Middle East and Southeast Asia – here many of the sites I worked at are no longer available for exploration and analysis due to urban expansion coverage and destruction, willful destruction by looters, political unrest and wars that make travel to key areas now impossible. The natural destructive forces of nature and man have obliterated many traces of ancient water systems – but these constraints are business as usual for archaeologists and in one way or the other, work will continue to bring forward the accomplishments of my brother engineers in centuries past.

Appendix A

A brief exposition of the use of nondimensional groups is presented. While commonly used for engineering applications, the use for archaeological analysis in the present text is unique. Early in the 20th century (in the pre-computer age) as industrialization took hold in major western countries, there was a need to have available reliable engineering formulas serving as the basis for reliable engineering designs. The challenge was to first list all the variables (p1, . . ., pn) governing an engineering problem and then trying to relate these variables in a deterministic, functional form that proved useful for an engineering design. One method to achieve a functional form was to develop correlations based upon isolating one variable (p1) and running tests over ranges of the remaining variables (p2, . . ., pn) to determine "weighting constants C2, . . ., Cn" for each of the p2, . . ., pn variables in the form p1 = f (p2•C2, . . ., pn•Cn) – this method clearly was time consuming, costly and limited in applicability. It was noticed by Buckingham (1914, 1915) that all physical laws must hold in all possible systems of units (English, metric, SI, etc.) and that for physical systems, three primary quantities are mass [M], length [L] and time [T] and that units for all physical variables can be expressed in these terms.

The insight was to develop a methodology to combine individual p1, . . ., pn variables each with combinations of units involving mass, length and time into a reduced number of nondimensional groups of variables – this would reduce the test range based on a fewer number of nondimensional groups. A nondimensional group is a combination of variables each with its appropriate units, such that all units cancel out and the group reduces to a number (which has no units). A simple demonstration of this concept is offered by the familiar Einstein formula $E = mc^2$ where E is energy, m is mass and c is the speed of light. Here dividing each part of this equation by mc^2, the result is $E/mc^2 = 1$ indicating that the units of E, m and c^2 must cancel in this equation as unity (1) has no units. As the units of E are $[M]\,[L^2]\,[T^{-2}]$ (or mass times velocity squared) and the unit of mass is [M] and the units of light velocity squared (c^2) are $[L^2]\,[T^{-2}]$, then the units of E/mc^2 are $[M]\,[L^2][T^{-2}]/[M]\,[L^2][T^{-2}]$ which cancel to produce a number – which in this case is (1) unity. The immediate utility of this method for experimenters is that fewer test variables in the form of nondimensional groups can be used rather than individual variables with their appropriate units. Here variables any unit system can be used (metric, English,

SI, etc.) to produce nondimensional groups. As many of the nondimensional groups have physical significance, their use in both theoretical and test applications is widespread (Murphy 1950); in fluid dynamics and heat transfer, Reynolds, Prandtl, Rayleigh, Nusselt, Knudsen, Froude and Mach numbers, among others, are widespread nondimensional groups with physical meaning appearing frequently in the literature. For the present application, only land, labor, water, technology and the food production rate are considered in the text; later applications can be extended to include further socioeconomic, political economy variables to widen the scope of similitude methodology in analyzing intensification development.

To illustrate the development of nondimensional group concepts, for example, velocity v has dimensions $[L]\,[T]^{-1}$, force has units $[M][L][T]^{-2}$, and energy (force times distance) has units $[M]\,[L^2]\,[T]^{-2}$. When, for example, two physical quantities with dimensions $[M]^a[L]^b[T]^b$ and $[M]^{a'}\,[L]^{b'}\,[T]^{c'}$ are multiplied together, the result is $[M]^{a+a'}\,[L]^{b+b'}\,[T]^{c+c'}$. If an equation representing some scientific law is to work in any unit system, its two sides must scale in the same way and their dimensions must be the same. If for example, the time period of a pendulum P is related to the mass m of the bob, the length L of the string and the gravitational constant g then $P \sim m^a\,L^b\,g^c$. In terms of consistent units on each side of the equation, then matching units $[T]^1 = [M]^a\,[L]^{b+c}\,[T]^{-2c}$ on each equation side leads to a = 0, b + c = 0 and c = -1/2, b = 1/2 so that $T \sim (L/g)^{1/2}$ – which is the well-known relationship obtainable from Newton's laws.

Relevant to the present application, extension of the preceding discussion lies in the concept that all equations representing physical laws must be expressible in a way that *both sides are dimensionless*. This means that if prime quantities are combined in such a way that their dimensions cancel to form nondimensional groups, then functions of these groups automatically satisfy physical laws that hold for all unit systems. This involves an extension of the preceding methodology where, for an example case, primary quantities are expressed in an arbitrary functional form as $s = f\{g, v, t, m, L, \rho, \mu\}$ where ρ is density $[ML^{-3}]$, μ is viscosity $[ML^{-1}T^{-1}]$, g is the gravitational constant $[LT^{-2}]$, m is mass [M]. T is time [T], v is velocity $[LT^{-1}]$ and s has length [L] dimensions. Then, in terms of dimensions of these primary variables,

$$L \sim [LT^{-2}]^{c1}\,[LT^{-1}]^{c2}\,[T]^{c3}\,[M]^{c4}\,[L]^{c5}\,[ML^{-3}]^{c6}\,[ML^{-1}T^{-1}]^{c7}$$

Collecting the exponents of [M], [L] and [T] leads to:

$$[M]: 0 = c4 + c6 + c7;\ [L]: 1 = c1 + c2 + c5 - 3c6 - c7;\ [T]: 0 = -2c1 - c2 + c3 - c7.$$

As there are seven unknowns (c1, . . ., c7) and only three equations involving [M], [L] and [T], three unknowns c1, c2, c7 may be expressed in terms of the remaining four unknowns c3, c4, c5, c6. One possible combination is c1= c3 + 2c4 + c5 – c6–1, c2 = – c3–3c4–2c5 + 3c6 + 2 and c7 = – c4 – c6 so that

$$s \sim f\{(g^{\,c3 + 2c4 + c5 - c6 - 1})\,(v^{\,-c3 - 3c4 - 2c5 + 3c6 + 2})\,t^{c3}\,m^{c4}\,L^{c5}\,\rho^{c6}\,\mu^{-c4 - c6}\}.$$

Now grouping together all terms with c3, c4, c5 and c6 exponents, there results

$$s \sim (v^2/g)\,(g\,t/v)^{c3}\,(g^2\,m/v^3\mu)^{c4}\,(gL/v^2)^{c5}\,(\rho v^3/g\mu)^{c6}$$

where now all parenthesis (. . .) terms are dimensionless and c3, c4, c5 and c6 can have any value as powers of nondimensional groups remain nondimensional. The appropriate c3, c4, c5 and c6 constants can be determined by experimental testing. This procedure is codified as the Buckingham Pi-Theorem (1914, 1915) and is the basis of the generation of equations derived from the basic equation as developed in the text. Details of this development are presented in Ortloff 2009:134–137.

Appendix B

Petra site numbers mentioned in Chapter 8

1 Zurraba, M reservoirs
2 Petra Rest House
3 Park entrance
4 Tomb of the Obelisks
5 Dijn Monument
6 Obelisk Tomb and Triclinium
7 Siq Entrance elevated arch remnants
8 Flood bypass tunnel and dam
9 Eagle Monument
10 Siq passageway
11 Treasury (El Kazneh)
12 High Place Sacrifice Center
13 Dual Obelisks
14 Lion Fountain Monument
15 Garden Tomb
16 Roman Soldier Tomb
17 Renaissance Tomb
18 Broken Pediment Tomb
19 Roman Theater
20 Uneishu Tomb
21 Royal Tombs (62, 63, 64, 65)
22 SextusFlorentinus Tomb
23 Carmine Façade
24 House of Dorotheus
25 Colonnade Street
26 Winged Lions Temple
27 Pharaoh's Column
28 Great Temple
29 Q'asar al Bint
30 Museum
31 Quarry
32 Lion Triclinium
33 El Dier (Monastery)
34 468 Monument
35 North City Wall
36 Turkamaniya Tomb
37 Armor Tomb
38 Outer SiqWadi Drainage
39 Aqueduct
40 Al Wu'aira Crusader Castle
41 Byzantine Tower
42 North Nymphaeum
43 Paradeisos, Temenos Gate area
44 WadiMataha Major Dam
45 Bridge Abutment
46 WadiThughra Tombs
47 Royal Tombs
48 Jebel el Khubtha High Place
49 El Hubtar Necropolis
50 Block Tombs
51 Royal Tombs
52 Obelisk Tomb
53 Columbarium
54 Conway Tower
55 Tomb Complex
56 Convent Tomb
57 Tomb Complex
58 Pilgrim's Spring
59 Jebel Ma'Aiserat High Place
60 Snake Monument
61 Zhanthur Mansion
62 Palace Tomb
63 Corinthian Tomb
64 Silk Tomb
65 Urn Tomb
66 Rectangular Platform/temple
67 South Nymphaeum

Supplemental Petra Site Details

—	pipeline
- - -	buried channel
B	dual pipeline supplied basin area
d	catchment dam
c	cistern
s	spring
D	multi-level platform
A	Kubtha aqueduct
T	Water distribution tank
-d	major dam

Bibliography

Abramowitz, M., I. Stegun 1964 *Handbook of Mathematical Functions*, National Bureau of Standards, Applied Mathematics Series 55. Government Printing Office, Washington, DC.

Abulnagy, B. 2002 *Slurry Systems Handbook*, Vol. 3, No. 3, McGraw-Hill, New York, pp. 183–210.

Akasheh, T. S. 2002 Ancient and Modern Water Management at Petra. *Near Eastern Archaeology* 65(4).

Albarracin-Jordan, J. 1996 Tiwanaku Settlement Systems: The Integration of Nested Hierarchies in the Lower Tiwanaku Valley. *Latin American Antiquity* 3(3):183–210.

Angelakis, A., L. Mays, D. Koutsoyiannis, N. Manassis 2012 *Evolution of Water Supplies Throughout Millennia*, IWA Publishing, London, pp. 231–243.

Avila, J. 1986 *Sistemas Hidráulicos Incas*, Lluvia Editores, Lima.

Baba, A. et al. 2018 Development of Water Dams and Water Harvesting Throughout History in Different Civilizations. *International Journal of Hydrology* 2(2).

Bain, A., S. Bonnington 1970 *The Hydraulic Transport of Solids by Pipeline*, Pergammon Press, New York.

Bakhmeteff, B. 1932 *Hydraulics of Open Channels*, McGraw Hill Book Company, New York.

Bandelier, A. 1911 The Ruins at Tiahuanaco. *Proceedings of the American Antiquarian Society* 21, Part 1.

Bandy, M. 2013 Tiwanaku Origins and Early Development: The Political and Moral Economy of a Hospitality State. In *Visions of Tiwanaku*, A. Vranich, C. Stanish, eds., Cotsen Institute of Archaeology Monograph 78, Los Angeles, pp. 135–150.

Barrera, A. 2008 *Bandurria*, Servicios Gráficos Jackeline, Huara, Peru.

Bartlett, F. 1958 *Thinking*, G. Allen and Unwin, London.

Barwald, A. 2000 *History and Theory in Anthropology*, Cambridge University Press, Cambridge.

Bastien, J. 1985 *Mountain of the Condor: Metaphor and Ritual in an Andean Ayllu,* Waveland Press, Prospect Heights, IL.

Baudin, L. 1961 *A Socialist Empire: The Inkas of Peru*. D. van Nostrand, Inc., Princeton, NJ.

Bauer, B. 1996 The Legitimization of the Inka State in Myth and Ritual. *American Anthropologist* 98(2):327–337.

Bauer, B. 1998 *Ancient Cuzco: Heartland of the Inka*, University of Texas Press, Austin.

Bauer, B., A. Covey 2004 The Development of the Inka State (AD 1000–1400). In *Ancient Cuzco: Heartland of the Inka*, B. Bauer, ed., University of Texas Press, Austin.

Bawden, G. 1999 *The Moche*, Blackwell Publishers, Oxford.

Bawden, G., R. M. Reycraft 2000 Environmental Disaster and the Archaeology of Human Response. In *Maxwell Museum of Anthropology Papers*, No. 7, University of New Mexico Press, Albuquerque, NM.

Bawden, G., R. M. Reycraft 2009 Exploration of Punctuated Equilibrium and Culture Change in the Archaeology of Andean Ethnogenesis (Chapter 12). In *Andean Civilization*, J. Marcus, P. R. Williams, eds., UCLA Cotsen Institute of Archaeology Press, Los Angeles, pp. 195–210.

Bear, J. 1972 *Dynamics of Fluids in Porous Media,* Dover Publications, New York.

Beckers, B., B. Schott, S. Tsukamoto, M. Frechen 2013 Age Determination of Petra's Engineered Landscape-Optically Stimulated Luminescence (OSL) and Radiocarbon Ages of Runoff Terrace Systems in the Eastern Highlands of Jordan. *Journal of Archaeological Science* 40(1):333–348.

Bedal, L-A. 2002 *The Petra Pool Complex: A Hellenistic Paradeisos in the Nabataean Capital*, Gorgias Dissertations, Near Eastern Studies 4. Gorgias Press, NJ.

Bedal, L-A. 2018 From Urban Oasis to Desert Hinterland: The Decline of Petra's Water System. The Case of the Petra Garden and Pool Complex. In *Water and Power in Past Societies*, Proceedings of the 8th IEMA Conference, Chapter 7, E. Holt, ed., IEMA Proceedings, SUNY Press, Albany, NY, pp. 131–158.

Bedal, L-A., K. Gleason, J. Schryver 2007 The Petra Garden and Pool Complex. *Annual of the Department of Antiquities of Jordan* 51.

Bellwald, U. 2006 *Petra Hydrology Uncovered*, Petra National Trust Publication, Amman, Jordan.

Bellwald, U. 2007a *Cultivated Landscapes of Native Amazonia and the Andes*, Oxford University Press, Oxford.

Bellwald, U. 2007b The Hydraulic Infrastructure of Petra – A Model for Water Strategies in Arid Lands. In *Cura Aquarum in Jordanien.* Proceedings of the 13th International Conference on the History of Water Management and Hydraulic Engineering in the Mediterranean Region, C. Ohlig, ed., Amman, Jordan.

Bellwald, U., M. al-Huneidi, A. Salihi, R. Naser 2002 *The Petra Siq: Nabataean Hydrology Uncovered*, Petra National Trust Publication, Amman, Jordan.

Bennett, W. C. 1934 Excavations at Tiahuanaco. *Anthropological Papers of the American Museum of Natural History* 34(3):354–359.

Bennett, W. C. (translator) 1961 *Frontinus: Stratagems and the Aqueducts of Rome*, Harvard University Press, Cambridge, MA.

Bentley, N. 2013 The Tiwanaku of A. F. Bandelier. In *Advances in Titicaca Basin Archaeology* 2, A. Vranich, A. Levine, eds., Cotsen Institute of Archaeology, University of California, Los Angeles, pp. 117–126.

Benyus, J. 1997 *Biomimicry: Inspiration Inspired by Nature*, William Morrow Publishers, New York.

Bermann, M. P. 1994 Domestic Life and Vertical Integration in the Tiwanaku Heartland. *Latin American Antiquity* 8(2):93–112.

Bermann, M. P. 1997 Domestic Life and Vertical Integration in the Tiwanaku Heartland. *Latin American Antiquity* 8(2):93–112.

Betancourt, P. 2008 Minoan Trade. In *The Aegean Bronze Age*, C. Shelmerdine, ed., Cambridge University Press, Cambridge, pp. 209–229.

Betanzos, Juan de 1996 [1551–1557] *Narrative of the Inkas*, University of Texas Press, Austin, TX.

Binford, M., M. Brenner, B. Leyden 1996 Paleoecology and Tiwanaku Agroecosystems. In *Tiwanaku and Its Hinterland: Archaeology and Paleoecology of an Andean Civilization*, Vol. 1, A. L. Kolata, ed., Smithsonian Institution Press, Washington, DC, pp. 89–108.

Binford, M., A. L. Kolata, M. Brenner 1997 Climate Variation and the Rise and Fall of an Andean Civilization. *Quaternary Research* 47:235–248.

Borsch, L., M. Sanger 2017 An Introduction to Anarchism in Archaeology. *SAA Archaeological Record* 17(1):9–16.

Boserup, E. 1965 *The Conditions of Agricultural Growth*, Aldine Publishers, Chicago, IL.

Bray, T. 2013 Water, Ritual, and Power in the Inka Empire. *Latin American Antiquity* 24(2):164–190.

Bray, T. 2015a *Explorations of the Sacred in the PreColumbian Andes*, University Press of Colorado, Boulder, CO.

Bray, T. 2015b *The Archaeology of Wak'as: Explorations of the Sacred in the Pre-Columbian Andes*, University Press of Colorado.

Browman, D. 1997 Political Institutional Factors Contributing to the Integration of the Tiwanaku State. In *Emergence and Change in Early Urban Societies*, Chapter 9, L. Manzanilla, ed., Plenum Press, New York.

Browning, I. 1982 *Petra*, Chatto and Windus, Ltd. Publishers, London.

Brundage, B. 1967 *Lords of Cuzco*, University of Oklahoma Press, Norman.

Bruno, M. 1999 *Sacred Springs: Preliminary Investigation of the Choquepacha Spring / Fountain, Tiwanaku, Bolivia*. Paper presented at the 69th Annual Meeting of the Society for American Archaeology, Philadelphia, PA.

Bruno, M. 2014 Beyond Raised Fields: Exploring Farming Practices and Processes in the Ancient Lake Titicaca Basin on the Andes. *American Anthropologist* 116:1–16.

Buckingham, E. 1914 On Physically Similar Systems: Illustrations of the Use of Dimensional Equations. *Physical Review* 4(4):345–376.

Buckingham, E. 1915 The Principles of Similitude. *Nature* 96(2406):396–397.

Cachot, R. C. 1955 *El Culto Agua en el Antigo Peru: la Paccha Elemento Cultural Pan-Andino*, Museo National de Antropología y Arqueología Publicacions.

Carballo, R., P. Roscoe, G. Feinman 2014 Cooperation and Collective Action in the Cultural Evolution of Complex Societies. *Journal of Archaeological Method and Theory* 21(1):98–133.

Carneiro, R. 1970 A Theory on the Origin of the State. *Science* 169:733–738.

Carneiro, R. 1981 The Chiefdom: Precursor of the State. In *The Transition to Statehood in the New World*, G. Jones, R. Kautz, eds., Cambridge University Press, Cambridge.

Carney, H., M. Binford, A. L. Kolata 1996 Nutrient Fluxes and Retention in Andean Raised-Field Agriculture: Implications for Long-Term Sustainability. In *Tiwanaku and Its Hinterland: Archaeology and Paleoecology of an Andean Civilization*, Vol. 1, A. L. Kolata, ed., Smithsonian Press, Washington, DC, pp. 169–179.

Chen, P. 2018 *Marine-Based Subsistence and its Social Implications in the Late Preceramic an Initial Period: A Different Pattern from Huaca Negra, North Coast of Peru*. Paper presented at the 58th Institute of Andean Studies Conference, Berkeley, CA.

Chepstow-Lusty, K., K. Bennett 1996 4000 Years of Human Impact and Vegetation Change in the Central Peruvian Andes. *Antiquity* 70:824–833.

Chepstow-Lusty, K., K. Bennett, V. Switsur, A. Kendall 1996 4000 Years of Human Impact Change in the Central Peruvian Andes- with Event Parallel in the Maya Record? *Antiquity*, December 1996 Issue. doi:10.1017/S0003598X0008409X.

Childe, G. V. 1950 The Urban Revolution. *The Town Planning Review* 21:13–17.

Clark, N., H. Ammann 2015 *Damp Indoor Spaces and Health, Ch.5: Human Health Effects in Damp Indoor Environment*, National Academies Press, Washington, DC.

Clement, C., M. E. Moseley 1991 The Spring-Fed Irrigation System of Carrizal, Peru: A Case Study of the Hypothesis of Agrarian Collapse. *Journal of Field Archaeology* 18(4):425–443.

Cline, E., Yasur-Landau 2013 Minoan Frescoes at Tel Kabri. *Biblical Archaeology Review* 39(4):37–44.

Coats, H. (ed.) 1984 *The Grand Canal of China*, South China Morning Post, Ltd., New China News, Hong Kong.

Cobo, B. 1990 [1653] *Inka Religion and Customs*, R. Hamilton, translation editor, University of Texas Press, Austin, TX.

Cohen, M. 1977 *Food Crisis in Prehistory*, Yale University Press, New Haven, CT.

Cohen, R. 1978 Introduction. In *Origins of the State: The Anthropology of Political Evolution*, R. Cohen, E. Service, eds., Institute for the Study of Human Issues, Philadelphia, pp. 1–20.

Collier, G. 1982 In the Shadow of Empire: New Directions in Mesoamerican and Andean Ethnohistory. In *The Inka and Aztec States* 1400–1800, G. Collier, R. Rosaldo, J. Wirth, eds., Academic Press, New York.

Collingwood, R., J. Meyers 1936 *Roman Britain and the English Settlements*, Clarendon Press, Oxford.

Combey, A. 2017 *Les Barrages de Haute Vallée de Nepeña Stratégiés de Gestion de L'eau el Impacts Sociaux*. MS Thesis, University of Paris, Pantheon-Sorbonne.

Contreras, D. 2000 Landscape and Environment: Insights from the Prehispanic Central Andes. *Journal of Archaeological Research* 18:241–238.

Cooke, K. 1979 Mathematical Approaches to Culture Change. In *Transformations- Mathematical Approaches to Culture Change*, C. Renfrew, K. Cooke, eds., Academic Press, New York, pp. 45–81.

Couture, N. 2002 *The Construction of Power: Monumental Space and an Elite Residence in Tiwanaku, Bolivia*. Ph.D. Dissertation, Department of Anthropology, University of Chicago.

Couture, N., K. Sampeck 2003 Putuni: A History of Palace Architecture in Tiwanaku. In *Tiwanaku and Its Hinterland: Archaeology and Paleoecology of an Andean Civilization*, Vol. 2, A. L. Kolata, ed., Smithsonian Institution Press, Washington, DC, pp. 226–263.

Coward, E. 1979 Principles of Social Organization in an Indigenous Irrigation System. *Human Organization* 38(1):28–36.

Craig, N., C. Aldenderfer, P. Rigsby, L. Baker, L. Blanco 2011 Geologic Constraints on Rain-fed Qocha Reservoir Agricultural Infrastructure, Northern Lake Titicaca Basin, Peru. *Journal of Archaeological Science* 38:2897–2907.

Creamer, W., A. Ruiz 2007 Archaeological Investigation of Late Archaic sites (3000–1800 BC) in the Pativilca Valley, Peru. *Fieldiana Anthropology* 40, Field Museum of Chicago Publication.

Créqui-Montfort, G. 1906 Fouilles de la Mission Scientifique Française à Tiahuanaco. Ses Recherches Archéologiques et Ethnographiques en Bolivie, au Chile et dans la République Argentine. *Proceedings of the 14th International Congress of Americanists*, Stuttgart, Germany, pp. 531–550.

Crumley, C. 2017 Assembling Conceptual Tools to Examine the Moral and Political Structures of the Past. *SAA Archaeological Record* 17(1):22–27.

D'Altroy, T. 2003 *The Inkas*, The Blackwell Publishing Company, Oxford.

D'Andrade, R. 2004 *The Development of Cognitive Anthropology*, Cambridge University Press, Cambridge.

Deeds, E., J. Kus, M. E. Moseley, F. Nials, C. R. Ortloff, L. Pippin, S. Pozorski, T. Pozorski 1978 Un Studio de Irrigation Prehispanica en Pampa Esperanza, Valle de Moche. In *III Congress Peruano, el Hombre y la Cultura Andina*, Actas y Trabajos, Lima, pp. 217–234.

Defner, M. 1921 Late Minoan Water Treatment Device. *Archaeological Newspaper* 8, Iraklion, Greece (in Greek).

Denevan, W. 2001 *Cultivated Landscapes of Native Amazonia and the Andes*, Oxford University Press, Oxford.

DiFenizio, C. 1916 Sulla Portada degli Acquedotti Romani e Determinazione della Quinaria, *Giornale del Genio Civile* 14:227–381.

Dillehay, T. 2001 Town and Country in Late Moche Times: A View from Two Northern Valleys. In *Moche Art and Archaeology in Ancient Peru*, J. Pillsbury, ed., Yale University Press, New Haven, CT, pp. 259–283.

Dillehay, T. 2011 *From Foraging to Farming in the Andes*, Cambridge University Press, Cambridge.

Dillehay, T. (ed.) 2017 Foundational Understandings. In *Where the Land Meets the Sea*, University of Texas Press, Austin, pp. 15–28.

Dillehay, T., H. Eling, J. Rossen 2005 Preceramic Irrigation Canals in the Peruvian Andes. *Proceedings of the National Academy of Science* 102(47):17241–17244.

Dillehay, T., J. Rossen, P. Netherly 1997 The Nanchoc Tradition: The Beginnings of Andean Civilization. *American Scientist* 85(1):46–85.

Doig, F. K. 1973 *Manual de Arqueologia Peruana*, Ediciones Peisa, Lima.

Downing, T., M. Gibson (eds.) 1974 *Irrigation's Impact on Society*, University of Arizona Press, Tucson, AZ.

Du, P., A. Koenig 2012 History of Water Supply in Pre-modern China. In *Evolution of Water Supply Through the Millennia*, A. Angelakis, L. Mays, D. Koutsoyiannis, N. Mamassis, eds., IWA Publishing, London, pp. 169–226.

Durant, W., A. Durant 1968 *The Lessons of History*, Simon and Schuster, New York.

Earle, T., T. D'Altroy 1989 The Political Economy of the Inka Empire: The Archaeology of Power and Finance. In *Archaeological Thought in America*, C. Lamberg-Karlovsky, ed., Cambridge University Press, Cambridge, pp. 183–204.

Eeckhout, P. 2019 *Pachacamac: Pilgrimages and Power in Ancient Peru*. World Archaeology, No. 92, Current Publishing, London, pp. 24–30.

Eling, H. 1986 *Pre-hispanic Irrigation Sources and Systems in the Jequetepeque Valley, Northern Peru*, Andean Archaeology: Papers in Memory of Clifford Evans, Institute of Archaeology, University of California, Los Angeles, pp. 130–148.

Erickson, C. 2000 Lake Titicaca Basin: A Precolumbian Built Landscape. In *Imperfect Balance: Landscape Transformations in the Precolumbian Americas*, D. Lentz, ed., Columbia University Press, New York, pp. 311–356.

Ertsen, M., J. van der Spek 2009 Modeling an Irrigation Ditch Opens up the World: Hydrology and Hydraulics of an Ancient Irrigation System. *Physics and Chemistry of the Earth* 34:176–191.

Evans, A. J. 1903 *Knossos Excavations*. Annual of the British School at Athens, No.9, London.

Fagan, B. 2008 *The Great Warming*, Bloomsbury Press, New York.

Fagan, B. 2011 *A History of Water and Humankind*, Bloomsbury Press, New York.

Feinman, G., J. Neitzel 1984 Too Many Types: An Overview of Sedentary Prestate Societies in the Americas. In *Advances in Archaeological Method and Theory*, Vol. 7, M. Schiffer, ed., Academic Press, New York, pp. 39–102.

Feldman, R. 1985 Preceramic Corporate Architecture: Evidence for Development of Non-egalitarian Social Systems in Peru. In *Early Ceremonial Architecture in the Andes*, C. Donnan, ed., Dumbarton Oaks Publications, Washington, DC, pp. 71–92.

Flores Ochoa, J. 1987 Cultivation in the Qocha of the South Andean Puna. In *Arid Land Use Strategies and Risk Management in the Andes: A Regional Anthropological Perspective*, D. Browman, ed., Westview Press, Colorado Springs, pp. 271–296.

Flores Ochoa, J., M. Bologna, D. Urzagast 2011 A Mathematical Model of the Andean Tiwanaku Civilization Collapse: Climate Variation. *Journal of Theoretical Biology* 291:29–32.

Floris Cohen, H. 1994 *The Scientific Revolution: A Historiographical Inquiry*, University of Chicago Press, Chicago, pp. 35–39.

Flow Science FLOW-3D 2018 User Manual V. 9.3. Flow Science, Inc. Santa Fe, New Mexico.

Freeman, K. 1950 *Greek City-States*, Norton & Company, Inc., New York.

Freeze, R. A., J. A. Cherry 1979 *Groundwater*, Prentice-Hall, Inc., Englewood Cliffs, NJ.

Gallo, I. 2014 Roman Water Supply Systems: New Approach. In *De Aquaductu Atque Aqua Urbium Lyciae Pamphyliae Pisidiae*, Peeters Publishing, Leuven, Belgium (Antalya Conference Proceedings).

Giddens, A. 1979 *Central Problems in Social Theory*, The Macmillan Company, London.

Giddens, A. 1984 *The Constitution of a Society*, University of California Press, Berkeley, CA.

Gilbert, A. 2017 *Encyclopedia of Geoarchaeology*, Springer Reference. ISBN 978-94-007-4827-4829.

Gillet, H., H. Mowbray 2008 *Dujiangyan: In Harmony with Nature*, Matric International Publishing House, Beijing.

Gleik, P. 1996 Basic Water Requirements for Human Activities. *Water International* 21:83–92.

Glowacki, M., M. Malpass 2003 Water, *Huacas* and Ancestor Worship: Traces of a Wari Sacred Landscape. *Latin American Antiquity* 14:431–449.

Glueck, N. 1965 *Deities and Dolphins*, Strauss and Cutahy Publishers, New York.

Goldstein, P. 2005 *Andean Diaspora: The Tiwanaku Colonies and the Origins of South American Empire*, University of Florida Press, Gainesville, FL.

Gorokhovich, Y., L. Mays, L. Ullman 2011 A Survey of Ancient Minoan Water Technology. Water Science and Technology, Water Supply. *IWA* 14:308–399.

Graffam, G. 1992 Beyond State Collapse: Rural History, Raised Fields and Pastoralism in the South Andes. *American Anthropologist* 94(4):882–904.

Graham, J. 1987 *The Palaces of Crete*, Princeton University Press, Princeton, NJ.

Green, M. 1997 *Dictionary of Celtic Myth and Legend*, Thames and Hudson Publishers, London.

Greider, T., A. Mendoza, C. E. Smith, R. Malina 1998 *La Galgada, Peru*, University of Texas Press, Austin, TX.

Griffin, J. 1990 Virgil. In *The Roman World*, J. Boardman, J. Griffin, O. Murray, eds., Oxford University Press, Oxford.

Guzzo, M., E. Schneider 2002 *Petra*, University of Chicago Press, Chicago.

Haas, J., W. Creamer 2004 Cultural Transformations in the Central Andes in the Late Archaic. In *Andean Archaeology*, H. Silverman, ed., Blackwell Publishers, London.

Haas, J., W. Creamer 2006 Crucible of Andean Civilization: The Peruvian Coast from 3000 to 1800 BC. *Current Anthropology* 47(5):745–775.

Haas, J., W. Creamer, A. Ruiz 2009 Dating the Late Archaic Occupation of the Norte Chico Region of Peru. *Nature* 432:1020–1023.

Hale, W. 1965 *The Horizon Book of Ancient Greece*, American Heritage Publishing Company, New York.

Hammond, P. 1973 *The Nabataeans: Their History, Culture and Archaeology: Studies in Mediterranean Archaeology* 37, Astrom Publishers, Gothenburg, Germany.

Hampson, N. 1968 *The Enlightenment*, Harmondsworth Publishers, London.

Harrington, L. 1974 *The Grand Canal of China*, Bailey Brothers and Swinfen, Ltd., London.

Hastorf, C. 1993 *Agriculture and the Onset of Political Inequality Before the Inka*, Cambridge University Press, Cambridge, pp. 1–8.

Hauck, G., Novak, R. 1988 Water Flow in the Castellum at Nimes. *American Journal of Archaeology* 92(3):393–407.

Haut, B., D. Viviers 2012 Water Supply in the Middle East During Roman and Byzantine Periods. In *Evolution of Water Supply Through the Millennia*, N. Angelakis, L. Mays, D. Koutsoyiannis, N. Mamassis, eds., IWA Publishing, London, pp. 319–350.

Henderson, F. 1966 *Open Channel Flow*, New York, the Macmillan Company.

Henderson, M. 2012 The Ancient Raised Fields of the Taraco Region of the Northern Lake Titicaca Region. In *Advances in Titicaca Basin Archaeology*-III, A. Vranich, E. A. Clarich, C. Stanish, eds., Memoirs of the Museum of Anthropology, No. 15, Ann Arbor, MI.

Herschell, C. 1973 *The Water Supply of the City of Rome*, New England Water Works Association, Boston, MA.

Hidalgo, C., P. Hansmann 2009 The Building Blocks of Economic Complexity. *Proceedings of the National Academy of Science* 106(26):10570–10575.

Higgins, R. 1973 *The Archaeology of Minoan Crete*, The Bodley Head Publishers, London.

Hobbes, T. 1991 [1658, 1642] *Man and Citizen* (De Homme and De Cive), B. Gert, ed., Hackett Publishing, Indianapolis, IN.

Hobbes, T. 1996 [1651] *Leviathan*, R. Tuck, ed., Cambridge University Press, Cambridge.

Hodge, A. T. 2011 *Roman Aqueducts and Water Supply*, MPG Books Group, London.

Hunt, R. 1988 Size and Structure of Authority in Canal Irrigation Systems. *Journal of Anthropological Research* 44:335–355.

Hvistndahl, M. 2008 China's Three Gorges Dam: An Environmental Catastrophe? *Scientific American*, March 2008 Issue.

Hyslop, J. 1990 *Inka Settlement Planning*, University of Texas Press, Austin.

Isbell, B. 1985 *To Defend Ourselves: Ecology and Rituals in an Andean Village*, Waveland Press, Prospect Heights, IL.

Janusek, J. 1999 Craft and Local Power: Embedded Specialization in Tiwanaku Cities. *Latin American Antiquity* 10(2):107–131.

Janusek, J. 2003 The Changing Face of Tiwanaku Residential Life: State and Social Identity in an Andean City. In *Tiwanaku and Its Hinterland: Archaeology and Paleoecology of an Andean Civilization*, Vol. 2, A. L. Kolata, ed., Smithsonian Institution Press, Washington, DC, pp. 264–295.

Janusek, J. 2004 *Identity and Power in the Ancient Andes*, Routledge Press, London.

Janusek, J. 2008 *Ancient Tiwanaku*, Cambridge University Press, Cambridge.

Janusek, J., H. Earnest 2009 Excavations in the Putuni: The 1988 Season. In *Tiwanaku and its Hinterland*, Vol. 2, Report Submitted to the National Science Foundation and the National Endowment for the Humanities, A. L. Kolata, ed., University of Chicago Press, Chicago, pp. 236–246.

Janusek, J., A. L. Kolata 2004 Top-down or Bottom-up Rural Settlement and Raised-Field Agriculture in the Lake Titicaca Basin, Bolivia. *Journal of Anthropological Archaeology* 23:404–430.

Jardine, L. 1999 *Ingenious Pursuits: Building the Scientific Revolution*, London Publishers.

Joukowsky, M. 1998 *Petra Great Temple, 1993–1997*, Vol. 1, Brown University Press, RI, pp. 265–270.

Joukowsky, M. 2004 *The Petra Great Temple Water Supply*. The 9th International Conference on the History and Archaeology of Jordan- Cultural Interaction throughout the Ages. Department of Antiquities, Al-Hussein Bin Talal University Publication, Abstracts 25.

Kaulicke, P., R. Kondo, T. Kusuda, J. Zapata 2003 *Aqua, Ancestros, y Archaeologia del Pasaje*. Boletin de Pontificia Universidad Catolica del Peru 7:27–56.

Kendall, A. 1985 *Aspects of Inka Architecture- Description, Function and Chronology*, 2 Vols., British Archaeological Reports, International Series 242, Oxford University Press.

Kerner, S., R. Dann, P. Bangsgaard (eds.) 2015 *Climate and Ancient Societies*, Meseum Tusculanum Press, Copenhagen.

Knighton, D. 1998 *Fluvial Forms and Processes: A New Perspective*, Hodder Education, Part of Hachette Livre Publishers, London.

Koestler, A. 1964 *The Act of Creation: The Evolution of Ideas*, The Macmillan Company, New York, pp. 224–254.

Kolata, A. L. 1986 Agricultural Foundations of the Tiwanaku State. *American Antiquity* 51(4):748–762.

Kolata, A. L. 1991 The Technology and Organization of Agricultural Production in the Tiwanaku State. *Latin American Antiquity* 2:99–125.

Kolata, A. L. 1993 *The Tiwanaku: Portrait of an Andean Civilization*, Blackwell Publishers, Oxford.

Kolata, A. L. 1996a *Tiwanaku and its Hinterland: Archaeology and Paleoecology of an Andean Civilization*, A. L. Kolata, ed., Smithsonian Institution Press, Washington, DC.

Kolata, A. L. 1996b Theoretical Orientations and Implications of the Proyecto Wila Jawira Research Program. In *Tiwanaku and Its Hinterland: Archaeology and Paleoecology of an Andean Civilization*, Vol. 1, A. L. Kolata, ed., Smithsonian Institution Press, Washington, DC, pp. 265–279.

Kolata, A. L. 2003 *Tiwanaku Ceremonial Architecture and Urban Organization*. In *Tiwanaku and Its Hinterland: Archaeology and Paleoecology of an Andean Civilization*, Vol. 2, A. L. Kolata, ed., Smithsonian Institution Press, Washington, DC, pp. 175–201.

Kolata, A. L., C. R. Ortloff 1989a Thermal Analysis of the Tiwanaku Raised-Field Systems in the Lake Titicaca Basin of Bolivia. *Journal of Archaeological Science* 16:233–263.

Kolata, A. L., C. R. Ortloff 1989b Tiwanaku Raised Field Agriculture in the Titicaca Basin of Bolivia. In *Tiwanaku and Its Hinterland: Archaeology and Paleoecology of an Andean Civilization*, Vol. 1, A. L. Kolata, ed., Smithsonian Institution Press, Washington, DC, pp. 109–152.

Kolata, A. L., C. R. Ortloff 1996 Agroecological Perspectives on the Decline of the Tiwanaku State. In *Tiwanaku and Its Hinterland: Archaeology and Paleoecology of an Andean Civilization*, Vol. 1, A. L. Kolata, ed., Smithsonian Institution Press, Washington, DC, pp. 181–201.

Kosok, P. 1965 *Life, Land and Water in Ancient Peru*, Long Island University Press, New York.

Kurin, D. 2016 The Bioarchaeology of Social Collapse and Regeneration in Ancient Peru. In *Bioarchaeology and Social Theory*, D. Martin, ed., Springer Reference Publications. ISBN 13:978-3319284026. doi:10.1007/978-3-310-28404-0_3.

Lamberg Karlovsky, C. 2016 Irrigation Among the Shaykhs and Kings. In *Ancient Irrigation Systems of the Aral Sea Area: The History, Origin, and Development of Irrigated Agriculture*, B. Andrianov, ed., Oxbow Press Books, pp. 23–53.

Lane, K. 2009 Engineered Highlands: The Social Organization of Water in the Ancient North-central Andes (AD 1000–1480), *World Archaeology* 41:1, 169–190.

Lane, K. 2017 Pukios and Aqueducts in the Central Andes of South America. In *Underground Aqueducts Handbook*, A. Angelakis, E. Chiotis, S. Eslamian, H. Weingartner, eds., CRC Press, Boca Raton, FL, pp. 465–475.

Langlie, B. 2018 Building Ecological Resistance: Late Intermediate Period Farming in the South-Central Highland Andes (CE 1100–1450). *Journal of Anthropological Archaeology*, June 2018 Issue. doi:10.1016/j.jaa.2018.06.005.

Lanning, E. 1967 *Peru Before the Inkas*, Prentice Hall Publishers, Englewood Cliffs, NJ.

Laudan, L. 1981 *Science and Hypothesis: Historical Essays on Scientific Methodology*, D. Reidel Publishers, Dordrecht, Germany.

Laureano, P. 2002 *The Water Atlas, Traditional Knowledge to Combat Desertification*, Laia Libros, Barcelona.

Lepenieas, W. 1988 *Between Literature and Science: The Rise of Sociology*, Cambridge University Press, Cambridge.

Lepenieas, W. 2010 *Auguste Comte, die Macht der Zeichen*, Hanser Verlag, München.

LeVine, T. (ed.) 1992 *Inka Storage Systems*, University of Oklahoma Press, Norman, OK.

Levy-Bruhl, L. 1899 *La Philosophie d'Auguste Comte*, Alcan Publishers, Paris.

Lewis, M. 2001 *Surveying Instruments of Greece and Rome*, Cambridge University Press, Cambridge.

Liverani, M. 2006 *Uruk: The First City*, Z. Bahrani, M. van de Mieroop, eds., Equinox Publishers, London.

Lobell, J. 2015 The Minoans of Crete. *Archaeology* 68(3):28–35.

Logan, A., C. Hastorff, D. Parsall 2012 "Let's Drink Together": Early Ceremonial Use of Maize in the Titicaca Basin. *Latin American Antiquity* 23(3):235–258.

Lugt, J. 1983 *Vortex Flow in Nature and Technology*, John Wiley & Sons, New York.

Lumbreras, L. 1974 *The Peoples and Cultures of Ancient Peru*, Smithsonian Institution Press, Washington, DC.

Lumbreras, L. 1987 Childe and the Urban Revolution: The Central Andean Experience. In *Studies in the Neolithic and Urban Revolutions*, L. Manzanilla, ed., B. A. R. International Series 349, Archaeopress, Oxford, pp. 327–344.

MacCormack, S. 1991 *Religion in the Andes*, Princeton University Press, Princeton, NJ.

Macherey, P. 1989 *Comte, la Philosophie et les Sciences*, Presses Universitaires de France, Paris.

Magnuszewski, P., K. Krolikowska, A. Koch, et al. 2018 *Water* 10(3):346.

Manners, R., F. Migilligan, P. Goldstein 2003 Floodplain Development, El Niño, and Cultural Consequences in a Hyperarid Andean Environment. *Annals of the Association of American Geographers* 97:229–249.

Manzanilla, L. 1997 The Impact of Climate Change in Past Civilizations: A Revisionist Agenda for Further Investigators. *Quaternary International* 43/44:153–159.

Marcus, J., C. Stanish (eds.) 2006 *Agricultural Strategies*, Cotsen Institute of Archaeology, University of California, Los Angeles, pp. 1–13.

Martini, P., R. Drusiani 2012 History of Water Supply of Rome as a Paradigm of Water Services Development in Italy. In *Evolution of Water Supply Through the Millennia*, N. Angelakis, L. Mays, D. Koutsoyiannis, N. Mamassis, eds., IWA Publishing, London, pp. 443–465.

Mauricio, A. 2018 Reassessing the Impact of El Niño at the End of the Early Intermediate Period from the Perspective of the Lima Culture. *Journal of Andean Archaeology (Ñawpa Pacha)* 38(2):203–231.

Mays, L. 2007 *Ancient Water Technologies*, Figure1.15, pp. 21–22. Springer Publishers.

Mays, L. 2010 Nabataean City of Petra's Water Supply, Figure1.15, pp. 21–22; and The Mycenaeans, pp. 10–14 Both in *Ancient Water Technologies*, Springer Publishers.

McAndrews, J., J. Albarracin-Jordan, M. P. Bermann 1997 Regional Settlement Patterns of the Tiwanaku Valley of Bolivia. *Journal of Field Archaeology* 24:67–83.

McEwan, G., LeVine, T. 1991 Investigations of the Pikillacta Site: A Provincial Huari Administrative Structure in the Valley of Cuzco. In *Huari Administrative Structure Prehistoric Monumental Architecture and State Government*, W. Isbell, G. McEwan, eds., Dumbarton Oaks Research Library and Collection, Washington, DC, pp. 93–119.

McKenzie, J. 1990 *The Archaeology of Petra*, Oxford University Press, Oxford.

Means, P. 1931 *Ancient Civilizations of the Andes*, C. Scribner's Sons Publishers, New York.

Meddens, F., C. McEwan, C. Pomacanchari 2010 Inca "Stone Ancestors" in Context at a High Altitude Usnu Platform. *Latin American Antiquity* 21(2):173–194.

Menninger, K. 1970 *Number Words and Number Symbols: A Cultural History of Numbers*, The MIT Press, MA, pp. 252–255.

Miller, M., I. Kendall, J. Capriles, M. Bruno, R. Evershed, C. Hastorf 2014 The Trouble with Maize and Fish: Refining Our Understanding of the Diets of Lake Titicaca's Inhabitants Using Multiple Stable Isotope Methods (1500 BC–AD 1100). Paper presented at the 59th Institute of Andean Studies Conference, Berkeley, CA.

Mitchell, W. 1973 The Hydraulic Hypothesis: A Reappraisal. *Current Anthropology* 14(57):532–534.

Mithin, S. 2012 *Thirst: Water and Power in the Ancient World*, Harvard University Press, Cambridge, MA.

Moore, J. 1988a Cultural Responses to Environmental Catastrophes: Post-El Niño Subsistence on the Pre-historic North Coast of Peru. *Latin American Antiquity* 2(1):21–47.

Moore, J. 1988b Prehistoric Raised Field Agriculture in the Casma Valley, Peru. *Journal of Field Archaeology* 15:265–276.

Moore, J. 2005 *Cultural Landscapes in the Ancient Andes*, University Press of Florida.

Morris, C. 1982 The Infrastructure of Inka Control in the Peruvian Central Highlands. In *The Inka and Aztec States 1400–1800*. Anthropology and History, G. Collier, R. Rosaldo, J. Wirth, eds., Academic Press, New York.

Morris, C. 1992 Foreword. In *Inka Storage Systems*, University of Oklahoma Press, Norman, AZ.

Morris, C. 1998 Inka Strategies of Incorporation and Governance. In *Archaic States*, G. Feinman, J. Marcus, eds., School of American Research Press, Santa Fe, NM, pp. 293–309.

Morris, H., J. Wiggert 1972 *Open Channel Hydraulics*, The Ronald Press, New York.

Morrison, K. 1994 The Intensification of Production: Archaeological Approaches. *Journal of Archaeological Method and* Theory 1:111–159.

Moseley, M. E. 1975 *The Maritime Foundations of Andean Civilizations*, Cummings Publishing Company, Menlo Park, CA.

Moseley, M. E. 1983 Patterns of Settlement and Preservation in the Viru and Moche Valleys. In *Prehistoric Settlement Patterns*, University of New Mexico Press, pp. 423–442.

Moseley, M. E. 1993 Geophysical Change on the North Coast of Peru. Preliminary Report Distributed to Caral Project Members (R. Shady, D Sandweiss, D. Keefer, C. Ortloff).

Moseley, M. E. 1999 Convergent Catastrophes: Past Patterns and Future Implications of Collateral Natural Disasters in the Andes. In *The Angry Earth: Disasters in Anthropological Perspective*, A. Oliver-Smith, A. Hoffman, eds., *Journal of Field Archaeology* 18(4):425–443. Routledge Press, pp. 59–71.

Moseley, M. E. 2008 *The Inkas and Their Ancestors*, Thames & Hudson, New York.

Moseley, M. E., C. Clement, J. Tapia, D. Satterlee 1998 *El Collapso Agrario de la Subregion de Moquegua- Moquegua, Los Primeros Doce Mil Años*, K. Wise, ed., Museo Contisuyo Publicaciones, Lima, pp. 9–16.

Moseley, M. E., E. Deeds 1982 The Land in Front of Chan Chan: Agrarian Expansion, Reform, and Collapse in the Moche Valley. In *Chan Chan: Andean Desert City*, M. E. Moseley, K. Day, eds., University of New Mexico Press, Albuquerque, NM, pp. 25–33.

Moseley, M. E., J. Tapia, D. Satterlee, J. Richardson 1992 Flood Events, El Niño Events, and Tectonic Events. In *Paleo-Enso Records*, International Symposium Extended Abstracts, L. Ortlieb, J. Machare, eds., OSTROM, Lima, pp. 207–212.

Moseley, M. E., D. Wagner, J. Richardson 1992 Space Shuttle Imagery of a Recent Catastrophic Change Along the Arid Andean Coast. In *Paleoshorelines and Prehistory: An Investigation of Method*, L. Johnson, M. Stright, eds., CRC Press, Boca Raton, FL, pp. 215–236.

Moseley, M. E., G. R. Willey 1973 Áspero, Peru: A Reexamination of the Site and Its Implications. *American Antiquity* 38(4):452–468.

Murphy, G. 1950 *Similitude in Engineering*, The Ronald Press, New York.

Murra, J. 1960 Rite and Crop in the Inka State. In *Culture in History*, S. Diamond, ed., Columbia University Press, New York, pp. 393–340.

Murra, J. 1980a *The Economic Organization of the Inka State*, JAI Press, Greenwich, CT.

Murra, J. 1980b *Formaciones Economicas y Poliicas del Mundo Andino*, Instituto de Estudios Peruanos, Lima. Organization of the Inka State, JAI Press Greenwich, CT.

Murra, J. 1982 The Mit'a Obligations of Ethnic Groups to the Inka State. In *The Inka and Aztec States 400–1800: Anthropology and History*, G. Collier, R. Rosaldo, J. Wirth, eds., Academic Press, New York.

Needham, J. 1971 *Science and Civilization in China*, Vol. 4: Physics and Physical Technology, Cambridge University Press, Cambridge.

Netherly, P. 1984 The Management of the Late Andean Irrigation System on the North Coast of Peru. *American Antiquity* 49(2):227–254.

Nicolis, G., I. Prigogine 1989 *Exploring Complexity*, W. H. Freeman and Company, New York, pp. 238–242.

Nielsen, M. 1952 *Pressure Vibrations in Pipe Systems*, NYT Nordisk Forlag Arnold Busck, Copenhagen.

Nikolic, M. 2008 *Cross-Disciplinary Investigation of Ancient Long-Distance Water Pipelines*. Unpublished Ph.D. dissertation, Department of Greek and Roman Studies, University of Victoria, Canada.

Niles, F., E. Deeds, M. E. Moseley, S. Pozorski, T. Pozorski, R. Feldman 1979 El Niño: Catastrophic Flooding of Coastal Peru. *Field Museum Bulletin* 50:4–14.

Nocquett, J., J. Villegas-Lanza, M. Chlieh, P. Mothes, F. Rolandone, P. Jarrin, D. Cisneros 2014 Motion of Continental Slivers and Creeping Subduction of the Northern Andes. *Nature Geoscience* 7:287–329.

Oleson, J. 1995 The Origins and Design of Nabataean Water-Supply Systems. In *Studies in the History and Archaeology of Jordan*, Vol. 5, Department of Antiquities, Amman, Jordan.

Oleson, J. 2002 Nabataean Water Supply: Irrigation and Agriculture. In *Proceedings of the International Conference on the World of Herod and the Nabataeans*, Vol. 2, K. Politis, ed., Franz Steiner Verlag, Stuttgart.

Oleson, J. 2007 Nabataean Water Supply, Irrigation and Agriculture. In *World of the Nabataeans*, Vol. 2, International Conference: The World of the Herods and Nabataeans, K. Politis, ed., Oriens et Occidens 15, Franz Steiner Verlag, Munich.

ONERN 1976 *Mapa Ecologico del Peru: Guia Explicativa*, Oficina Nacional de Evaluation de Recursos Naturales, Lima, Peru.

Ortlieb, L., M. Fournier, J. Macharé 1993 Beach Ridge Series in Northern Peru: Chronology, Correlation and Relationship with Major Holocene El Niño Events. *Bulletin de Français d'Etudes Andines* 22:121–212.

Ortloff, C. R. 1993 Chimu Hydraulics Technology and Statecraft on the North-coast of Peru, AD 1000–1470. In *Economic Aspects of Water Management in the Prehispanic*

New World, V. Scarborough, B. Isaacs, eds., Economic Anthropology, Supplement 7, JAI Press, Greenwich, CT, pp. 327–367.

Ortloff, C. R. 1995 Surveying and Hydraulic Engineering of the Pre-Columbian Chimu State, A.D. 900–1450. *Cambridge Archaeological Journal* 5(1):55–74.

Ortloff, C. R. 1996 Engineering Aspects of Tiwanaku Groundwater Controlled Agriculture. In *Tiwanaku and Its Hinterland: Archaeology and Paleoecology of an Andean Civilization*, Vol. 1, A. L. Kolata, ed., Smithsonian Institution Press, Washington, DC, pp. 153–168.

Ortloff, C. R. 1997 Engineering Aspects of Groundwater Controlled Agriculture in the PreColumbian Tiwanaku State of Bolivia in the Period 400–1000 AD. In *Tiwanaku and its Hinterland: Archaeology and Paleoecology of an Andean Civilization*, Vol. 1, A. L. Kolata, ed., Smithsonian Institution Press, Washington, DC, pp. 109–152.

Ortloff, C. R. 2003 The Water Supply and Distribution System of the Nabataean City of Petra (Jordan), 300 BC–AD 300. *Cambridge Archaeological Journal* 15(1).

Ortloff, C. R. 2009 *Water Engineering in the Ancient World: Archaeological and Climate Perspectives on Societies of Ancient South America, the Middle East and South-East Asia*, Oxford University Press, Oxford.

Ortloff, C. R. 2014a *Hydraulic Engineering in Ancient Peru and Bolivia*. Encyclopaedia of the History of Science, Technology and Medicine in Non-Western Cultures, Springer Publications, Heidelberg.

Ortloff, C. R. 2014b Groundwater Management at the 300 BCE–CE 1100 Precolumbian City of Tiwanaku (Bolivia). *Hydrology: Current Research* 5(2). http://dx.doi.org/10.4172/2157-7587.1000168. (Internet open access).

Ortloff, C. R. 2014c Water Engineering at Petra (Jordan): Recreating the Decision Process underlying Hydraulic Engineering at the Wadi Mataha Pipeline. *Journal of Archaeological Science* 44:91–97.

Ortloff, C. R. 2014d *Three Hydraulic Engineering Masterpieces at 100 BC–AD 300 Nabataean Petra*, (Antalya, Turkey Conference Proceedings. In *De Aquaductu Atque Aqua Urbium Lyciae Phamphyliae Pisidiae: The Legacy of Julius Sextus Frontinus*, G. Wiplinger, ed., Peeters Publishing, Leuven, Belgium. ISBN 978-90-429-3361-3.

Ortloff, C. R. 2016a New Discoveries and Perspectives on Water Management and State Structure at AD 300–1100 Tiwanaku's Urban Center (Bolivia). *MOJ Civil Engineering* 1(3). doi:10.15406/mojce.2016.01.00014.

Ortloff, C. R. 2016b Similitude in Archaeology: Examining Agricultural System Science in Precolumbian Civilizations of Peru and Bolivia. *Hydrology: Current Research* 7(3). doi:10.4172/2157-7587.1000259.

Ortloff, C. R. 2018 The Pont du Garde Aqueduct and Castellum: Insight into Roman Hydraulic Engineering Practice. *Journal of Archaeological Science: Reports* 20:808–817.

Ortloff, C. R. 2019 Tipon: Insight into Inka Hydraulic Engineering Practice. *Article Among Others Submitted for Book Publication.*

Ortloff, C. R., D. Crouch 1998 Hydraulic Analysis of a Self-Cleaning Drainage Outlet at the Hellenistic City of Priene. *Journal of Archaeological Science* 25:1211–1220.

Ortloff, C. R., D. Crouch 2002 The Urban Water Supply and Distribution System of the Ionian City of Ephesos in the Roman Imperial Period. *Journal of Archaeological Science* 28:843–860.

Ortloff, C. R., R. Feldman, M. E. Moseley 1985 Hydraulic Engineering and Historical Aspects of the Pre-Columbian Intravalley Canal System of the Moche Valley, Peru. *Journal of Field Archaeology* 12:77–98.

Ortloff, C. R., J. Janusek 2014 Water Management and Hydrological Engineering at 300 BCE–1100 CE Precolumbian Tiwanaku (Bolivia). *Journal of Archaeological Science* 44(2):91–97.

Ortloff, C. R., J. Janusek 2015 *Water Management at BCE 300–CE 1100 Tiwanaku (Bolivia): The Perimeter Canal and its Hydrological Features*, Encyclopaedia of the History of Science, Technology and Medicine in Non-Western Civilizations, Springer Publications.

Ortloff, C. R, A. L. Kolata 1989 Hydraulic Engineering of Tiwanaku Aqueducts at Lukurmata and Pajchiri. *Journal of Archaeological Science* 16(5):513–535.

Ortloff, C. R., A. L. Kolata 1993 Climate and Collapse: Agro-ecological Perspectives on the Decline of the Tiwanaku State. *Journal of Archaeological Science* 16:513–535.

Ortloff, C. R., M. E. Moseley 2009 Climate, Agricultural Strategies and Sustainability in PreColumbian Andean Societies. *Andean Past* 9:1–27.

Ortloff, C. R., M. E. Moseley 2012 2600–1800 BCE Caral: Environmental Change at a Late Archaic Period Site in North Central Coast Peru. *Journal of Andean Archaeology (Ñawpa Pacha)* 12(2):189–206.

Ortloff, C. R., M. E. Moseley, R. Feldman 1982 Hydraulic Engineering Aspects of the Chimu Chicama-Moche Intervalley Canal. *American Antiquity* 47:572–595.

Ossorio, F., D. Porter 2009 *Petra- Splendors of the Nabataean Civilization Highlights*, White Star Publishers, Vercelli, Italy.

Outram, D. 1995 *The Enlightenment*, Cambridge University Press, Cambridge.

Papapostolou, J. 1981 *Crete*, Clio Editions, Athens.

Parcak, S., C. Tuttle 2016 Hiding in Plain Sight: The Discovery of a New Monumental Structure at Petra, Jordan, Usung WorldView-1 Satellite Imagery. *Bulletin of the American Schools of Oriental Research* 375:35–51.

Patterson, T. 1991 *The Inka Empire: The Formation and Disintegration of Pre-Capitalist State*, Berg Publishers, New York.

Paulson, A. 1996 Environment and Empire: Climatic Factors in Prehistoric Andean Culture Change. *World Archaeology* 8:121–132.

Payne, R. 1959 *The Canal Builders*, The Macmillan Company, New York.

Peck, R., W. Hanson, T. Thornton 1974 *Foundation Engineering*, John Wiley and Sons, New York.

Perfettini, H., P. Avouac, H. Tavera, A. Kosit, et al. 2010 Seismic and Aseismic Slip on the Central Peru Megathrust. *Nature* 465:78–81.

Plowman, D., S. Solansky, T. Beck, l. Baker, M. Kulkani 2007 *The Leadership Quarterly*, 341–356. doi:10:1016/jleaqua2007.04.004.

Podany, A. 2010 *Brotherhood of Kings: How International Relations Shaped the Ancient Near East*, Oxford University Press, Oxford.

Podany, A., J. Sasson 2008 Texts, Trade and Travelers. In *Beyond Babylon: Art, Trade and Diplomacy in the Second Millennium B.C.*, J. Aruz, ed., Yale University Press, New Haven, CT.

Poma, J. 2018 *El Aqua de los Ancestros: Algunas Notas sobre el Systema de Riego Prehispanico Huiru Catac*. Presentation at Ponencias Desarrolladas del 1 Coloquio del Museo de Julio G. Tello de Paracas. Publicacion disponible en http://repositorio cultura.gob.pe/.

Ponce Sangines, C. 1961 *Informe de Labores. La Paz: Centro de Investigaciones Arqueologicas en Tiwanaku Tiwanaku: Espacio, Tiempo y Cultura: Ensayo de Sintesis Arqueológica*, Academía Nacional de Ciencias Publication, La Paz, Bolivia.

Ponce Sanguines, C. 2009 *Descripción Sumaria del Templete Semisuterraneo de Tiwanaku*, La Paz (Bolivia), Juventud Publicaciones, La Paz, Bolivia.

Porter, B. 2013 *Complex Communities: The Archaeology of Early Iron Age West-Central Jordan*, The University of Arizona Press, Tucson, AZ.

Posnansky, A. 1945 *Tihuanacu: The Cradle of American Man*, Vols. I & II., J. Augustin Publishers, New York.

Poston, T., I. Stewart 1978 *Catastrophe Theory and its Applications*, Dover Publications, New York.

Prentiss, A. 2019 Evolutionary Archaeology at Middle Age: Reflections from the Handbook of Evolutionary Research in Archaeology. *The Society of American Archaeology, SAA Record* 19(5):35–38.

Pruett, J. 2016 Marsh Creation in Coastal Louisiana. *ECO*, September 2016 Issue.

Pulgar Vidal, J. 2012 *Geografía del Peru*, Novena Edición, Editorial PEISA, Lima.

Quilter, J. 2014 *The Ancient Central Andes*, Routledge Press, New York, pp. 274–275.

Rasmussen, C. 2017 *A Comparative Analysis of Roman Water Systems in Pompeii and Nîmes*. MS Thesis, University of Arizona.

Reinhard, J. 1992 An Archaeological Investigation of the Inka Ceremonial Platforms on the Volcano Copispo, Central Chile. In *Ancient America: Contributions to New World Archaeology*, N. Saunders, ed., Oxbow Monographs 24, Oxford.

Renfrew, C. 1979 System Collapse as Social Transformation: Catastrophe and Anastrophe in Early State Societies. In *Transformations- Mathematical Approaches to Culture Change*, C. Renfrew, K. Cooke, eds., Academic Press, New York.

Renfrew, C., P. Bahn 2000 *Archaeology*, Thames & Hudson, Ltd., London.

Reycraft, R. 2000 Long-term Human Response to El Niño in South Coastal Peru circa A.D. 1400. In *Environmental Disaster and the Archaeology of Human Response*, G. Bawden, R. Reycraft, eds., Maxwell Museum of Anthropology and Archaeological Papers, No.7, Albuquerque, NM, pp. 99–120.

Richardson, J. 1983 The Chira Beach Ridges, Sea Level Change, and the Origins of Maritime Economies on the Peruvian Coast. *Annals of the Carnegie Museum* 52:265–275.

Rivera Casanovas, C. 1994 *Ch'iji Jawira: Evidencias sobre la Producción de Cerámica en Tiwanaku*. Licentiatura Thesis, Universidad Mayor de San Andrés, Peru.

Rivera Casanovas, C. 2003 Ch'iji Jawira: A Case of Ceramic Specialization in the Tiwankau Urban Periphery. In *Tiwanaku and Its Hinterland: Archaeology and Paleoecology of an Andean Civilization*, Vol. 2, A. L. Kolata, ed., Smithsonian Institution Press, Washington, DC, pp. 296–315.

Rogers, S., D. Sandweiss, K. Maasch, D. F. Belknap, P. Agouris 2004 Coastal Change and Beach Ridges Along the Northwest Coast of Peru: Image and GIS Analysis of the Chira, Piura and Colán Beach Ridge Plains. *Journal of Coastal Research* 20(4):1102–1125.

Romer, P. 1986 Increasing Returns and Long-Run Growth. *Journal of Political Economy* 94(5):1002–1037.

Romey, K. 2019 An Unthinkable Sacrifice. *National Geographic*, February Issue.

Romilly, J. 1991 *The Rise and Fall of States According to Greek Authors*, The University of Michigan Press, Ann Arbor, MI.

Rosenzweig, M., Marston, J. 2016 Archaeologies of Empire and Environment. *Journal of Anthropological Archaeology* 52:87–102.

Rostworowski, M. 1999 *History of the Inca Realm*, Cambridge University Press. ISBN 0-521-63759-7.

Rouse, H. 1978 *Elementary Mechanics of Fluids*, Dover Publications, New York.

Routledge, B. 2013 *Archaeology and State Theory*, Bloomsbury Academic Press, London.

Rowe, J. 1946 Inca Culture at the Time of the Spanish Conquest. In *Handbook of South American Indians 2: The Andean Civilization*, J. Steward, ed., Bureau of American Government Printing Office, Washington, DC, pp. 183–330.

Rowe, J. 1979 An Account of the Shrines of Ancient Cuzco. *Journal of the Institue of Andean Studies (Ñawpa Pacha)* 17:2–80.

Sabersky, R., A. Acosta, E. Hauptman 1971 *Fluid Flow*, The Macmillan Company, New York.

Sage, M. 2011 *Roman Conquests: Gaul*, Pen and Sword Books, Ltd., London.
Sanders, W., B. Price 1968 *Mesoamerica: The Evolution of a Civilization*, Random House Press, New York.
Sandweiss, D. 1986 The Beach Ridges at Santa, Peru: El Niño, Uplift and Prehistory. *Geoarchaeology* 1:17–28.
Sandweiss, D., K. Maasch, R. Burger, J. Richardson, H. Rollins, A. Clement 2001 Variation in Holocene El Niño Frequencies: Climate Records and Cultural Consequences in Ancient Peru. *Geology* 29:603–606.
Sandweiss, D., J. Quilter 2009 *El Ñino, Catastrophism and Culture Change in Ancient America*, Dumbarton Oaks Research Library and Collection, Washington, DC.
Sandweiss, D., J. Richardson, E. Reitz, H. Rollin, K. Maasch 1996 Geoarchaeological Evidence from Peru for a 5000 BP Onset of El Niño. *Science* 273:1531–1533.
Sandweiss, D., R. Shady-Solis, M. E. Moseley, D. Keefer, C. R. Ortloff 2009 Environmental Change and Economic Development in Coastal Peru Between 5,800 and 3,600 Years Ago. *Proceedings of the National Academy of Sciences* 106(5):1359–1363.
Sarminto, P. 2007 [1572] *The History of the Inkas*, University of Texas Press, Austin.
Sarton, G. 1931 *Introduction to the History of Science*, Williams & Wilkins Company, Washington, DC.
Scarborough, V. 2004 Intensification and the Political Economy- a Contextual Overview. In *Agricultural Strategies*, J. Markus, C. Stanish, eds., Cotsen Institute of Archaeology Publication, Los Angeles, pp. 401–418.
Scarborough, V. 2015 Human Niches, Abandonment Cycling, and Climate. In *Water History, Springer Science + Business Media*, Dordrecht Publishers, Germany. doi:10.1007/S12685-015-0147-5.
Schmid, S. 2007 *Die Wasserversorgung des Wadi Farasa Ost in Petra*. Cura Aquarum in Jordanien. Proceedings of the 13th International Conference on the History of Water Management and Hydraulic Engineering in the Mediterranean Region. DWhG Publication, Band 12, Berlin, pp. 95–117.
Schmid, S., P. Bienkowski, Z. Ziema, B. Kolb 2012 The Palaces of the Nabataean Kings at Petra. In *The Nabataeans in Focus: Current Archaeological Research at Petra*, L. Nehme, L. Wadeson, eds., Archaeopress, Oxford.
Schreiber, K., J. Rojas 2007 *Irrigation and Society in the Peruvian Desert: The Pukios of Nazca*, Lexington Books, New York.
Shady, R. 2000 Sustento Socioeconómico del Estado Pristino de Supe-Peru: Las evidencia de Caral-Supe. *Arqueología y Sociedad* 13:49.
Shady, R. 2001 *La Ciudad Sagrada de Caral- Origines de la Civilización Andina*, Museo de Archaeología, Universidad Nacional de San Marcos Publicaciónes, Lima.
Shady, R. 2004 *Caral- La Ciudad del Fuego Sagrado*, Interbank Publishers, Lima.
Shady, R. 2007 *The Social and Cultural Values of Caral-Supe, the Oldest Civilization in Peru and America and its Role in Integral and Sustainable Development*, Proyecto Especial Arqueologico Caral-Supe, Instituto Nacional de Cultura Publicación No.4:1–69, Lima.
Shady, R. 2009 Caral-Supe: Y Su Entorno Natural y Social en Los Origenes de la Civilización. In *Andean Civilization*, J. Marcus, P. Williams, eds., Cotsen Institute of Archaeology 63, Los Angeles, pp. 99–117.
Shady, R., W. Creamer, A. Ruiz 2004 Dating the Late Archaic Occupation of the Norte Chico Region of Peru. *Nature* 432:1020–1023.
Shady, R., J. Haas, W. Creamer 2002 Dating Caral, a Preceramic Site in the Supe Valley on the Central Coast of Peru. *Science* 292:723–726.

Shady, R., C. Leyva 2003 *La Ciudad Sagrada de Caral- los Origines de la Civilización Andina y la Formación del Estado Pristina*, Proyecto Especial Arqueológico Caral-Supe, Instituto National de Cultura Publicación, Lima.

Shady, R., E. Quispe, P. Novoa, M. Machacuay 2014 *Vichama, Civilización Agropesquera de Végueta, Huara: La Ideologia de Nuestros Ancestros, 3800 Ãnos de Arte Mural*, Impreso en Servicios Graphicos JMD S.R.L., Lima.

Shalaeva, V. 2014 *Symbolism and Mythology of the Ancients: An Outline of Georg Friedrich Creuzer's Argument*, National Research University Higher School of Economics, Moscow.

Shapin, S. 1996 *The Scientific Revolution*, University of Chicago Press, Chicago.

Sharratt, N. 2019 Tiwanaku's Legacy: A Chronological Reassessment of the Terminal Middle Horizon in the Moquegua Valley, Peru. *Latin American Antiquity* 30(3):529–549.

Sherbondy, J. 1992 Water Ideology in Inka Ethnogenesis. In *Andean Cosmologies Through Time: Persistence and Emergence*, R. Dover, K. Siebold, J. McDowell, eds., Indiana University Press, pp. 46–56.

Shimada, L., C. Schaff, L. Thompson, E. Moseley-Thompson 1991 Cultural Impacts of Severe Droughts in the Prehistoric Andes: Application of a 1,500-year Ice Core Precipitation Record. *World Archaeology* 22(3):247–270.

Skelland, A. 1967 *Newtonian Flow and Heat Transfer*, John Wiley & Sons, New York.

Sloan, K. (ed.) 2003 *Enlightenment: Discovering the World of the Eighteenth Century*, British Museum Publications, London.

Smith, A. 2003 *The Political Landscape: Constellations of Authority in Early Complex Polities*, University of California Press, Berkeley, CA.

Spencer, C. 1993 Human Agency, Biased Transmission, and the Cultural Evolution of Chiefly Authority. *Journal of Anthropological Archaeology* 12:41–74.

Spier, F. 2016 Early State Formation from a Big History Point of View. In *Eurasia at the Dawn of History*, M. Fernandez-Gotz, D. Krausse, eds., Cambridge University Press, Cambridge.

Spooner, B. (ed.) 1972 *Population Growth: Anthropological Implications*, MIT Press, Cambridge, MA.

Squire, G. 1877 *Peru: Incidents of Travel and Exploration in the Land of the Inkas*, Harper & Brothers Publishers, Franklin Square, New York.

Stanish, C. 2001 The Origin of State Societies in South America. *Annual Review of Anthropology* 30:41–64.

Stanish, C. 2006 Prehispanic Agricultural Strategies of Intensification in the Titicaca Basin of Peru and Bolivia. In *Agricultural Strategies*, J. Marcus, C. Stanish, eds., Cotsen Institute of Archaeology, University of California Publishers.

Stanish, C. 2017 *The Evolution of Human Cooperation: Ritual and Social Complexity in Stateless Societies*, Cambridge University Press, Cambridge.

Steward, J. (ed.) 1955 *Irrigation Civilizations: A Comparative Study*, Pan American Union Press, Washington, DC.

Thompson, L. G., M. Davis, E. Mosley-Thompson 1994 Glacial Records of Global Climate: A 1500-year Tropical Ice Core Record of Climate. *Human Ecology* 22(1):83–95.

Thompson, L. G., E. Moseley-Thompson 1989 One-half Millennium of Tropical Climate Variability as Recorded in the Stratigraphy of the Quelccaya Ice Cap, Peru. In *Aspects of Climate Variability in the Pacific and Western Americas*, D. Peterson, ed., *Geophysical Union Monograph* 55(17), American Geophysical Union, Washington, DC, No. 445.

Thompson, L. G., E. Mosley-Thompson, M. Davis 1995 Late Glacial Stage and Holocene Tropical Ice Core Records from Huascaran, Peru. *Science* 269:46–50.

Toynbee, A. 1972 *A Study of History*, Weathervane Books, New York, pp. 348–350.

Tseropoulos, G., Y. Dimalopoulos, J. Tsamopoulos, G. Lyberalos 2013 On the Flow Characteristics of the Conical Minoan Pipes Used in Water Supply Sytems via CFD Simulations. *Journal of Archaeological Science* 40:2057–2068.

Tung, T. 2012 *Violence, Ritual and the Wari Empire*, University of Florida Press.

Tung, T., N. Vang, B. Culleton, D. Kennett 2017 *Dietary Inequality and Indiscriminant Violence: A Social Bioarchaeological Study of Community Health During Times of Climate Change and Wari State Decline*. Paper presented at the Institute of Andean Studies, Berkley, CA.

Urban, T., S. Alcock, C. Tuttle 2012 Virtual Discoveries at a Wonder of the World: Geophysical Investigations and Ancient Plumbing at Petra, Jordan. *Antiquity* 86:331.

Urton, G. 1999 *Inka Myths*, University of Texas Press, Austin.

Vallières, C. 2012 *Taste of Tiwanaku: Daily Life in an Ancient Andean Urban Center as Seen Through Cuisine*. Unpublished Ph.D. Thesis, Department of Anthropology, McGill University, Montreal.

Van Deman, E. 1934 *The Building of Roman Aqueducts*, Series: Carnegie Institution of Washington Publication, Washington, DC.

Ven Te Chow 1959 *Open-Channel Hydraulics*, McGraw-Hill Book Company, New York.

Villafana, J. 1986 *Sistemas Hidráulicos Incas*, Lluvia Editores, Lima.

Webster, L., R. Hughs 2008 The Mystery of Minoan Tapered Pipes. *Journal of Hydraulic Research* 47(2):27–29.

Weis, L. 2014 Das Wasser der Nabataer: Zwishen Lebensotwendigkeit und Lexus: The Northwestern Project. In *Conference Proceedings: De Aquaductu Atque Aqua Urbium Lyciae Phamphyliae Pisidiae: The Legacy of Julius Sextus Frontinus*, G. Wiplinger, ed., Peeters Publishing, Leuven, Belgium.

Weismantel, M. 2018 Cuini Raya Superhero-Ontologies of Water on Peru's North Coast. In *Powerful Places in the Ancient Andes*, J. Jennings, E. Swenson, eds., University of New Mexico Press, Albuquerque, NM.

Weiss, H. 2000 Beyond the Younger-Dryas: Collapse as an Adaptation of Abrupt Climate Change in Ancient West Asia and the Eastern Mediterranean. In *Environmental Disaster and the Archaeology of Human Response*, G. Bawden, R. Reycraft, eds., Maxwell Museum of Anthropology Paper No.7:75–98, Albuquerque, NM.

Weiss, Z. 2014 *Public Spectacles in Roman and Late Antique Palestine*, Harvard University Press, Cambridge, MA.

Wells, L. 1992 Holocene Landscape Change on the Santa Delta, Peru: Impact on Archaeological Site Distributions. *Holocene* 2:193–204.

Wells, L. 1996 The Santa Beach Ridge Complex: Sea Level and Progradational History of an Open Gravel Coast in Central Peru. *Journal of Coastal Research* 12:1–17.

West, M. 1971 Early Watertable Farming on the North Coast of Peru. *American Antiquity* 44(1):138–144.

White, K. 1984 *Greek and Roman Technology*, Cornell University Press, Ithica, New York.

Wieczorek, A., Wright, M. 2012 History of Agricultural Biotechnology: How Crop Development has Evolved. *Nature Education Knowledge* 3(10).

Willets, R. 1977 *The Civilization of Ancient Crete*, Berkeley, CA: The University of California Press.

Williams, P. 2002 Rethinking Disaster-induced Collapse in the Demise of Andean Highland States. *World Archaeology* 33(3).

Williams, P. 2006 Agricultural Innovation, Intensification, and Sociopolitical Development. In *Agricultural Strategies*, J. Marcus, C. Stanish, eds., The Cotsen Institute, University of California Publishers, Los Angeles.

Wittfogel, K. 1955 Developmental Aspects of Hydraulic Societies. In *Irrigation Civilizations: A Comparative Study*, J. Steward, et al., eds., Social Science Monographs, Pan American Union, Washington, DC.

Wittfogel, K. 1956 *The Hydraulic Civilizations*, University of Chicago Press, Chicago.

Wittfogel, K. 1957a Oriental Despotism: A Comparative Study of Total Power, Water Management in the Ancient World. *Philosophical Transactions of the Royal Society*.

Wittfogel, K. 1957b *Oriental Despotism*, Yale University Press, New Haven, CT.

Woodward, S., C. Posey 1941 *Hydraulics of Steady Flow in Open Channels*, John Wiley and Sons, London.

Woolf, G. (ed.) 2003 *Cambridge Illustrated History: The Roman World*, Cambridge University Press, Cambridge.

Wright, H. 1977 Recent Research on the Origin of the State. *American Review of Anthropology* 6:379–397.

Wright, K., A. Gilbaja Oviedo, G. McEwan, R. Miksad, R. Wright *2016 Incamisana: Engineering an Inca Water Temple*, American Society of Civil Engineers Press, Reston, VA. ISBN 978-0-7844-1079-0.

Wright, K., G. McEwan, R. Wright 2006 *Tipon: Water Engineering Masterpiece of the Inca Empire*, American Society of Civil Engineers Press, Reston, VA. ISBN 0-7844-0851-3.

Wright, K., R. Wright, A. Zegarra, G. McEwan 2011 *Moray: Inka Engineering Mystery*, American Society of Civil Engineers Publication, Reston, VA. ISBN 978-0-7844-1079-0.

Zuidema, T. 1990 *Inka Civilization at Cuzco*, University of Texas Press, Austin.

Index

Note: Page numbers in *italic* indicate a figure and page numbers in **bold** indicate a table.